Biotransformations

Volume 1

Biotransformations

*A survey of the biotransformations
of drugs and chemicals in animals*

Volume 1

Edited by

D. R. Hawkins

Huntingdon Research Centre Ltd

ROYAL
SOCIETY OF
CHEMISTRY

British Library Cataloguing in Publication Data

Biotransformations: a survey of the
 biotransformations of drugs and chemicals in
 animals.
 Vol. 1
 1. Animals. Effects of drugs
 I. Hawkins, D. R.
 591.19

 ISBN 0-85186-157-1

Published by The Royal Society of Chemistry,
Burlington House, London W1V 0BN

Set by Unicus Graphics Ltd, Horsham, West Sussex
and printed in Great Britain
by Whitstable Litho Ltd, Whitstable, Kent

Preface

Investigation of biotransformation processes has been a rapidly expanding aspect of xenobiotic metabolism. This has been brought about by the increasing realization of the important influence of biotransformation on pharmacological activity and toxicity and the parallel rapid advances in technology in analytical chemistry and in particular physico-chemical techniques such as mass spectrometry.

In drug discovery and development extrapolation of activity from *in vitro* pharmacological screens to animals *in vivo* and finally to man can be improved by consideration of the impact of biotransformation on the compound in different systems. Consideration of the structures of compounds shown to be active in *in vitro* screens would indicate compounds likely to be inactive *in vivo* due to rapid biotransformation. Structural modifications can then be introduced to inhibit metabolism at key positions in the molecule.

Species differences in toxicity are commonplace and are often attributable to differences in biotransformation pathways. An understanding of the mechanisms of toxicity may help to extrapolate the relevance of findings in laboratory animals to human exposure.

Prediction of biotransformation pathways would be an ideal goal but one which we are far from achieving. However, a key to knowing how close we can get to this goal is to ensure that full use can be made of available knowledge. This series has been devised to bring together all current information on biotransformations in a readily accessible form. It is hoped that this will provide a valuable database which will increase awareness of patterns in species differences and influences of chemical structure on biotransformation with a view to increasing the ability to make predictions for new compounds.

The series will cover biotransformations of chemical entities whether they are pharmaceuticals, agrochemicals, food additives, or environmental or industrial chemicals in vertebrates of the animal kingdom, which includes mammals, birds, fish, amphibians, and reptiles. This first volume broadly covers the literature for the calendar year 1987 although it has not been possible to include material in late issues of some journals. This will be included in the second volume which will cover 1988 literature. We have attempted to include all relevant literature but realize that this goal may not have been achieved. The Editor would be grateful to receive copies of omitted papers which could be included in a subsequent volume.

Arrangement of Material and Access

An overview chapter has been prepared which contains highlights such as novel biotransformation, mechanisms of toxicity, and notable species differences. The abstracts are arranged according to compound class, although there may be cases where allocation to one or another class is somewhat subjective. It has been considered valuable to be able to access information on the biotransformation of compounds with similar structural features. For this purpose the concept of key functional groups has been developed. Selection and naming of these groups has evolved during preparation of the material. For each compound functional groups have been selected where biotransformation has been shown to occur but in addition groups have also been included where biotransformation has not taken place. A list of the functional groups is provided which may be referred to before proceeding to the corresponding index. Two other indexes have been included containing compound names and types of biotransformation processes respectively.

In the precis for each compound certain key information has been included when available. Where radiolabelled compounds have been used the position(s) of labelling have been indicated on the structure. Comments on the source of metabolites and information on the quantitative importance of individual metabolites such as percentage material in the sample or percentage administered dose are given where possible. Also, in order to provide a perspective on the criteria for identification, the procedures used for separation and isolation of metabolites and structural assignments such as chromatographic and physico-chemical techniques and use of reference compounds have been discussed.

The Editor

Contents

Contributors

P. Batten *ICI Central Toxicology Laboratory, Alderley Park, Macclesfield, Cheshire, SK10 4TJ, UK*

A. Bell *Glaxo Group Research, Ware, Herts., SG12 0DJ, UK*

P. Farmer *MRC Toxicology Unit, Medical Research Council Laboratories, Woodmansterne Road, Carshalton, Surrey, SM5 4EF, UK*

D. R. Hawkins *Huntingdon Research Centre, Huntingdon, Cambs., PE18 6ES, UK*

D. Kirkpatrick *Huntingdon Research Centre, Huntingdon, Cambs., PE18 6ES, UK*

M. Prout *ICI Central Toxicology Laboratory, Alderley Park, Macclesfield, Cheshire, SK10 4TJ, UK*

Key Functional Groups

R may be any unspecified group including H. Where two or more R groups are indicated these may be the same or different groups. Where aromatic rings or other cyclic systems are shown they may also contain substituents when they are not specified as part of the key functional group.

Acetal		Alkoxyphenyl	$RCH_2O-\langle\bigcirc\rangle$
Acetamide	$CH_3\overset{O}{\overset{\|}{C}}NHR$	Alkyl alcohol	RCH_2OH
N-Acetyl aryl amine	$CH_3\overset{O}{\overset{\|}{C}}NH-\langle\bigcirc\rangle$	sec-Alkyl alcohol	$\overset{R}{\underset{R}{>}}CH-OH$
		Alkyl aldehyde	RCH_2CHO
Acetylthio	$CH_3\overset{O}{\overset{\|}{C}}SR$	Alkyl amide	$R\overset{O}{\overset{\|}{C}}NHCH_2R$
Alanine	$\underset{H_2N\overset{\|}{C}HCO_2H}{\overset{CH_3}{}}$	Alkyl amine	RCH_2NH_2
tert-Alcohol	$\underset{R}{\overset{R}{R}}C-OH$	Alkyl tert-amine	$\overset{R}{\underset{R}{>}}N-R$
Alkadiene	$>C=CH-CH=C<$	sec-Alkyl amine	$\overset{R}{\underset{R}{>}}CH-NHR$
Alkane	$CH_3(CH_2)_nR$	tert-Alkyl amine	$\underset{R}{\overset{R}{R}}C-NHR$
iso-Alkane	$\overset{RCH_2}{\underset{RCH_2}{>}}CH-$	Alkylamino	RCH_2NH-
Alkene	$>C=C<$	iso-Alkylamino	$\overset{R}{\underset{R}{>}}CHNH-$

Alkyl aryl amide	$RC(=O)-N(CH_2R)$(phenyl)	N-Alkyl cycloalkylamine	$(CH_2)_n - NHCH_2R$
Alkyl aryl amine	$R-N(CH_2R)$(phenyl)	Alkylcyclohexane	$RCH_2 -$ (cyclohexane)
iso-Alkyl aryl amine	$R-N-CH(R)(R)$(phenyl)	Alkyl ester	$RCOCH_2R$ ($C=O$)
sec-Alkyl aryl amine	$R-N-CH_2CH(R)(R)$(phenyl)	iso-Alkyl ester	$RCOCH(R)(R)$ ($C=O$)
Alkyl aryl ether	RCH_2O-(phenyl)	Alkyl ether	RCH_2OR
		Alkyl hydrazine	RCH_2NHNHR
Alkyl aryl sulphoxide	$RCH_2S(\rightarrow O)-$(phenyl)	N-Alkyl imide	$R-C(=O)-N(CH_2R)-C(=O)-R$
Alkyl aryl thioether	RCH_2S-(phenyl)	Alkyl ketone	$RCH_2\overset{O}{\underset{\parallel}{C}}R$
Alkyl carboxamide	$RCH_2\overset{O}{\underset{\parallel}{C}}NHR$	Alkylphenyl	$RCH_2 -$(phenyl)
Alkyl carboxylic acid	$RCH_2\overset{O}{\underset{\parallel}{C}}OH$	iso-Alkylphenyl	$(R)(R)CH -$(phenyl)
iso-Alkyl carboxylic acid	$(R)(R)CH\overset{O}{\underset{\parallel}{C}}OH$	N-Alkylpiperazine	$RN(\text{piperazine})NCH_2R$
		Alkyl quaternary ammonium	$>\overset{+}{N}-CH_2R$
		Alkyl sulphate	RCH_2OSO_3H

x

Alkyl sulphonate	RCH_2OSO_2R

Anthraquinone

Alkyl sulphoxide	$RCH_2\overset{O}{\underset{\uparrow}{S}}R$

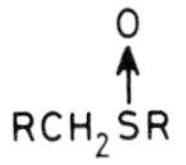

Alkyl thioether	RCH_2SR

Aryl acetic acid

$-CH_2CO_2H$

Allyl amine	$RCH{=}CH{-}NH_2$

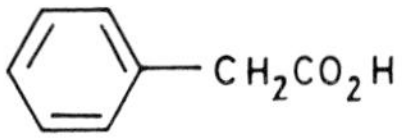

Aryl aldehyde

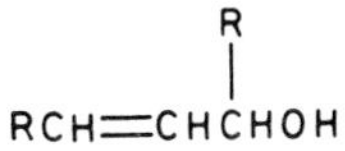

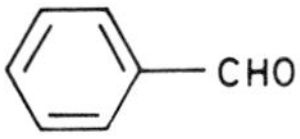

$-CHO$

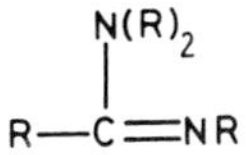

Allylic alcohol	$RCH{=}CHCHOH$ with R

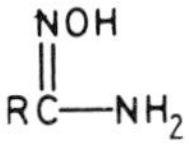

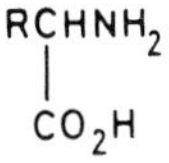

Arylalkene

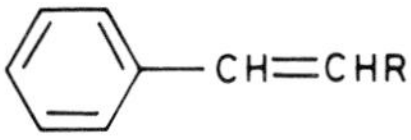

$-CH{=}CHR$

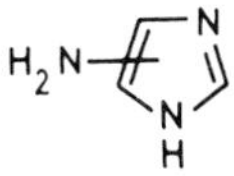

Amidine	$R{-}\overset{N(R)_2}{\underset{}{C}}{=}NR$

Arylalkyl

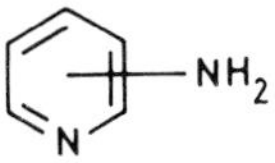

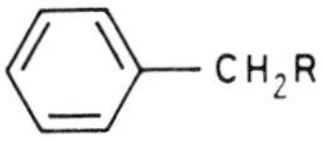

$-CH_2R$

Amidoxime	$R\overset{NOH}{\underset{}{C}}{-}NH_2$

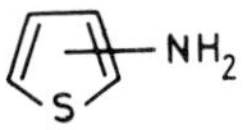

Aryl tert-alkyl

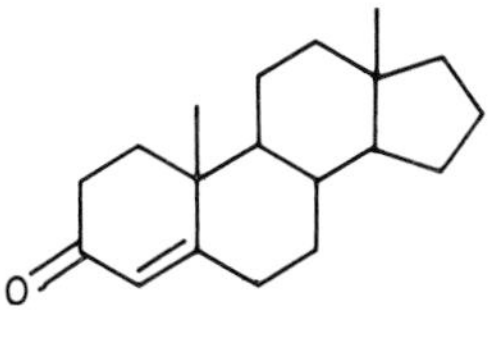

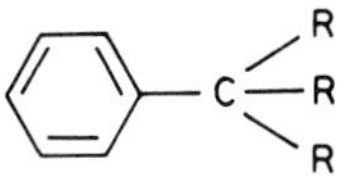

$-C\overset{R}{\underset{R}{<}}R$

Amino acid	$\overset{RCHNH_2}{\underset{CO_2H}{|}}$

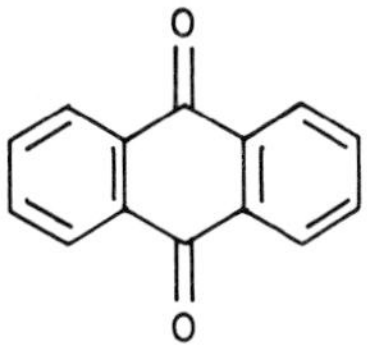

Aryl amide

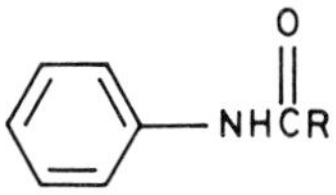

$-NH\overset{O}{\overset{\|}{C}}R$

Aminoimidazole	

Aryl amine

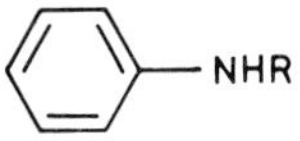

$-NHR$

Aminopyridine	$-NH_2$

Aryl amino acid

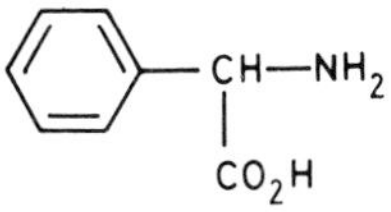

$-\overset{CH{-}NH_2}{\underset{CO_2H}{|}}$

Aminothiophene	$-NH_2$

Aryl carboxamide

$-\overset{O}{\overset{\|}{C}}NHR$

Androsten-3-one	

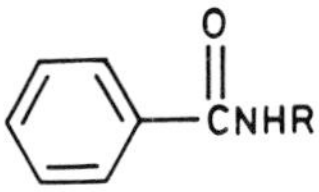

Aryl carboxylate

$-\overset{O}{\overset{\|}{C}}OR$

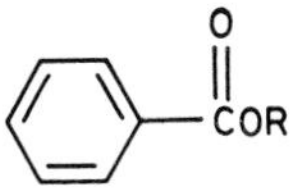

Aryl carboxylic acid	![benzoic acid structure]	Aryl thioether	—SR
Aryl ester		Azobenzene	—N=NR
Aryl ether	—OR	Barbiturate	
Arylethylene	—CH=CH$_2$	Benzanthracene	
Aryl hydrazine	—NHNH$_2$	Benzhydrol	
Arylhydroxymethyl	—CH$_2$OH	Benzhydryl	
N-Arylimine	—N=R	Benzidine	—NH—NH—
Aryl ketone		Benzimidazole	
Arylmethyl	—CH$_3$	Benzofuran	
N-Arylnitrosamine	—NNO R	Benzo[a]pyrene	
Arylnitroso	—NO		
Aryl propionic acid	—CHCO$_2$H (CH$_3$)		

Benzo[c]phenanthrene		Bromophenyl	Br—⬡
Benzodiazepine		iso-Butyl	CH₃\CHCH₂— / CH₃
Benzyl	⬡—CH₂R	tert-Butyl	CH₃\ CH₃—C— / CH₃
Benzyl alcohol	⬡—CH₂OH	Chiral carbon	R¹—C*—R³ (R², R⁴)
Benzyl amine	⬡—CH₂N(R, R)		
		Chloroacetamide	ClCH₂CNHR (O)
Benzyl bromide	⬡—CH₂Br		
		Chloroacetyl	ClCH₂CR (O)
Benzyl ether	⬡—CH₂OR	Chloroalkane	ClCH₂R
		Chloroalkene	ClCH=CHR
Benzyl nitrile	⬡—CH₂CN		
		Chloroalkyl	ClCH(R, R)
Biphenyl	⬡—⬡		
		Chlorobenzoyl	Cl—⬡—C(O)
Bromoacetyl	BrCH₂CR (O)		
		Chlorobiphenyl	⬡—⬡ (Cl)ₙ
Bromoalkane	BrCH₂R		
Bromoalkyl	BrCH(R, R)	N-Chloroethyl	ClCH₂CH₂N<

Chlorophenyl

N-Cyclopropylmethyl $R-NCH_2-\triangleleft$

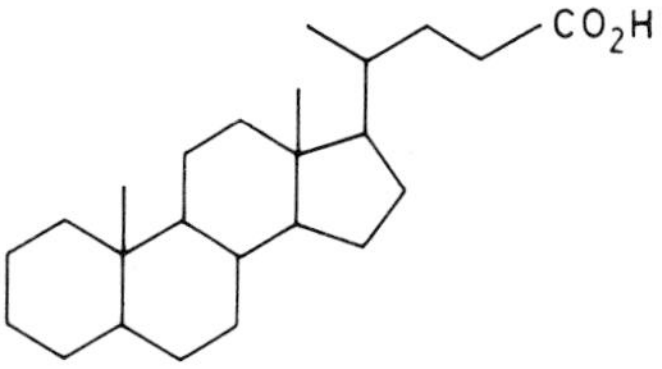

Cholanic acid

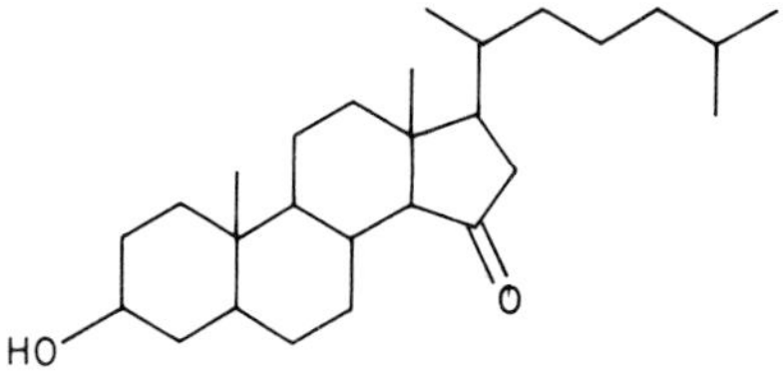

CO_2H

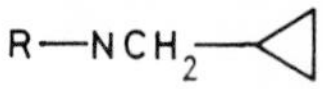

Cytidine

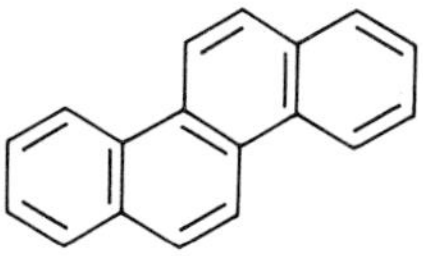

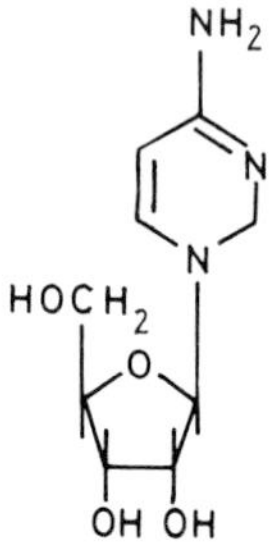

$HOCH_2$

NH_2

$OH \; OH$

Cholestenone

HO

Cytosine

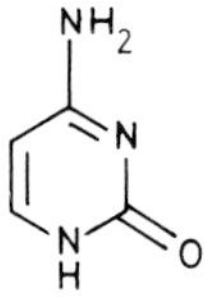

NH_2

Chrysene

Dialkyl amine

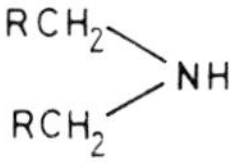

RCH_2
RCH_2
NH

Coumarin

Dialkylamino

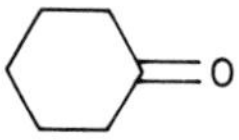

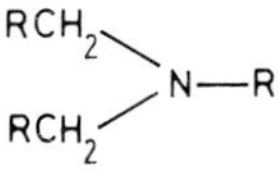

RCH_2
RCH_2
$N-R$

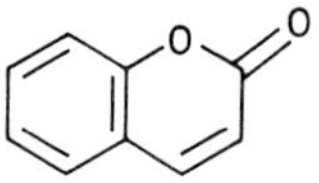

Cycloalkane/cycloalkyl $(CH_2)_n$

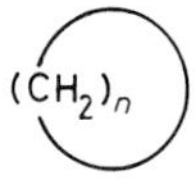

Dialkylaminoalkyl

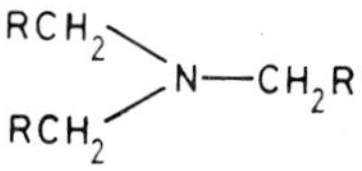

RCH_2
RCH_2
$N-CH_2R$

Cyclohexane/cyclohexyl

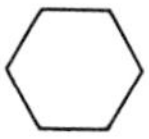

Dialkyl aryl amine

RCH_2
RCH_2
N

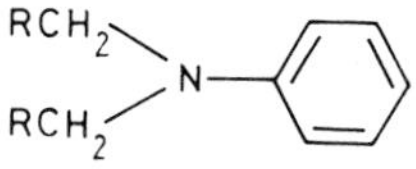

Dialkyl ether RCH_2OCH_2R

Cyclohexanone

Dialkylisoxazole

R R
N O

Cyclopentenone

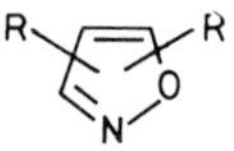

Dialkyl thioether RCH_2SCH_2R

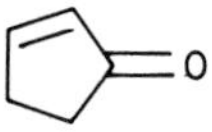

Cyclopropyl carboxylate

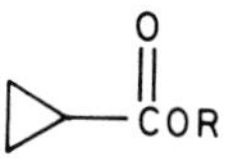

O
COR

Diaryl thioether

S

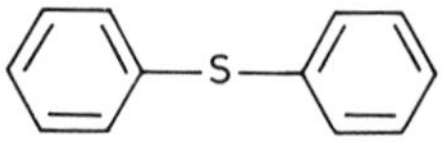

Dibenzofuran		Ethynyl	$R—C{\equiv}C—R$
		Fatty acid (saturated)	$R(CH_2)_nCO_2H$
Dichloroalkene		Fatty acid (unsaturated)	$R(CH{=}CH)_nCH_2CO_2H$
Digitoxigenin	—	Fluorene	
Dihydropyran		Fluoroacetyl	$FCH_2\overset{O}{\overset{\|}{C}}—R$
Dihydropyridine		Fluoroalkyl	$FCH_2CH{<}^R_R$
Dihydroquinoline		Fluorocytosine	
Dimethylamide		Fluorouracil	
Dimethylaminoalkyl		Formamide	$RNHCHO$
Dimethylcyclopropyl		Furan	
Disulphide	$R—S—S—R$	Furfural	
Dithiocarbamate		Glutathione	$HOOCCHCH_2CH_2CO—NHCHCO—NHCH_2COOH$
Dithiolane		Glycolamide	$R\overset{OH}{\overset{\|}{C}HCONHR}$
Epoxide			

Glycoside

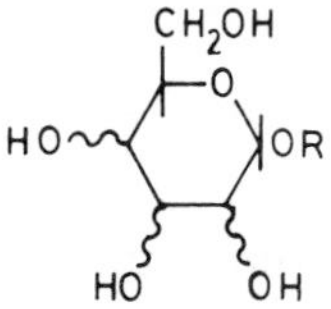

Guanidine $R{-}NHC({=}NH)NH_2$

Hexose

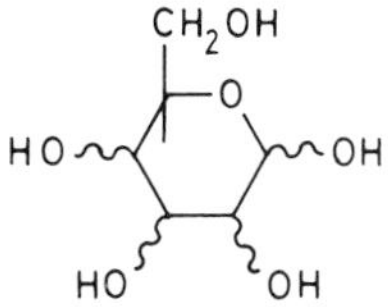

Hydantoin

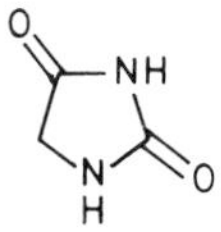

Hydrazide $RC({=}O)NHNH_2$

Hydrazine $RNHNH_2$

Hydroperoxide R_2CHOOH

N-Hydroxy $R_2N{-}OH$

N-Hydroxy sulphate $R_2N{-}OSO_3H$

Imidazole

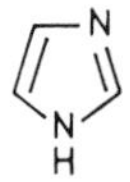

Imidazoline

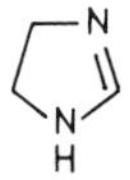

Iminecarboxylate $={NC(=O)OR}$

Indane

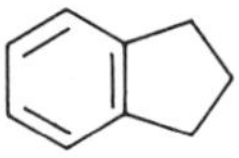

Indene

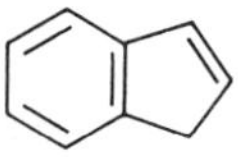

Indole

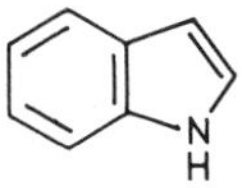

Iodoalkyl ICH_2R

Isoprenoid —

Isoquinoline

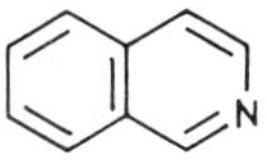

Lactam

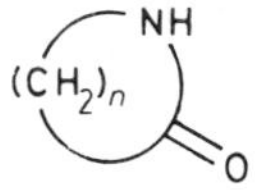

Lactone

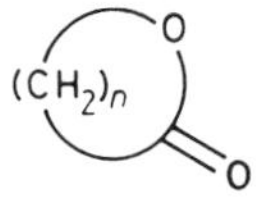

Leucine $(H_3C)_2CHCH_2CH(NH_2)COOH$

Methanesulphonate CH_3SO_2OR

o-Methoxyphenol

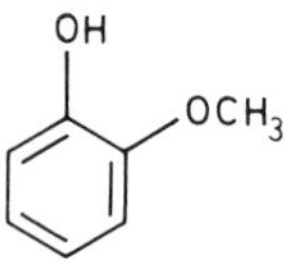

Methoxyphenyl — OCH₃ (methoxyphenyl structure)

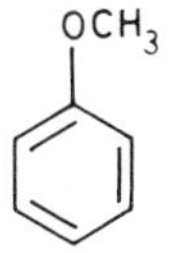

Methoxyphenyl — phenyl ring with OCH₃

Methoxyphenyl

OCH₃ on benzene ring

N-Methylnitrosamine — CH₃NNO / R

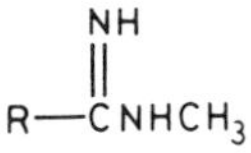

Methylphenyl — H₃C—phenyl

N-Methyl alkyl amine — R, R, N—CH₃

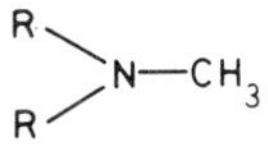

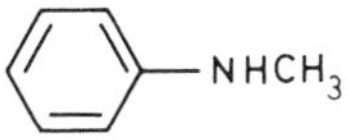

Methyl amide — RCNHCH₃ (O)

Methylphosphinoyl — CH₃P—OR (OH)

N-Methylamidine — R—CNHCH₃ (NH)

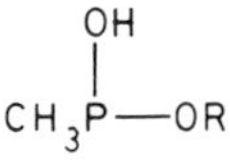

O-Methylphosphorodithioate — CH₃O, CH₃O, P—SR (S)

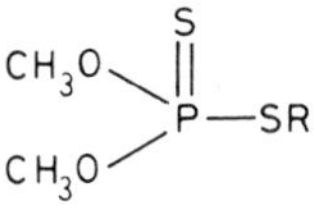

Methylamino — CH₃NHR

O-Methylphosphorothioate — CH₃O, CH₃O, P—SR (O(S)) (O)

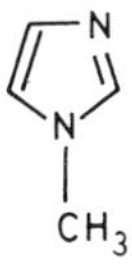

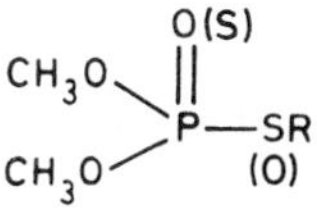

N-Methyl aryl amine — phenyl—NHCH₃

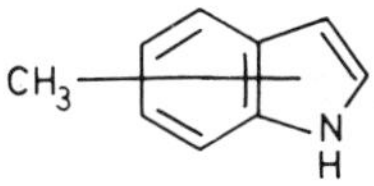

N-Methylpiperidine — piperidine N—CH₃

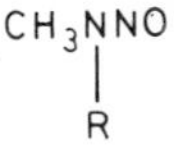

Methyl carbamate — RNHCOCH₃ (O)

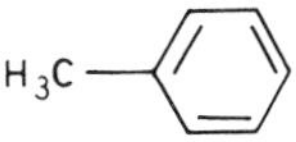

N-Methylpurine — purine structure, R=H or CH₃

Methyl carboxylate — CH₃COR (O)

Methylpyrimidine — pyrimidine with NHR and CH₃

Methylcyclohexene — cyclohexene—CH₃

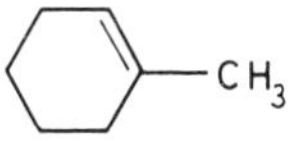

Methyl ether — CH₃OR

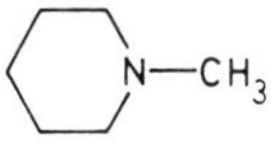

N-Methylpyrrolidine — pyrrolidine N—CH₃

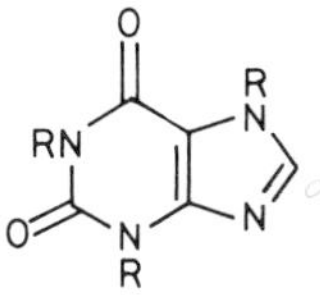

N-Methylimidazole — imidazole N—CH₃

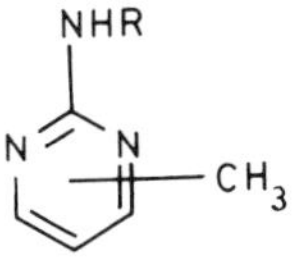

Methylindole — CH₃—indole (N-H)

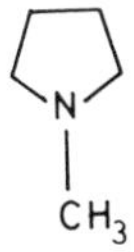

N-Methylpyrrolidone — pyrrolidone N—CH₃, O

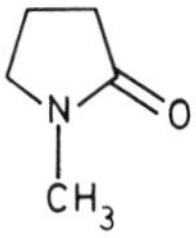

Methylquinoxaline		Nitrofuran	
Methyl sulphonate	CH_3OSO_2R	Nitroimidazole	
Methylsulphinyl		Nitrophenyl	
N-Methyltetrahydropyridine		Nitrosamine	
		Nucleoside	—
Methyl thioether	CH_3SR	Octyl	$CH_3(CH_2)_7-$
Monoclonal antibody	—		
Morphinan		Oestradiol	
Morpholine		Oestren-3-one	
Naphthalene		Oestrone	
Naphthaquinone		Organoarsenic	$(R)_3As$
Nitrile	RCN	Oxazole	

Oxazolidinone	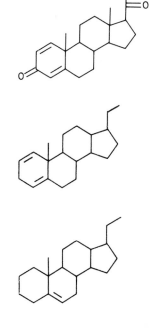	Phthalazine	

Oxazolidinone

Phthalazine

Oxime $RCH={\!=\!}NOH$

Oximino $R{=\!=}N{-\!}OR$

Phthalidyl

Pentyl $CH_3(CH_2)_4{-}$

Peptide

Piperazine $R{-}N\underset{}{}N{-}R$

Piperidine NR

Phenol

Piperidinedione

Phenoxybenzyl CH_2R

Polycyclic aromatic   —

Polycyclic aromatic amine   —

N-Phenyl

Phenylethyl CH_2CH_2R

Prednisolone

Phosphoramide

Pregnadiene

Phosphoramidothioate

Phosphorothioate

Pregnene

Prochiral carbon		Pyrrole	
Prostanoid	—	Pyrrolidine	
Pteridine			
		Pyrrolidine amide	
Pyrazine		Pyrrolizidine	
Pyrazolin-5-one		Quinoline	
Pyrazole		Quinonedi-imine	
Pyrene			
		Quinoneimine	
Pyridazine			
Pyridine		Steroid	—
		Sulphonamide	RSO_2NHR
Pyridinium		Sulphonic acid	RSO_3H
		N-Sulphonoxy	
Pyrimidine		Sydnone	

Terpene	—	Triazole	
Tetrahydroindazole		Trichothecene	
Tetrahydropyridine		Trifluoromethylphenyl	F_3C—
Thiazole		Uracil	
Thienylalkyl	$-CH_2R$		
Thiocarbamate	$RNHC S^- X^+$	Ureide	$RNHCNH_2$
Thiocarbonyl		Vinca alkaloid	—
Thiophene		Xanthine	

An Overview

The purpose of this chapter is to highlight biotransformations included in this volume which may be of particular interest. These include novel biotransformations, stereoselective and stereospecific biotransformations, and examples where mechanisms of toxicity have been attributed to specific biotransformations.

Novel biotransformations include new pathways or those where few examples are known and also where combinations of known pathways combine to produce novel types of metabolites for particular compounds. The latter would include metabolites where the character of the parent compound had been markedly changed by the introduction of diverse functional groups or formation of new ring systems. There is now a greater realization that stereoselective metabolism means that compounds consisting of mixtures of enantiomers or diastereoisomers cannot be considered as single entities. Similarly biotransformation is often stereospecific when it results in the introduction of an asymmetric centre. Both stereoselectivity and stereospecificity may have important consequences for pharmacological activity and toxicity. Since there may be species differences in stereochemical aspects of biotransformation these should be considered when extrapolating from animals to man. An understanding of the mechanisms of toxicity observed in animals is often important in extrapolation of the relevance to man and in making risk assessments for human exposure.

1 NOVEL PATHWAYS

1.1 Oxidative

Investigations with o-, m-, and p-dinitrobenzenes in rats have shown that biotransformation proceeds via the expected pathways, namely nitro-group reduction and glutathione displacement of a nitro-group (p. 178). In addition some ring hydroxylation occurs with the resulting phenols being excreted as sulphates. Thus, two sulphates (1) and (2) were produced from the 1,2-isomer, (1) being a major urine metabolite (17% dose). The 1,4-isomer formed an *ortho*-amino-sulphate (3; 33% dose) but the 1,3-isomer gave an unusual sulphate (4; 6% dose) resulting from loss of the nitro-group. It is possible that this metabolite is formed via a 3,4-epoxide with the nitro-group being lost during a rearrangement. Similar loss of other substituents, particularly halogens, during aromatic hydroxylation has been reported. Other metabolites with retention and migration of the substituent are also often found, this process being commonly referred to as the NIH shift.

(1)   (2)

(3)

(4)

The oxidation of benz[a]pyrene (5) by lipid hydroperoxides has been studied in liposomes derived from rat liver (p. 68). Three quinones were identified, the 1,6-, 3,6-, and 6,12-quinone (6). It was proposed that the mechanism involves direct attack of a lipid peroxy radical followed by autoxidation.

(5)   (6)

Hydroxylation on an alkyl carbon adjacent to a heteroatom is a process which leads to O-, N-, and S-dealkylation since the intermediates are usually unstable and spontaneously liberate the alkyl group as an aldehyde or ketone. However, when an N-alkyl group is part of an amide function the hydroxyalkyl intermediate may be more stable and can sometimes be isolated as a metabolite. Compounds of this type have been isolated as metabolites of the benzodiazepine precursor (7) (p. 201). Hydroxymethyl derivatives of a primary (9) and secondary (8) amide function in the benzodiazepine were identified. The minor metabolite (9) decomposed during isolation and was identified after formation of its acetyl derivative.

(7)

(8) R = CH$_3$
(9) R = H

A similar type of intermediate metabolite of a tertiary amine has now also been identified (p. 000). The benzylic alcohol (11) of tripelennamine (10) was

2

identified as a major metabolite in rat urine. The corresponding *N*-debenzylated metabolite was also found in smaller amounts.

(10)　　　　　　　　　　　　(11)

N-Dealkylation results in formation of an intermediate aldehyde which is usually rapidly oxidized to the corresponding carboxylic acid. However, certain aldehydes may also be reduced to alcohols systemically. Thus, both the acid (13) and alcohol (14) have been identified as metabolites of chlorophenoxamine (12) arising from an *N*-dealkylation pathway (p. 229).

(12)　　　　　　　　(13)　　　　　　　　(14)

Quaternary ammonium compounds are polar ionic compounds with high water solubility. Methylation of a tertiary amine could provide a prodrug derivative if *in vivo* N-demethylation of the quaternary compound occurs. After intravenous administration of naltrexonium methiodide (15) to rats, naltrexone (16) could be detected in plasma and brain (p. 266). The major component in plasma was the parent compound, indicating that the extent of this biotransformation was probably very low.

(15)　　　　　　　　　　　　　　(16)

Conversion of carboxylic acid esters into the acids is one of the most common biotransformation processes which is usually assumed to involve the well-characterized esterases. However, it is now realized that esters can be cleaved by an alternative oxidative pathway involving cytochrome P-450. Evidence for the pathway has been demonstrated for de-esterification of the diethyl ester (17) by rat liver microsomes (p. 305). It was shown that the monoester (19) was formed with liberation of acetaldehyde. Experiments using the compound with deuterium-labelled ethyl groups showed a high kinetic isotope effect indicating C–H cleavage as a rate-limiting step. Hence it is most likely that de-esterification involves formation of the hydroxylated intermediate (18).

3

Studies with metoprolol (20) in man has demonstrated the array of metabolites which can be formed by oxidative degradation of an aryl methoxyethyl function (p. 244). Benzylic hydroxylation to (21) was followed by O-demethylation (22) and oxidation to a glycolic acid derivative (23). A benzyl alcohol (24) was formed by further degradation of the glycolic acid. In all these metabolites the alkylaminopropanol side-chain was intact although this group is normally subject to extensive metabolism. The metabolites were identified after formation of novel oxazolidinone derivatives (25) by reaction with phosgene. This derivatization would be applicable to other drugs containing a similar side-chain.

Although chloramphenicol (26) is a long-established drug and various studies on its metabolism have been reported a new major human metabolite has now been identified (p. 238). This metabolite, representing 13% of a 500 mg oral dose, was identified as the oxamic acid derivative (27) formed by oxidation of the dichloromethyl group.

The butadiene (28) can be oxidized to epoxides which are known to be mutagenic. After inhalation exposure of rats the two monoepoxides (29) and (31) and the diepoxide (30) were identified as metabolites (p. 103). Fat

4

appeared to be a reservoir for these epoxides and the diepoxide was also detected in a range of other tissues.

$$CH_2=C-CH=CH_2 \qquad \overset{O}{CH_2-C-CH=CH_2} \qquad \overset{O}{CH_2-C-CH-CH_2} \qquad CH_2=C-\overset{O}{CH-CH_2}$$

$$\underset{CH_3}{|} \qquad\qquad \underset{CH_3}{|} \qquad\qquad \underset{CH_3}{|} \qquad\qquad \underset{CH_3}{|}$$

$$(28) \qquad\qquad\quad (29) \qquad\qquad\quad\quad (30) \qquad\qquad\quad\quad (31)$$

Canavine (32) is an unusual amino acid possessing a guanidine group linked to homoserine via an oxime-type function. Studies in rats have shown that both urea and guanidine are formed as metabolites, with L-canaline (33) and L-homoserine (34) respectively released as the amino acid group (p. 439). These two pathways correspond to cleavage of the imino or N–O bonds in the oximino function. It is also of interest that guanidine was excreted in urine partly unchanged and partly as the N-methyl derivative (35).

$$\overset{NH_2}{\underset{|}{}} \qquad \overset{NH_2}{\underset{|}{}}$$
$$H_2N-C=N-OCH_2CH_2CHCO_2H$$
$$(32)$$

$$\overset{NH_2}{\underset{|}{}} \qquad\qquad\qquad \overset{NH_2}{\underset{|}{}}$$
$$H_2N-C=O \qquad\qquad HOCH_2CH_2CHCO_2H$$
$$(34)$$

$$\overset{+}{\underset{NH_2}{\underset{|}{}}} \qquad\qquad \overset{+}{\underset{NH_2}{\underset{|}{}}} \qquad\qquad \overset{NH_2}{\underset{|}{}}$$
$$H_2NOCH_2CH_2CHCO_2H \qquad H_2N-C=NH \longrightarrow CH_3HN-C=NH$$
$$(33) \qquad\qquad\qquad\qquad\qquad\qquad\qquad\qquad (35)$$

N-Formylamines have been reported as metabolites of aromatic amines produced by enzymic formylation. The N-formyl function may also be formed by oxidative degradation of an amidine group. Sulfamidine (36) is extensively metabolized in rats, one of the terminal metabolites being 5-chloroanthranilic acid (p. 410). An intermediate metabolite was the N-formyltoluidine (37), which was a urinary metabolite and was also formed in a rat liver supernatant.

$$Cl\langle\bigcirc\rangle\underset{CH_3}{\overset{N=CH-N}{}}\overset{CH_3}{\underset{CH_2SCH_3}{}} \qquad\qquad Cl\langle\bigcirc\rangle\underset{CH_3}{\overset{NHCHO}{}}$$
$$(36) \qquad\qquad\qquad\qquad\qquad (37)$$

A dihydroquinoline contains a secondary aromatic amine function but there are few examples in the literature on the biotransformation of this type of compound. One of the major metabolites of the trimethylquinoline (38) in rat urine was identified as a sulphate of a dihydroxylated compound (p. 340). This metabolite (39) was formed by N-hydroxylation and hydroxylation in the aromatic ring.

5

(38)

(39)

1.2 Reductive

Nitroaromatic compounds are frequently reduced to the corresponding aryl amines but studies with aristolochic acids in several animal species have shown that reductive displacement of a nitro-group can occur (p. 99). Minor metabolites of the acids (40; R = H or OCH_3) have been identified where the nitro-group has been replaced by hydrogen. This process is attributed to the action of intestinal bacteria.

(40)

A novel metabolite of the allylamine (41) in rats has been identified as a propionic acid derivative (43) (p. 278). The proposed mechanism involved deamination and oxidation of the resulting aldehyde to a carboxylic acid (42) followed by reduction of the alkene function. It is of interest that formation and reduction of an α,β-unsaturated carboxylic acid has been postulated as the mechanism for the stereospecific interconversion of enantiomeric 2-aryl-propionic acids. In this investigation the enantiomeric composition of the propionic acid metabolite was not determined.

(41)

(42)

(43)

Oximes may be expected to be metabolized to aldehydes and ketones although there are few reported examples. *In vitro* studies have been carried out with three classes of compound containing the oxime function, namely an aldoxime, ketoxime, and amidoxime (p. 406). Rabbit liver cytosol was shown to metabolize salicylaldoxime (44) to the aldehyde, *d*-camphor oxime (45) to the ketone (46; R = O) and imine (46; R = NH), and benzamidoxime (47) to benzamidine (48). The mechanism proposed was reduction of the oximes by liver oxidase to ketoimines followed by hydrolysis of the latter to the oxo-compound and ammonia.

6

(44)   (45)   (46)

(47)   (48)

The common pathways for biotransformation of epoxides include hydrolysis and conjugation. The mycotoxin deoxynivalenol (49) contains an exocyclic epoxide attached to a fused-ring system, and after an oral dose to rats a metabolite (50) has been identified with an alkene function formed by reduction of the epoxide (p. 257). The metabolite was present in urine (about 3% dose) and in faeces. It is possible that this metabolite is formed in the intestinal tract and then absorbed.

(49)   (50)

1.3 Ring Cleavage

8-Methoxypsoralen (51) undergoes two interesting biotransformations which may have some toxicological significance (p. 372). One rat metabolite has been identified resulting from oxidative cleavage of the furan ring (52). This metabolite may be formed via epoxidation of the furan double bond, a process which has been implicated in the toxicity of other furans. Of further interest is the functionalization of the central aromatic ring which undergoes *para*-hydroxylation and *O*-demethylation to give a hydroquinone (53) excreted as a sulphate conjugate in urine. The hydroquinone was readily oxidized to the quinone (54), and these two compounds may exist in equilibrium *in vivo*.

(51)   (52)

(53)   (54)

7

Another example of oxidative cleavage of a furan ring has been reported for the experimental drug (55) (p. 290). Two metabolites, an alcohol (56) and a carboxylic acid (57), were detected in the urine of rat, rabbit, and dog where together they accounted for 22%, 13%, and 2% of an administered oral dose respectively. The nature of these metabolites suggests that they would be formed by hydroxylation at the unsubstituted α-carbon.

(55)　　　　　　　　　　　(56)　　　　　　　　　　　(57)

Compounds containing an imidazoline ring can undergo extensive oxidative degradation to form a series of different functional groups, which is illustrated in studies with midaglizole (58) in dogs (p. 287). One of the primary metabolites was the imidazole (59) believed to be formed by dehydration of a mono-hydroxylated intermediate. The only other metabolite containing the intact ring was the diketone (60), while four ring-opened metabolites were identified containing amino acid (61), amide (62) and (63), and amidine (64) functions.

(58)　　　　　　　　　　　(59)

(60)　　　　　　　　　(61)　　　　　　　　　(62)

(63)　　　　　　(64)

1.4 Cyclization

Novel tetrahydrothiophene metabolites of busulfan (65) have been identified and evidence has been obtained that glutathione conjugation is the primary pathway involved (p. 147). In an experiment with a perfused rat liver a single major metabolite in bile was identified as the thiophenium compound (66). Formation of this metabolite was inhibited in the presence of ethacrynic acid, a glutathione-*S*-transferase inhibitor. The novel metabolite is presumably formed by nucleophilic displacement of a methanesulphonyl group by glutathione followed by further internal displacement of the second group to give a sulphonium ion. It was shown that at physiological pH the metabolite decomposed to tetrahydrothiophene (67).

$$CH_3SO_2O(CH_2CH_2)_2OSO_2CH_3 \longrightarrow CH_3SO_2O(CH_2CH_2)_2SG$$

(65)

(66)   (67)

GS = glutathionyl

Recent studies on the biotransformation of cyclosporin A (p. 446) represent one of the few extensive investigations carried out on a complex peptide although it is likely that many more products of this type will be investigated in the future. The complex structure of these molecules will bring many new challenges to those involved in biotransformation studies. Several metabolites of cyclosporin A have been identified from rat, rabbit, and man. All identified metabolites retained the intact cyclic oligopeptide structure and the main processes involved hydroxylation at the γ-position in three of the *N*-methyl-leucines and *N*-demethylation (68). One unusual metabolite was identified which involved an internal cyclization of the hydroxyalkene side-chain to give a tetrahydrofuran (69). The mechanism is unknown but must be initiated by facilitated nucleophilic attack of the hydroxy-group on the alkene function.

(68) R = H or CH$_3$

(69)

Two new metabolites of chloroquine (70) have been isolated from human urine which contain a pyrrolidinone (73) and dihydropyrrole (74) function respectively (p. 342). These metabolites would be formed by oxidative cleavage of the diethylamino-group and cyclization of the resulting aldehyde (71)

9

or derived carboxylic acid (72). However, cyclization of these metabolites could occur during their isolation and mass spectral analysis.

(70)   (71)   (72)

(74)   (73)

(R)-$(+)$-Pulegone (75) is a hepatotoxin and studies in mice have shown that there was an increase in toxicity in animals pretreated with phenobarbital and that the compound labelled with deuterium in the allylic methyls was much less toxic (p. 109). These results suggest that biotransformation is important for the toxicity of pulegone, and a major metabolite formed on incubation with mice liver microsomes and believed to be the proximate hepatotoxin has been identified as the menthofuran (76). Since the furan oxygen was derived from molecular oxygen, the proposed route of formation involved oxidation of an allylic methyl and intramolecular cyclization.

(75)   (76)

Formation of a novel cyclized metabolite of chlorpheniramine (77) has been demonstrated on incubation with rat liver microsomes (p. 219). Oxidative N-dealkylation of the side-chain occurs which presumably results in initial formation of the aldehyde (78); this was not identified but evidence for its formation was shown by identification of the corresponding carboxylic acid and alcohol and by trapping experiments with methoxylamine. A novel metabolite (79) containing an indolizine ring was also identified which was postulated as being formed by spontaneous cyclization of the aldehyde.

(77)   (78)   (79)

10

1.5 Conjugates

There have been occasional reports of urea adducts as urine metabolites but their mechanism of formation and whether they are true metabolites or artefacts is unclear. A ureide (80; R=NHCONH$_2$) of etodolac (80; R=H) has now been reported as a minor urine metabolite in mouse, rat, dog, and man (p. 371). It is likely that this component is formed from a metabolite with a functional group at the benzylic carbon.

(80)

A new novel metabolite of 5-fluorouracil has been identified in samples of bile from patients administered intravenous doses (p. 314). 2-Fluoroalanine (82) is a known metabolite derived by hydrolytic cleavage of the ring. The novel metabolite was shown to be a conjugate (83) of 2-fluoroalanine with cholic acid and represents the first example of a pathway where an exogenous amino acid competes with endogenous glycine and taurine to produce an abnormal bile acid.

(81)   (82)   (83)

N-Acetylation is a commonly observed biotransformation pathway for aromatic amines but a similar pathway for primary and secondary aliphatic amines has now been reported. Both the mono-desmethyl and di-desmethyl metabolites of doxylamine (84) were isolated from human urine but in addition the corresponding *N*-acetyl metabolites (85) and (86) have also been identified (p. 231). However, the quantitative importance of these metabolites was not reported.

(84)   (85)   (86)

N-Methylation of primary amines is not a common pathway but since it leads to metabolites with greater lipophilicity it may produce pharmacologically active metabolites of drugs. The thiophene carboxylate (87) is rapidly hydrolysed to the acid which is a major urine metabolite in rat and dog (p.

11

277). In addition the *N*-methyl derivative (88) was identified as a dog urine metabolite representing about 7% of the dose.

(87)   (88)

Although *N*-formyl derivatives as metabolites of primary and secondary amines are known there are sufficiently few examples for it to be considered a novel process. *N*-Formyl metabolites derived from 1-nitropyrene (89) and 2-nitrofluorene (90) have now been reported (p. 92) in both rat and rabbit. It was demonstrated that liver cytosol from a variety of species catalysed formation of these metabolites in the presence of *N*-formyl-L-kynurenine.

(89)   (90)

Three different glutathione conjugates of melphalen (91) have been identified after incubation with immobilized glutathione-*S*-transferases from cynomolgus monkey liver (p. 202). Two of these were formed by displacement of one or both chlorines in the dichloroethyl function. The third was more unusual and formed by displacement of the nitrogen mustard group (92). It has been proposed that this was formed via an aziridinium ion (93) derived by an internal nucleophilic displacement of one chlorine.

(91)   (92)

(93)

Conjugation of substituted acetic acids is the most important biotransformation for this class of compound but the type of conjugate varies according to structure and also between species. 3,4-Dichlorobenzyloxyacetic acid is unusual in that the taurine conjugate (94) was the major rat urine metabolite (60% of the dose) (p. 196). The glycine conjugate (95), which is more commonly encountered, represented only 2% of the dose and there was no evidence for a glucuronide.

Cl₂C₆H₃—CH₂OCH₂C(=O)NHCH₂CH₂SO₃H

(94)

Cl₂C₆H₃—CH₂OCH₂C(=O)NHCH₂CO₂H

(95)

Owing to the indirect procedures frequently used to identify conjugates the evidence for structural assignments is not unambiguous. In particular the existence of diconjugates would not usually be distinguished. By the use of two different derivatizations before and after conjugate hydrolysis the occurrence of mono- and di-sulphate conjugates of norethindrone (96) has been detected. The disulphate of oestranediol (97) was a major metabolite in milk samples taken from a woman receiving daily 1 mg doses of norethindrone.

(96)

(97)

1.6 Miscellaneous

4-Nitrosophenetol (98) was rapidly metabolized when incubated with human red blood cells (p. 397). While reduction of the nitroso-group to give 4-phenetidine was the major pathway two dimers, 4-ethoxy-4′-nitrosodiphenylamine (99) and the corresponding 4′-amino derivative, were formed. The mechanism of formation is thought to involve nucleophilic *para*-substitution of an activated 4-phenetidine.

C_2H_5O—〈〉—NO

(98)

C_2H_5O—〈〉—NH—〈〉—NO

(99)

Owing to the potential toxicity of nitrosamines there is an increasing interest in studying the fate of these compounds. Nitrosocimetidine (100; R = NO) is a potentially toxic metabolite of cimetidine (100; R = H) and it has been found that the whole blood of a number of species is capable of denitrosation (p. 421). Cimetidine is the main product formed and the denitrosation pathways appears to be dependent on haemoglobin cysteine.

CH_3—N—C(=N·CN)NHCH₂CH₂—S—CH₂—[imidazole]—CH₃

(100)

Denitrosation of *N*-nitrosodimethylamine has been shown to occur in rat liver microsomes, the two main products being methylamine and nitrite (p. 420). It appears that dimethylamine is probably not an intermediate since demethylation to methylamine is not a significant pathway. Formation of diphenylamine from the *N*-nitroso derivative has also been demonstrated in

13

mouse liver microsomes (p. 433) but there was no indication of the mechanism of formation.

Investigations on the mechanisms for the biotransformation of N-nitroso-N-methylaniline (101) have been carried out using rat liver and oesophageal S9 fractions (p. 431). Aniline was the major metabolite with N-methylaniline as a minor component. Evidence was obtained that the mechanism of formation of these metabolites involved an initial denitrosation to give a nitroxyl radical which was rapidly oxidized to nitrite. N-Methylaniline was demethylated to aniline and formaldehyde. A parallel pathway involved α-hydroxylation and demethylation to give the toxic reactive intermediate benzenediazonium hydroxide (102).

(101)   (102)

Introduction of unsaturation into an alkyl chain is not a common process but its occurrence could be of toxicological significance. The isoalkyl carboxylic acid valproic acid (103) is metabolized by hydroxylation at the three alkyl carbons but a novel metabolite Δ^4-VPA (104) has now been identified (p. 143) formed by microsomes from phenobarbital-treated rats. Its formation was inhibited by cytochrome P-450 inhibitors. The mechanism of formation is unknown but it was not produced from either the 4-hydroxy (105) or 5-hydroxy (106) metabolites. It is of particular interest that Δ^4-VPA is a potent hepatotoxin since there are occasional incidents of liver damage associated with valproic acid therapy.

(103)   (104)   (105)   (106)

Formation of alkenes from alkanes is a process which introduces a functional group and changes the nature of the compound. Thus the alkene function may be oxidized to an epoxide which may be of toxicological significance. Metabolites (108) of both mitometh (107; R = CH$_3$) and lysodren (107; R = H) resulting from dehydrochlorination have been identified in the rat and guinea pig (p. 122).

(107)   (108)

The ability of a compound to participate in a reversible biotransformation process may make an appreciable contribution to its pharmacokinetic profile. Oxidation and reduction processes provide well known examples as in the interconversion of sulphides and sulphoxides and alcohols and ketones.

14

Another example is with the carcinogen 2-acetylaminofluorene (109) which is metabolized to the N-sulphonoxyl derivative (110) (p. 87). It has been shown that conversion of this metabolite into acetylaminofluorene can occur in a one-step non-enzymatic reaction catalysed by rat liver cytosol. Subsequent N-hydroxylation and sulphation would lead to metabolic recycling.

(110)

(109)

Isomazole (111) is metabolized to the corresponding sulphone and sulphide with both metabolites contributing to pharmacological activity. [^{18}O]Isomazole has been used to investigate the extent of reversible metabolism of the sulphide to the sulphoxide (p. 351). Measurement of plasma concentrations of [^{18}O]- and [^{16}O]-isomazole in rats showed that there was a progressive decrease in the proportion of ^{18}O retained in the circulating isomazole from 95% at 30 minutes to 47% at 6 hours. Thus, a large proportion of the isomazole in plasma was produced by recycling of the sulphide metabolite.

(111)

Studies with a vinca alkaloid monoclonal antibody conjugate represent one of the first with a compound of this type (p. 269). It was demonstrated that the alkaloid which was linked to the antibody by a succinic acid diester bridge was released by hydrolysis after intravenous administration to animals. An N-oxide metabolite of the unconjugated alkaloid was detected in plasma at later times.

2 STEREOSPECIFIC/STEREOSELECTIVE BIOTRANSFORMATION

A major metabolite of caprolactam (112) in rats is 6-amino-γ-caprolactone (114) which equilibrates with the amino acid (113) under acidic conditions (p. 327). The lactone metabolite was shown to be optically active and since administration of 6-aminohexanoic acid did not yield this metabolite it was concluded that caprolactam undergoes stereospecific hydroxylation followed by hydrolysis and rearrangement to the lactone.

(112)   (113)   (114)

The inversion of configuration of enantiomeric arylpropionic acids has been established but the novelty of this process and its pharmacological importance for drugs containing this function make further investigations on its occurrence worthy of mention. Investigations in rabbits with ketoprofen (115) and fenoprofen (116) have shown that there was some conversion of the R- into the S-enantiometer but no conversion of S into R (pp. 193, 194). A mean of 9% of a dose of R-ketoprofen was converted into the S-enantiomer and a mean of 73% of a dose of R-fenoprofen was converted into S-fenoprofen.

(115)   (116)

For compounds where an sp^3 carbon carries two identical functional groups, biotransformation of one of these groups introduces an asymmetric carbon into the molecule and the metabolite may exist as a pair of enantiomers. Aromatic hydroxylation of phenytoin (117) is a major pathway in man. Analysis of the metabolite (118) isolated from urine has shown that it consists of 95% of the S-isomer.

(117)   (118)

Biotransformation of terodiline (119) by rat liver microsomes involves aromatic and benzylic hydroxylation although the relative importance of these pathways is different for the R/S enantiomers (p. 218). The ring-hydroxylated metabolite (120) was a major component from R-terodiline (53%) but was only a minor metabolite from the S-isomer (5%). Benzylic hydroxylation (121) was similar for both isomers, R (17%) and S (14%).

(119)   (120)   (121)

16

Budesonide (122) contains a $16\alpha,17\alpha$ unsymmetrical acetal group which is a mixture of two epimers due to the asymmetric carbon at C-22 in the acetal function. From *in vitro* studies with soluble liver enzyme preparations from rat, mouse, and man it has been shown that there are marked differences in the biotransformation of the epimers (p. 463). Thus, for the 22R-epimer, the major metabolites were those formed by loss of the acetal group (123) and hydroxylation in the B-ring (124) whereas the 22S-epimer gave (124) and a metabolite hydroxylated in the propyl side-chain. There were species differences in the rates and routes of biotransformation but man was more similar to the mouse than the rat. Evidence was also obtained that cleavage of the acetal group occurred by a two-step process involving hydroxylation (125) and hydrolysis of the resulting ester (126) to give the diol and butyric acid.

(122) (123) (124)

(125)

(126)

Two of the main biotransformation pathways for α-bromoisovalerylurea (127) are glutathione conjugation and hydrolysis of the amide function to give (128) and (129) respectively. The acid (129) is also conjugated with glutathione but in studies using rat hepatocytes it has been shown that there is a pronounced stereoselectivity for the two enantiomers (p. 119). The R-enantiomer showed a much greater rate of conjugation with glutathione whereas the S-enantiomer was substantially hydrolysed leading to formation of the conjugated acid (130).

(127)   (128)   GS = glutathionyl

(129)   (130)

17

Styrene (131) is known to be metabolized to the 7,8-oxide (132), which is a chiral compound, and phenylethylene glycol (133) is formed from this primary metabolite by the action of epoxide hydrolase. Measurement of the enantiomeric composition of the glycol excreted in urine of humans after occupational exposure to styrene showed an increase in the L/D ratio (maximum about 3) (p. 104). It was concluded that epoxide hydrolase favours formation of the L-glycol and that the D-epoxide would be likely to persist longer in the body.

(131)   (132)   (133)

The epoxidation of naphthalene has been studied *in vitro* using microsomal preparations from various species and tissues in order to obtain a biochemical explanation for tissue-selective toxicity. Thus in mouse lung preparations, the major target organ for acute toxicity in this species, there was evidence that formation of the 1R,2S-oxide (134) predominated since the glutathione conjugate (135) was a major metabolite (p. 45). In contrast there was no similar stereoselectivity observed with mouse liver and kidney preparations.

(134)   (135) GS = glutathionyl

Benzo[c]phenanthrene (136) is converted into 3,4- and 5,6-dihydrodiols by rat liver microsomes via hydration of the corresponding epoxides (p. 50). In the presence of an epoxide hydrolase inhibitor, the 5,6-epoxide (137) was shown to be enriched in the 5S,6R-enantiomer. The major dihydrodiol (138) was the 5S,6S-enantiomer, indicating that the 5S,6R-epoxide was exclusively hydrated to this compound. This result was confirmed using the purified epoxide enantiomers. However, the 5R,6S-enantiomer of (137) was hydrated to a mixture of the enantiomeric dihydrodiols.

(136)   (137)   (138)

Formation of a K-region epoxide at the 5,6-position in 12-methyl-benz[a]anthracene (139) is one of the important routes of biotransformation.

18

Formation of the 5S,6R-enantiomer (140) by rat liver microsomes predominates particularly in 3-methylcholanthrene-treated animals (p. 54). Conversion of the epoxide into a dihydrodiol also proceeds stereoselectively with a predominance of the 5S,6S-enantiomer (141) formed by hydration at C-6.

(139)   (140)   (141)

Analysis of dihydrodiol metabolites of 5-methyl and 6-methylchrysene formed by rat liver fractions has shown a high level of stereospecificity. Both the 1,2-diol (143) and 7,8-diol (144) of 5-methylchrysene (142) were shown to be exclusively of the R,R-configuration. It was demonstrated that this was not caused by a rapid metabolism of any S,S-isomer formed. A similar result was found with formation of the 1,2-diol of 6-methylchrysene. It is of interest that the R,R-diols were shown to be more tumourigenic than the S,S-enantiomers.

(142)   (143)   (144)

3 MECHANISMS OF TOXICITY

Trimethylpentane (145) is one of a group of alkanes which produce a specific type of nephrotoxicity in male but not female rats. Studies using a radio-labelled compound have shown that there is a selective retention of radio-activity in male kidneys (p. 34). The biotransformation pathways appeared to be qualitatively the same in males and females, involving hydroxylation of the secondary carbon and sequential oxidation of the terminal methyls. Three hydroxycarboxylic acids (146)–(148) appeared to be the terminal metabolites; however, the mechanism of toxicity and the involvement of a specific metabolite are unknown.

(145)   (146)

19

$$HO_2C-\underset{\underset{CH_3}{|}}{\overset{\overset{CH_3}{|}}{C}}-CH_2-\underset{}{\overset{\overset{CH_3}{|}}{CH}}CH_2OH$$

(147)

$$CH_3-\underset{\underset{CH_3}{|}}{\overset{\overset{CH_3}{|}}{C}}-CH_2-\underset{\underset{OH}{|}}{\overset{\overset{CH_3}{|}}{C}}-CO_2H$$

(148)

The biotransformation of tetrachloroethylene follows two major pathways, one being oxidative yielding trichloroacetic acid and oxalic acid as metabolites and the other involving conjugation with glutathione. The latter is believed to be responsible for the nephrotoxicity of tetrachloroethylene. The cysteine conjugate (149) is converted into dichloroacetic (150) and pyruvic (151) acids as the terminal metabolites. Formation of these metabolites is believed to involve cleavage of the cysteine conjugate to a toxic thiol intermediate.

(149)

(150)

(151)

MPTP (152) is a compound which destroys specific neurons in the brain producing clinical symptoms similar to those of Parkinson's disease. The toxicity has been attributed to oxidative biotransformation to the pyridinium compound (154). Studies with rat liver hepatocytes have shown that MPTP is converted into (153) and (154) and that the latter accumulates in cells (p. 300). Exposure of cells to (153) showed a more rapid accumulation of (154). The cytotoxicity was dose-dependent and was preceded by complete depletion of intracellular ATP.

(152)

(153)

(154)

Hydrazines are known to cause particular types of cell damage for which various mechanisms have been proposed. Formation of free radicals has been postulated, produced via N-acetylation and liberation from a reactive intermediate. Spin-trapping techniques have now been used to demonstrate formation of acetyl radicals from hydrazine and isoniazid (155) in the perfused rat liver (p. 387). Similarly, it was shown that the isopropyl radical was formed from iproniazid (156).

20

(155)   (156)

Free-radical intermediates have also been demonstrated in the biotransformation of phenelzine (157). Phenylacetaldehyde and benzaldehyde with their corresponding alcohols were identified as metabolites formed by rat liver microsomes (p. 389). Investigations of the mechanism of formation indicated that the 2-phenylethyl radical was the precursor of all the microsomal metabolites.

(157)

The interaction of isoniazid (155) with DNA has been investigated since there have been indications that it is carcinogenic in mice. Three DNA adducts have been detected in liver and lungs of mice administered oral doses of isoniazid (p. 388). One of these components, which had been identified from *in vitro* studies, was 4-deamino-4-isoniazidocytosine (158).

(158)

Studies with 3-methylindole (159) have shown that its pneumotoxicity is reduced by substitution of deuterium in the methyl group (p. 335). The deuterium-labelled compound also caused less depletion of pulmonary glutathione. The iminemethide (160) was proposed as a reactive intermediate.

(159)   (160)

The N-dealkylation of tetramethylbenzidine (161) that occurs in a peroxidase system is inhibited by glutathione. It has been shown that two diglutathionyl adducts (162) and (163) are formed in the presence of glutathione (p. 400). In the absence of glutathione, tetramethylbenzoquinone di-imine (164) was formed but was removed on addition of glutathione to give the same diglutathionyl adducts.

21

(161) (162)

(163) (164)

GS = glutathionyl

An apparent novel metabolite of *N*-methylformamide (165), a glutathione conjugate (166), has been isolated from mouse bile (p. 408). The corresponding mercapturic acid was detected as a urine metabolite. Methylamine was also found as a terminal metabolite and its formation together with that of the glutathione conjugate was subject to a primary kinetic isotope effect when the formyl hydrogen was replaced with deuterium. Thus, the mechanism involves a primary oxidative process to give an intermediate capable of reacting with glutathione. Methyl isocyanate was postulated as a possible intermediate.

(165)

(166) GS = glutathionyl

Clomiphene (167) is converted into several ring-hydroxylated metabolites on incubation with rat liver microsomes. The 4'-hydroxy metabolite (168) was shown to react readily with alcohols by displacement of the chlorine. It also reacted with γ-(*p*-nitrobenzyl)pyridine as a test for electrophilic activity. Thus, the chemical reactivity of this metabolite may indicate a potential mechanism for toxicity.

(167)

(168)

The nitrosamine (169) is a potential carcinogen present in tobacco smoke. Metabolic activation of this compound is thought to be an initiating stage in the carcinogenic process leading to a reactive species capable of alkylating

22

DNA and protein. Although reaction with protein is of less toxicological significance, measurement of the formation of these adducts could provide an index of carcinogen exposure. After administration of the nitrosamine to rats, about 0.1% of the dose was bound to haemoglobin (p. 427). Acid or alkaline hydrolysis released 10–15% of this material, of which the main component was identified as 4-hydroxy-1-(3-pyridyl)butan-1-one (172). It was postulated that the activation pathway involved α-hydroxylation to (170) followed by formation of the diazohydroxide (171), which reacts with nucleophilic sites in haemoglobin.

(169)   (170)

(172)   (171)

When α-hydroxylation occurs at the methylene adjacent to nitrogen this leads to formation of the methylating agent methyldiazohydroxide. The occurrence of this pathway has been investigated by measurement of O^6-methylguanine in the DNA from lung cells of rats after intraperitoneal doses of (169) (p. 429). Higher concentrations of the methylated adduct were found in Clara cells which are known for their localization of cytochrome P-450 activity. The extent of alkylation was about twice that obtained from equimolar doses of nitrosodimethylamine.

N-Methyl-4-aminoazobenzene (173) is a known carcinogen in rats and studies have been carried out to establish whether there is a correlation of covalent binding to DNA with the observed toxicity (p. 401). After administration of the tritiated compound in the diet, DNA was isolated from liver and, following enzymatic hydrolysis, radiolabelled nucleoside adducts were separated by HPLC. The two major adducts identified were those linked to either the N^2 or C-8 positions of guanine (174) and (175). In a parallel experiment using non-labelled material, sixteen of thirty animals developed hepatocellular carcinoma and it was postulated that formation of the N^2-guanine adduct provided the best correlation with tissue carcinogenic specificity.

(173)

(174)   (175)

Etoposide is a complex polycyclic glucoside which also contains a di-methoxyphenol. *O*-Demethylation of this group forms an *o*-hydroquinone. Formation of this metabolite (176) was demonstrated after incubation with mouse liver microsomes and evidence was also obtained for a semiquinone radical (177) as an intermediate in its oxidation to the quinone (178). It is postulated that the quinone or the free radical are reactive entities which bind covalently to microsomal proteins.

(176)   (177)   (178)

The carcinogenicity of polycyclic aromatic hydrocarbons is believed to involve, in many cases, covalent binding of a reactive intermediate to nucleic acid as an initiating event. Benzylic methyl groups can be oxidized to an alcohol which is capable of being activated to form an electrophilic species. This process has been investigated using the hydroxymethylbenzanthracene (179) where good evidence was obtained for formation of the sulphate ester as the reactive intermediate (p. 58). Hepatic nucleic acid was isolated from rats dosed with the benzanthracene and two nucleoside adducts were identified after enzymatic hydrolysis. The deoxyguanosine (180) and deoxyadenosine (181) adducts were formed in larger amounts when the sulphate ester was administered and they could be synthesized by reaction of the ester with the nucleosides.

(179)   (180)   (181)

X = Benzanthracenylmethyl

7,12-Dihydroxymethylbenz[*a*]anthracene has been shown to form the highly mutagenic 7-sulphate (182) in rat liver cytosol which can be inactivated by reaction with glutathione. Two purine adducts were isolated from

24

calf thymus DNA after incubation with the 7-sulphate and identified as the N^6-adenine and N^2-guanine adducts (p. 56).

CH_2OH

$CH_2OSO_3^-$

(182)

The covalent binding of 6-nitrochrysene (183) and the 6-amino- and 6-hydroxyamino-derivatives to DNA has been investigated with rat hepatocytes (p. 97). Three major DNA adducts were identified including the N-deoxyguanosine (184) and N-deoxyinosine (185) compounds. The latter was thought to arise from spontaneous oxidation of the corresponding deoxyadenosine adduct during sample preparation. Formation of this adduct and its oxidation is a new reaction involving N-hydroxylamines and DNA.

NO_2

(183)

(184)

(185)

4 MISCELLANEOUS

Benzarone (186) contains a phenolic function which is available for conjugation with glucuronic acid or sulphate and it might be expected that this would be an important elimination pathway for this drug. In rat and dog oral doses were well absorbed and excreted predominantly as a glucuronide conjugate of the parent (p. 352) in bile and urine. However, in man, Phase I biotransformation predominated with mainly the formation of two hydroxylated metabolites (187) and (188), which were ultimately excreted as glucuronides.

25

(186)

(187)

(188)

With the development of new techniques in mass spectrometry it is now a more feasible proposition to identify intact polar conjugates. This allows greater confidence in structural assignments and may also provide information on the position of conjugation when there are two or more functional groups available. In man, butolol (189) undergoes direct conjugation with glucuronic acid at the secondary alcohol function but also aromatic hydroxylation (p. 242). The resulting phenol was also eliminated as a conjugate with glucuronic acid at the aliphatic hydroxyl rather than the phenolic group. *In vitro* incubation of the phenol with glucuronyl transferase from rat, rabbit, and bovine liver microsomes resulted in formation of the phenolic glucuronide. It is possible that the regioselectivity of human glucuronyl transferase is different to that of animals.

(189)

^{19}F is an isotope which can be detected by NMR spectroscopy and some investigations on the biotransformation of flucytosine (190) in man have been performed using the ^{19}F-labelled compound (p. 313). By the use of ^{19}F NMR a glucuronide conjugate of flucytosine and a hydroxylated metabolite (191) were identified.

(190)

(191)

^{18}F s a short-lived radioisotope which can be useful for following the fate of organofluorine compounds and any fluoride released as a result of metabolism. The biotransformation of [^{18}F]fluoroacetate has been studied in mice and the presence of [^{18}F]fluoride in plasma demonstrated by chromatography (p. 135). It was proposed that a large proportion of the administered fluoroacetate was degraded to fluoride.

Hydrocarbons

n-Hexane, Cyclohexane

Use:	Solvent
Key functional groups:	Alkane, cycloalkane
Test system:	Human (occupational exposure)

Structure and biotransformation pathway

(1) (2) (3) (4) (5)

The urine of thirty female and ten male factory workers who were occupationally exposed to a mixture of hexanes was monitored for the presence of metabolites. Urine was collected from each individual at the end of a weekly shift. Hexane-2,5-dione (2) and γ-valerolactone (3) were the major metabolites of n-hexane that were detected and were present in urine from all forty workers. Hexan-2-ol (1) was also found but only in the urines of eleven workers and was a minor metabolite. γ-Valerolactone is probably not a true metabolite but an artefact caused by cyclization of a hydroxylated carboxylic acid metabolite during gas chromatographic analysis. Cyclohexanol (4) and cyclohexanone (5), which were formed from cyclohexane, were found in the urine of nine and seven workers respectively. Metabolites were identified by gas chromatography using a capillary column.

Reference

M. Governa, R. Calisti, G. Coppa, G. Tagliavento, A. Colombi, and W. Troni, Urinary excretion of 2,5-hexanedione and peripheral polyneuropathies in workers exposed to hexane, *J. Toxicol. Env. Health*, 1987, **20**, 219.

n-Hexane, Hexane-2,5-dione

Use/occurrence:	Solvent, gasoline constituent (n-hexane)
Key functional groups:	Alkane, alkyl ketone
Test system:	Rat (inhalation, 2000 p.p.m., for n-hexane, intraperitoneal, 200 mg kg^{-1}, for hexane-2,5-dione)

Structure and biotransformation pathway

$$CH_3CH_2CH_2CH_2CH_2CH_3 \longrightarrow CH_3CCH_2CHCHCH_3 \longleftarrow CH_3CCH_2CH_2CCH_3$$

(1) → (3) ← (2)

The metabolism of n-hexane (1) has been previously shown to lead to hexane-2,5-dione (2), which is a neurotoxic compound that is believed to act via the formation of pyrrole adducts with proteins. However, there was evidence for a urinary 'preformed hexane-2,5-dione' which could be converted into (2) by acid hydrolysis.

Rats were dosed intraperitoneally with (2) and the urine was collected. Glucuronide conjugates were hydrolysed with β-glucuronidase and the products analysed by GC–MS as their methoxime derivatives. In addition to derivatives of (2), three extra compounds were detected, and identified as isomers of the methoxime of 4,5-dihydroxyhexan-2-one (3). As there should be four enantiomeric pairs for this derivative it appears that they were incompletely resolved. Further derivatization of the hydroxy-groups by trimethylsilylation did, however, yield four products. $[^2H_{10}]$Hexane-2,5-dione was metabolized to $[^2H_9]$-(3).

Rats were exposed to (1) at 2000 p.p.m. for 8 hours, and (3) was again found after enzymic hydrolysis of the glucuronide fraction. Acidic hydrolysis of this fraction gave only (2) (identified as its methoxime derivative). Thus compound (3) appears to be identical to 'preformed hexane-2,5-dione'.

Reference

N. Fedtke and H. M. Bolt, 4,5-Dihydroxy-2-hexanone: a new metabolite of n-hexane and of 2,5-hexanedione in rat urine, *Biomed. Environ. Mass Spectrom.*, 1987, **14**, 563.

2,3,4-Trimethylpentane

Use/occurrence:	Gasoline constituent
Key functional groups:	Alkane
Test system:	Rat (oral)

Structure and biotransformation pathway

Rats were administered 2,3,4-trimethylpentane (1) by gavage (7×1 ml over 14 days). Urinary metabolites were deconjugated with β-glucuronidase and sulphatase, purified on ClinElut columns, and analysed by GC or GC–MS, either underivatized or as trimethylsilyl derivatives. The metabolites were identified by comparison with authentic compounds as 2,3,4-trimethylpentan-1-ol (2), 2,3,4-trimethylpentan-2-ol (3), and 2,3,4-trimethylpentan-1-oic acid (4). No attempts were made to determine the diastereoisomeric composition of the metabolites. It is postulated that the nephrotoxic effects of (1) may be due to the interaction of metabolites with $\alpha_2 u$ globulin.

References

K. O. Yu, C. T. Olson, D. W. Hobson, and M. P. Serve, Identification of urinary metabolites in rats exposed to the nephrotoxic agent 2,3,4-trimethylpentane, *Biomed. Environ. Mass Spectrom.*, 1987, **14**, 639.

C. T. Olson, D. W. Hobson, K. O. Yu, and M. P. Serve, The metabolism of 2,3,4-trimethylpentane in male Fischer-344 rats, *Toxicol. Lett.*, 1987, **37**, 199.

2,2,4-Trimethylpentane

Use/occurrence: Component of unleaded gasoline

Key functional groups: Isoalkane

Test system: Rat (oral, 4.4 mmol kg^{-1})

Structure and biotransformation pathway

The disposition of 2,2,4-trimethylpentane in male and female rats was studied with a view to understanding why 2,2,4-trimethylpentane produces nephrotoxicity in the male rat but not in the female rat. The radioactivity in the 0–48 hour urine of both male and female rats accounted for

34

approximately 30% of the dose. Both male and female rats metabolize 2,2,4-trimethylpentane via the same pathway and at a similar rate. Female rats, however, excrete more conjugates of 2,4,4-trimethylpentan-2-ol (1) in urine than the males. Metabolite concentrations were determined by GC, and conjugates were treated with sulphatase and β-glucuronidase enzymes. A selective retention of radioactivity was seen in the kidneys of the male rat. These observations were believed to support the hypothesis that a kidney-specific accumulation of a metabolite is responsible for the male rat nephrotoxicity. A concomitant accumulation in renal α_{2u}-globulin was also seen and a metabolite complex was proposed. All the metabolites were formed by hydroxylation of the secondary carbons (1,4,7,10) and sequential oxidation of terminal methyls $(2 \rightarrow 5 \rightarrow 8 \rightarrow 11)$

Reference

M. Charbonneau, E. A. Lock, J. Strasser, M. G. Cox, M. J. Turner, and J. S. Bus, 2,2,4-Trimethylpentane-induced nephrotoxicity. 1. Metabolic disposition of TMP in male and female Fischer-344 rats, *Toxicol. Appl. Pharmacol.*, 1987, **91**, 171.

t-Butylcyclohexane

Use/occurrence:	Model compound (fuel and solvent component)
Key functional groups:	Cyclohexane, t-butyl
Test system:	Male rats (oral, 0.8 g kg^{-1})

Structure and biotransformation pathway

Compound	R^1	R^2	R^3	R^4	R^5	R^6	Relative molar abundancy
(1)	CH_3	CH_3	H	H	H	H	—
(2)	CH_3	CH_3	H	H	H	OH	1.0
(3)	CH_3	CH_3	H	H	OH	H	53.8
(4)	CO_2H	CH_3	H	H	H	H	12.2
(5)	CH_2OH	CH_2OH	H	H	H	H	5.0
(6)	CH_3	CH_3	H	OH	H	OH	15.4
(7)	CH_3	CH_3	OH	H	H	OH	5.7
(8)	CH_3	CH_3	H	OH	OH	H	4.8

t-Butylcyclohexane (1) was chosen for investigation for toxicological reasons — it is known to produce renal damage in male rats, and for structural reasons — it contains primary, secondary, and tertiary aliphatic carbons which may be preferentially oxidized. Additionally, because of the proclivity of the t-butyl group to remain in the equatorial position of the cyclohexane chain conformation, attack at the other ring carbons would yield non-intraconvertible *cis-* and *trans-*isomers, giving further information on preferential sites of attack by oxidative enzymes. Urine, collected during 48 hours after dosing, was incubated with glucuronidase/sulphatase, and dichloromethane extracts were prepared and analysed by GC and GC–MS. Quantification was by GC, using dodecane as an internal standard, analytical standards of each identified metabolite (2)–(8) being available. The majority of oxidative metabolism of (1) occurred on the cyclohexane ring yielding compounds (2), (3), (6), (7), and (8), and the 4-position was the preferred site of attack. The major phase 1 metabolite was *trans-*4-t-butylcyclohexanol (3). Stereochemical aspects of the metabolism are discussed in the paper. Oxidation of the t-butyl group led to compounds (4) and (5). The hypothetical precursor of both of these metabolites, 2-methyl-2-

36

cyclohexylpropanol, was not detected, suggesting that, if formed, it was rapidly metabolized. It was suggested by analogy with known metabolism and toxicological properties of other hydrocarbons, including 2,2,4-trimethylpentane, that the nephrotoxicity of (1) was a consequence of its extensive oxidative metabolism.

Reference

G. M. Henningsen, K. O. Yu, R. A. Solomon, M. J. Ferry, I. Lopez, J. Roberts, and M. P. Servé, The metabolism of t-butylcyclohexane in Fischer-344 male rats with hyaline droplet nephropathy, *Toxicol. Lett.*, 1987, **39,** 313.

Dodecylcyclohexane

Use/occurrence:	Petroleum hydrocarbon
Key functional groups:	Alkylcyclohexane, cyclohexane
Test system:	Trout (*Salmo gairdneri*) (oral, 5 mg)

Structure and biotransformation pathway

The urinary duct of rainbow trout was cannulated and [G-^{3}H]dodecylcyclohexane (1) was administered orally. Fourteen per cent of the dose was excreted in urine over 72 hours and TLC analysis of a neutral extract (ethyl acetate) showed the presence of four radioactive components, which were compared with reference compounds. Cyclohexylacetic acid (2; 5% of urine radioactivity), the ethyl ester of 1-hydroxycyclohexylacetic acid (3; 15% of urine radioactivity), 3-hydroxycyclohexylacetic acid (4) and 4-hydroxycyclohexylacetic acid (5), were identified and the structures confirmed by GC–MS.

Glucuronide conjugates of phenylacetic acid (6) and cyclohexylacetic acid (2) were also isolated. More than 30% of the urinary radioactivity was identified as the taurine conjugates of these two acids. These results indicate that the metabolites (7) and (8) formed by hydroxylation of the cyclohexane ring, previously identified *in vitro*, are subject to further biotransformation before excretion occurs.

Reference

J.-P. Cravedi and J. Tulliez, Urinary metabolites of dodecylcyclohexane in *Salmo gairdneri*: evidence of aromatisation and taurine conjugation in trout, *Xenobiotica*, 1987, **17**, 1103.

n-Butylbenzene, n-Hexylbenzene, n-Octylbenzene

Use/occurrence:	Model compounds
Key functional groups:	Alkylphenyl
Test system:	Rainbow trout

Structures and biotransformation pathway

n-Butylbenzene

(1)

n-Hexylbenzene

(2)

(3)

(4)

n-Octylbenzene

(5)

(7)

(6)

(8)

Rainbow trout were dosed by mouth with either n-butylbenzene, n-hexylbenzene, or n-octylbenzene. Fish were killed 120 hours after dosing and a sample of bile was taken. The biles were solvent extracted and the aqueous residues subject to β-glucuronidase hydrolysis prior to further extraction. GC–MS analysis of the hydrolysed bile from rats dosed with n-butylbenzene indicated the presence of (1) as the major metabolite.

6-Phenylhexan-1-ol (2) was the major metabolite in the bile of fish dosed with either n-hexylbenzene or n-octylbenzene. The corresponding acid metabolite (3) was found in the bile of fish dosed with n-hexylbenzene as were metabolites (4) and (7). For n-octylbenzene metabolites (5), (6), and (8) were also found. Thus primary and secondary alcohols are formed rather than phenols, oxidation taking place at either end of the aliphatic side-chain.

Reference

J. Hellou and A. King, Metabolism of n-butyl-, n-hexyl-, and n-octyl-benzene in the bile of rainbow trout, *Bull. Environ. Contam. Toxicol.*, 1987, **39**, 182.

5-(2-Dodecylphenyl)-4,6-dithianonanedioic acid (SKF 102 081)

Use/occurrence:	Anti-inflammatory agent
Key functional groups:	Alkyl carboxylic acid, dialkyl thioether, alkylphenyl
Test system:	Guinea pig (oral, 130 mg kg^{-1}) *in vitro* liver and kidney microsomes

Structure and biotransformation pathway

40

Following intravenous administration of [^{14}C]-SKF 102 081 to guinea pigs, 85% of the dose was excreted in the bile in 1 hour. Only 6% of the radioactivity was associated with parent compound and at least 14 metabolites were present in bile. The major metabolite (5) accounted for 46% of the material present in bile, the precursor (ω) oxidation product (2) was also found in bile.

The ($\omega - 1$) oxidation products (7) and (8) accounted for 10% of the radioctivity present in bile and a further 16% was accounted for by the glucuronides of the parent (9) and (10). Other metabolites found were the ($\omega - 2$), ($\omega - 3$), ($\omega - 4$), and ($\omega - 5$) oxidation products (11), (12), (13), and (14) respectively. Metabolites were identified by mass spectrometry and NMR.

The use of the *in vitro* test system enabled a comparison of the metabolism in liver and kidneys of a wide range of species. Differential inhibition studies demonstrated that (ω) and ($\omega - 1$) hydroxylations were mediated by different isozymes of cytochrome P-450. There were quite marked differences between the species in the degree of these hydroxylations; in the human liver microsomes (ω) hydroxylation predominated, whereas in the guinea pig ($\omega - 1$) hydroxylation predominated. The activity of kidney microsomes was less than that of the liver microsomes in each species studied.

References

J. F. Newton, K. M. Straub, R. H. Dewey, C. D. Perchonok, T. B. Leonard, M. E. McCarthy, J. G. Gleason, and R. D. Eckardt, *In vitro* microsomal metabolism of the leukotriene receptor antagonist 5-(2-dodecylphenyl)-4,6-dithianonanedioic acid (SKF 102 081), *Drug Metab. Dispos.*, 1987, **15**, 161.

J. F. Newton, K. M. Straub, G. Y. Kuo, C. D. Perchonok, M. E. McCarthy, J. G. Gleason, and R. K. Lynn, *In vivo* metabolism of the leukotriene receptor antagonist 5-(2-dodecylphenyl)-4,6-dithianonanedioic acid (SKF 102 081) in the guinea pig, *Drug Metab. Dispos.*, 1987, **15**, 168.

Indane

Use/occurrence:	Coal tar
Key functional groups:	Indane
Test system:	Rat (oral)

Structure and biotransformation pathway

Rats were administered indane (1) by gavage (7×0.25 ml kg^{-1} over 14 days). Urine was collected and metabolites of (1) were isolated after deconjugation with β-glucuronidase and sulphatase by chromatography on Clin-Elut columns. Identification was by GC or GC–MS of the underivatized or trimethylsilylated metabolites and comparison with authentic materials. The main metabolites were 2- and 3-hydroxyindan-1-ones (2) and (3). Seven other less abundant metabolites identified were indan-1-ol (4), *cis*-indane-1,2-diol (5), indan-2-one (6), indan-1-one (7), indan-2-ol (8), indan-5-ol (9), and *trans*-indane-1,2-diol (10).

Reference

K. O. Yu, C. T. Olson, M. J. Ferry, and M. P. Serve, Gas chromatographic/mass spectrometric studies of the urinary metabolites of male rats given indane, *Biomed. Environ. Mass Spectrom.*, 1987, **14**, 649.

Ketamine, 6-Hydroxyketamine

Use:	Anaesthetic
Key functional groups:	Tertiary alkylamine, chlorophenyl, cyclohexanone
Test system:	Rat, rabbit, and human liver microsomes

Structure and biotransformation pathway

Norketamine (1) was the major metabolite found following incubation of ketamine with either rat or rabbit liver microsomes. Small amounts of Z-6-hydroxynorketamine (2) were also present. Phenobarbital-induced rat and rabbit liver microsomes effected greater rates of metabolism than the untreated microsomes. In the rat three additional metabolites were produced using the induced microsomes; these were 4- and 5-hydroxynorketamine (3) and (4) respectively and either Z- or E-6-hydroxyketamine (5). The same 4-hydroxynorketamine isomer (3) was also produced by induced rabbit liver microsomes together with a different isomer of 5-hydroxynorketamine than the one detected in the rat.

Incubation of either the Z- or E-forms of 6-hydroxyketamine (5) with control or phenobarbital-induced rat or rabbit liver microsomes produced only the corresponding demethylated metabolite (2). Six metabolites were identified after incubation of ketamine with human liver microsomes. These were norketamine (2), 4-hydroxynorketamine (3) (the same isomer as found in the rat and rabbit), both 5-hydroxynorketamine isomers (4), Z-6-hydroxynorketamine (2), and 4-hydroxyketamine (6). 5,6-Dehydro-norketamine was not detected, suggesting that it was found as an artefact in previous studies. Since both 6-hydroxy isomers were shown to be stable, the 5,6-dehydro product was probably formed by dehydration of the 5-hydroxy metabolite. Metabolites were identified by GC and mass spectral fragmentation characteristics.

Reference

T. F. Woolf and J. D. Adams, Biotransformation of ketamine, *Z*-6-hydroxyketamine, and *E*-6-hydroxyketamine by rat, rabbit, and human liver microsomal preparations, *Xenobiotica*, 1987, **17,** 839.

Naphthalene

Use/occurrence:	Industrial chemical
Key functional groups:	Naphthalene, aromatic hydrocarbon
Test system:	Mouse, rat, and hamster pulmonary, hepatic, and renal microsomal enzymes

Structure and biotransformation pathway

The objective of this study was to look for a biochemical basis for the known species and tissue-selective toxicity of naphthalene. Three isomeric hydroxyglutathionyldihydronaphthalene metabolites (1), (2), and (3), all with the *trans*-configuration, were identified as *in vitro* metabolites of [^{14}C]naphthalene. Structural characterization was based on FAB-MS, ^{1}H NMR, and chemical synthesis of the metabolites from the separate (1S,2R)- and (1R,2S)-naphthalene 1,2-oxide enantiomers (5) and (6). The other possible conjugate (4) could not be detected. Incubation of naphthalene, glutathione, and glutathione transferases (from mouse liver cytosol) with pulmonary, hepatic, or renal microsomes resulted in the formation of all three conjugates in all three species, but substantial differences in rates of

45

formation of individual conjugates were observed. In mouse lung preparations (the major target organ for acute toxicity in this species) the (1R,2S)-naphthalene 1,2-oxide isomer (6) was the predominant enantiomer (10:1 ratio) formed since the glutathione adduct (2) was the predominant conjugate. By contrast no such stereoselectivity was observed in mouse liver or kidney preparations, although the rate of metabolism in kidney preparations was low. Total rates of naphthalene metabolism by rat and hamster lung and rat liver preparations were lower than observed in the mouse. In lung preparations, in contrast to the mouse, the (1S,2R)-oxide isomer (5) was formed in slight excess as judged by the ratio of the glutathione adducts formed. This isomer was also apparently favoured, in an approximate 4:1 ratio in rat and hamster liver preparations, although for the hamster this result could not be relied upon, as the dihydrodiol (7) was the major metabolite. The view that the stereochemistry of epoxidation is related to tissue-selective injury resulting from naphthalene administration was supported by the data obtained with mouse lung preparations but additional studies would be needed to test the basis for recently reported renal toxicity in the mouse.

Reference

A. R. Buckpitt, N. Castagnoli Jr., S. D. Nelson, A. D. Jones, and L. S. Bahnson, Stereoselectivity of naphthalene epoxidation by mouse, rat, and hamster pulmonary, hepatic, and renal microsomal enzymes, *Drug Metab. Dispos.*, 1987, **15**, 491.

2-Isopropylnaphthalene

Use/occurrence:	Industrial chemical
Key functional groups:	Isoalkylphenyl, naphthalene
Test system:	Rabbit (oral, 100 mg kg^{-1})

Structure and biotransformation pathway

This study was carried out using non-radiolabelled (1). The identified metabolites, excluding (6), but including conjugates, accounted for about 29% dose in 0–24 hour urine. The major biotransformation pathway involved oxidation of the isopropyl side-chain to yield metabolites (2)–(5) of which (2) (6.5% dose) and (3) (16.0% dose) were the most important. Conjugates of (2) and (3), capable of hydrolysis by β-glucuronidase, were also present but in much lower amounts than free aglycones. Compounds (2)–(5) were identified by MS, IR, and chromatographic comparison with the same metabolites isolated from rat urine for which structural characterization had been reported previously. MS, NMR, and IR data were obtained and interpreted in terms of a dihydrodiol structure for metabolite (6) but were insufficient to distinguish between the alternative 5,6- and 7,8-substitution patterns. Metabolite (6) and its conjugate accounted for 1.6% dose. No quantitative data were presented for the two naphthol metabolites (7a) and (7b), and as only MS data were obtained the positions of substitution could not be assigned.

Reference

T. Honda, A. Fukada, M. Kiyozumi, and S. Kojima, Identification and determination of urinary metabolites of 2-isopropylnaphthalene in rabbits, *Eur. J. Drug Metab. Pharmacokinet.*, 1987, **12**, 11.

Polycyclic Aromatic Hydrocarbons

Benzo[c]phenanthrene

Use/occurrence:	Model compound
Key functional groups:	Polycyclic aromatic hydrocarbon
Test system:	Rat liver microsomes

Structure and biotransformation pathway

(2)   (1)   (3)

(4)   (5)

Metabolism of benzo[c]phenanthrene (1) by liver microsomes (control, phenobarbital induced, 3-methylcholanthrene induced, Aroclor 1254 induced) from rats (Sprague–Dawley or Long–Evans) yielded benzo[c]-phenanthrene-*trans*-3,4-dihydrodiol (4) and benzo[c]phenanthrene-*trans*-5,6-dihydrodiol (5), which were separated by HPLC. The major enantiomer of (5), determined by chiral stationary phase HPLC and exciton chirality CD spectra of its bis-p-N,N-dimethylaminobenzoate, was 5S,6S (78–86%). In the presence of an epoxide hydrolase inhibitor (3,3,3-trichloropropylene 1,2-oxide) the benzo[c]phenanthrene 5,6-epoxide (3) was isolable and shown to be enriched in the (5S,6R) enantiomer (58–72%). This suggested that (5S,6R)-(3) is exclusively enzymatically hydrated to (5S,6S)-(5) and this was confirmed using the purified enantiomers. However, the (5R,6S)-enantiomer of (3) is hydrated to a mixture of 32% (5S,6S)- and 68% (5R,6R)-(5). For the 3,4-epoxide (2) there was evidence that the favoured enantiomer was (3S,4R) although some racemization occurred.

Reference

S. K. Yang, M. Mushtaq, and H. B. Weems, Stereoselective formation and hydration of benzo[c]phenanthrene 3,4- and 5,6-epoxide enantiomers by rat liver micro-somal enzymes, *Arch. Biochem. Biophys.*, 1987, **255**, 48.

Benzo[c]phenanthrene

Use/occurrence:	Model compound
Key functional groups:	Polycyclic aromatic hydrocarbon
Test system:	Rat liver microsomes

Structure and biotransformation pathway

The weak carcinogen benzo[c]phenanthrene (1) is known to be metabolized (partially) to a *trans*-3,4-dihydrodiol (2) and (3), which is further metabolized to the 3,4-dihydrodiol 1,2-epoxide that is believed to be the active carcinogenic species.

An earlier report described the enantiomeric composition (3S,4S):(3R,4R), of the 3-4-dihydrodiol to the 11:89 (control rats), 20:80 (phenobarbital pretreated rats), and 6:94 (3-methylcholanthrene pretreated rats). This paper presents different results for this metabolite.

(1) was incubated with liver microsomes from rats (control, phenobarbital, 3-methylcholanthrene, and Aroclor 1254 treated) at a concentration of 80 μM. The *trans*-3,4-dihydrodiol isomers (2) and (3) were extracted and purified by normal phase HPLC. Their enantiomeric composition was determined by CD and chiral stationary phase HPLC. At low concentrations of microsomes there was a decrease in the amount of the (3R,4R)-isomer (2) formed, and variations were also seen when the incubation time was altered. No difference was seen between Sprague–Dawley and Long–Evans rats. Ratios of (3S,4S):(3R,4R) for control rats were in the range 64–71%, for phenobarbital treated rats 42–54%, for 3-methylcholanthrene treated rats 0–6%, and for Aroclor 1254 treated rats 1–10%. No explanation was given for the difference from the earlier reported values. Other metabolites that were identified included the *trans*-5,6-dihydrodiol, phenols of (1), and a phenol of *trans*-5,6-dihydrodiol.

Reference

M. Mushtaq and S. K. Yang, Stereoselective metabolism of benzo[c]phenanthrene to the procarcinogenic *trans*-3,4-dihydrodiol, *Carcinogenesis*, 1987, **8**, 705.

Benzo[c]phenanthrene

Use/occurrence:	Model compound
Key functional groups:	Polycyclic aromatic hydrocarbon
Test system:	Rat liver microsomes

Structure and biotransformation pathway

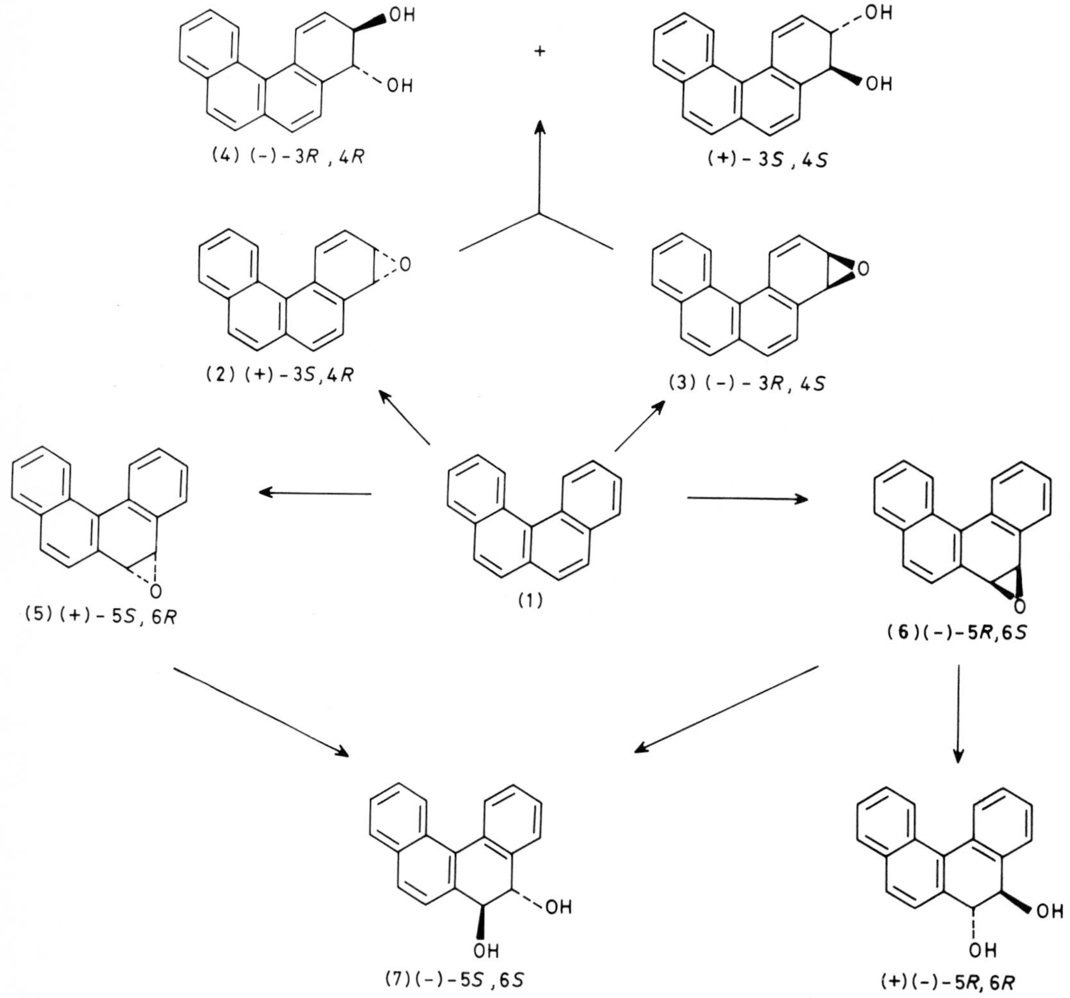

(4) (−)−3R , 4R

(+)− 3S , 4S

(2) (+)−3S,4R

(3) (−)− 3R, 4S

(5) (+)− 5S, 6R

(1)

(6)(−)−5R,6S

(7)(−)−5S ,6S

(+)(−)−5R, 6R

The stereoselective formation of the principal arene oxide metabolites of benzo[c]phenanthrene (1) by cytochrome P-450c from rat liver microsomes has been investigated. The enantiomeric composition of the two major metabolites 3,4- and 5,6-oxides [(2), (3) and (5), (6) respectively] was determined by trapping the oxides as thiolate adducts formed from N-acetyl-

L-cysteine. A total of eight non-enantiomeric adducts were formed from the two arene oxides and HPLC conditions were developed for their separation. Enantiomerically pure arene oxides were prepared as authentic reference compounds. Analysis of adducts from the 5,6-oxides indicated that the (+)-(5S,6R) (5) and (−)-(5R,6S) (6) enantiomers were formed in the ratio 76:24. Correspondingly, the (+)-(3S,4R)-oxide (2) was favoured over the (−)-(3R,4S)-oxide (3) in the ratio 90:10. The enantiomer composition of 3,4- and 5,6-dihydrodiols formed from hydrolysis of the epoxides was similar using microsomes from 3-methylcholanthrene treated rats and reconstituted cytochrome P-450c and epoxide hydrolase. There was pronounced stereoselectivity in the formation of dihydrodiols with the (+)-(3S,4R) (2) and (+)-(5S,6R) (5) oxides yielding 80–90% of the (−)-(3R,4R) (4) and (−)-(5S,6S) (7) dihydrodiols. A steric model for the catalytic binding site of cytochrome P-450c was also discussed.

Reference

P. J. van Bladeren, S. K. Balani, J. M. Sayer, D. R. Thakker, D. R. Boyd, D. E. Ryan, P. E. Thomas, W. Levin, and D. M. Jerina, Stereoselective formation of benzo[c]phenanthrene (+)-(3S,4R)- and (+)-(5S,6R)-oxides by cytochrome P-450c in a highly purified and reconstituted system, *Biochem. Biophys. Res. Commun.*, 1987, **145**, 160.

12-Methylbenz[*a*]anthracene

Use/occurrence:	Model compound
Key functional groups:	Benzanthracene, polycyclic aromatic hydrocarbon
Test system:	Rat liver microsomes

Structure and biotransformation pathway

The 5,6-epoxides (2) and (3) of 12-methylbenzanthracene (1) formed by rat liver microsomes in the presence of an epoxide hydrolase inhibitor were isolated by normal phase HPLC and their $(5S,6R):(5R,6S)$ enantiomer ratios were measured after separation by chiral stationary phase HPLC. With microsomes from untreated and phenobarbital treated animals formation of the $(5S,6R)$-enantiomer (2) was favoured (about 3:1) which was even more pronounced with microsomes from 3-methylcholanthrene treated animals (99:1). Absolute configurations of the epoxide enantiomers were determined by high resolution NMR and CD of the two isomeric methoxylation products derived from each of the two epoxides. The K-region *trans*-5,6-dihydrodiols (4) and (5) were analysed by chiral stationary phase HPLC which showed a predominance (about 19:1) of the $(5S,6S)$-enantiomer (4) compared with the $(5R,6R)$-enantiomer. Thus the $(5S,6R)$-epoxide (2) and the $(5S,6S)$-dihydrodiol (4) were the major enantiomers preferentially formed in the metabolism of the K-region 5,6-double bond.

The $(5S,6R)$-epoxide was hydrated predominantly at C-6 while the $(5R,6S)$-epoxide was hydrated nearly equally at both C-5 and C-6 to form an approximately equal mixture of the 5,6-dihydrodiols (4) and (5).

Reference

S. K. Yang, M. Mushtaq, H. B. Weems, D. W. Miller, and P. P. Fu, Stereoselective formation and hydration of 12-methylbenz[*a*]anthracene 5,6-epoxide enantiomers by rat liver microsomal enzymes, *Biochem. J.*, 1987, **245**, 191.

7,12-Dimethylbenz[*a*]anthracene

Use/occurrence:	Environmental carcinogen
Key functional groups:	Arylmethyl, benzanthracene, polycyclic aromatic hydrocarbon
Test system:	Mouse mammary cells, mouse mammary microsomes

Structure and biotransformation pathway

Diols (5,6-; 8,9-; 3,4-)

Phenols (2-; 3-; 4-)

Hydroxymethyls (7-; 12-)

(1)

7,12-Dimethylbenzanthracene (DMBA) (1) is a carcinogen with pronounced effects on the mammary gland in animals. The metabolism of DMBA by liver microsomal systems has been well studied and leads to a wide variety of diols, including diol epoxides which may be the active carcinogenic and mutagenic species. The metabolism of [³H]-DMBA has now been studied with mammary organs in culture and with mouse mammary microsomes. Microsomes produced DMBA phenols, but only trace amounts of diols (*i.e.* no significant quantities of the epoxides). Cultured mammary glands metabolized [³H]-DMBA (to the extent of 46% in the organ culture medium) to a mixture of DMBA 5,6-diol, 8,9-diol, 3,4-diol, 2-phenol, 3-phenol, 4-phenol, 12-hydroxymethyl, and 7-hydroxymethyl derivatives. About 95% of the DMBA-derived radioactivity was associated with the mammary fat pad and dispersion of this released DMBA, phenols, and diols. The metabolites were all analysed by reverse phase HPLC.

Reference

R. Menon, J. Bartley, S. Som, and M. R. Banerjee, Metabolism of 7,12-dimethyl-benz[*a*]anthracene by mouse mammary cells in serum-free organ culture medium, *Eur. J. Cancer Clin. Oncol.*, 1987, **23**, 395.

7,12-Dihydroxymethylbenz[*a*]anthracene

Use/occurrence:	Carcinogenic metabolite
Key functional groups:	Arylhydroxymethyl, benzanthracene
Test system:	Calf thymus DNA (*in vitro*)

Structure and biotransformation pathway

The present study examined the covalent binding of 7,12-dihydroxymethylbenz[*a*]anthracene 7-sulphate (2) to calf thymus DNA, inhibition of such binding by rat liver glutathione (GSH) transferases in the presence of GSH, and isolation and identification of the major carcinogen–DNA adducts by HPLC and field desorption MS.

7,12-Dihydroxymethylbenz[*a*]anthracene (1) has been shown to be regiospecifically conjugated in rat liver cytosol by sulphotransferase in the presence of 3′-phosphoadenosine 5′-phosphosulphate to form the highly mutagenic 7-sulphate (2). This metabolite can be inactivated by GSH transferases to give *S*-(12-hydroxymethylbenz[*a*]anthracen-7-yl)methyl glutathione (3). This occurs at a much greater rate than the sulphate conjugation.

There is much evidence to suggest that sulphate esters of carcinogenic alcohols react with nucleophiles mainly via an S_N2 reaction. Compound (2) bound covalently (in much higher ratios than with higher concentrations of hepatic cytosolic proteins) to the exocyclic amino-groups of DNA purine bases of calf thymus DNA.

7-Sulphate formation from DHBA, as well as covalent binding to DNA, was strongly inhibited by dehydroepiandrosterone sulphate, a hydroxysteroid sulphotransferase inhibitor, but not by pentachlorophenol or dichloronitrophenol, inhibitors of phenol sulphotransferases. DNA binding was also markedly inhibited in the presence of GSH and rat liver cytosolic GSH transferases.

Two purine adducts, which accounted for 70% of the total carcinogen covalently bound to nucleic acids, were isolated from DNA after hydrolysis. They were identified as N^6-(12-hydroxymethylbenz[*a*]anthracen-7-

yl)methyladenine and N^2-(12-hydroxymethylbenz[a]anthracen-7-yl)methyl-guanine. The ratio of adenine to guanine adducts was 1:2.5.

Reference

T. Watabe, A. Hiratsuka, and K. Ogura, Sulphotransferase-mediated covalent binding of the carcinogen 7,12-dihydroxymethylbenz[a]anthracene to calf thymus DNA and its inhibition by glutathione transferase, *Carcinogenesis*, 1987, **8,** 445.

7-Hydroxymethyl-12-methylbenz[*a*]anthracene

Use/occurrence:	Model compound
Key functional groups:	Arylhydroxymethyl, benzanthracene, polycyclic aromatic hydrocarbon
Test system:	Preweanling rats and mice (intraperitoneal, 0.25 mmol kg^{-1})

Structure and biotransformation pathway

Hepatic DNA and RNA were isolated from 12 day old female rats sacrificed 6 hours after an intraperitoneal dose of the [³H]benzanthracene (1). After enzymatic digestion radiolabelled adducts were analysed by HPLC. Two adducts were identified by co-chromatographic comparison with standards prepared by the reaction of deoxyguanosine and deoxyadenosine with the sulphate conjugate (2). The amounts of the N^2-deoxyguanosine (3) and N^6-deoxyadenosine (4) adducts represented 2–3 pmol and 0.6–0.7 pmol per mg of hepatic DNA respectively. There was little detectable adduct formation in the livers of male mice. Administration of equivalent doses of the sulphate (2) caused an eight- to ten-fold increase in the adduct formation in rat livers, and high levels of the deoxyguanosine adduct were detected in mouse liver (1100 pmol mg^{-1} DNA). The results provided strong evidence for the formation of a sulphate ester as an electrophilic reactant responsible for the covalent binding of the benzanthracene (1) to DNA. Mouse liver cytosol was shown to contain a much lower level of sulphotransferase activity

58

compared with rat liver cytosol which could explain the lower level of covalent binding in the mouse.

Reference

Y.-Y. Surh, C.-C. Lai, J. A. Miller, and E. C. Miller, Hepatic DNA and RNA adduct formation from the carcinogen 7-hydroxymethyl-12-methylbenz[*a*]anthracene and its electrophilic sulfuric acid ester metabolite in preweanling rats and mice, *Biochem. Biophys. Res. Commun.*, 1987, **144**, 576.

7-Ethylbenz[*a*]anthracene, 7α-Hydroxyethylbenz[*a*]anthracene, 7β-Hydroxyethylbenz[*a*]anthracene

Use/occurrence:	Experimental carcinogen
Key functional groups:	Arylalkyl, benzanthracene, polycyclic aromatic hydrocarbon
Test system:	Rat liver microsomes

Structure and biotransformation pathway

60

7-Ethylbenz[a]anthracene (1) is a less powerful carcinogen than 7-methylbenz[a]anthracene. The latter compound is thought to be activated via metabolism to a 3,4-diol 1,2-oxide. The metabolism of (1) has therefore now been studied in order to compare it with that of the methyl analogue. To assist with elucidating the possible involvement of alkyl group hydroxylation in the metabolism of (1) the metabolic profiles of α-hydroxyethyl- (2) and β-hydroxyethyl benz[a]anthracene (3) were also studied.

Compounds were incubated with liver microsomes from 3-methyl-cholanthrene pretreated rats. Products were extracted with ethyl acetate and chromatographed on silica TLC. The polyhydroxylated metabolites were further purified by reverse phase HPLC and identified by UV, fluorescence, mass, and NMR spectra. Some metabolites were also analysed by GC–MS as their trimethylsilyl derivatives. Synthetic procedures were used to prepare the dihydrodiols (7)–(11) from (1) and (4), (5), (6), and (12) from (2), for use as reference compounds for the metabolites.

Metabolism of (1) yielded three phenolic dihydrodiols (position of phenol group unknown and dihydrodiol in the 8-, 9-, 10-, and 11-ring positions), two isomers of the 8,9-dihydrodiol (4) of (2), the 3,4-dihydrodiol (5), the 10,11-dihydrodiol (6), and the 5,6-dihydrodiol (7), 8,9-dihydrodiol (8), 1,2-dihydrodiol (9), 3,4-dihydrodiol (10), and 10,11-dihydrodiol (11) of (1). No evidence was obtained for any β-hydroxyethyl metabolites.

Metabolism of (2) yielded the following dihydrodiols: 1,2- (12), 8,9- (two isomers) (4), 3,4- (5), and 10,11- (6). Three phenolic 8,9- or 10,11-dihydrodiols were also identified.

Compound (3) was metabolized to the 1,2- (13), 8,9-(14), 10,11- (15), and 3,4- (16) dihydrodiols. Again 10,11-dihydrodiols containing an additional phenol group (position unknown) were found. Other partially characterized metabolites were believed to be phenolic and diphenolic derivatives of (3) and 7-α,β-dihydroxyethylbenz[a]anthracene. It was concluded that (1) is metabolized by similar routes to 7-methylbenz[a]anthracene.

Reference

S. McKay, P. B. Farmer, P. D. Cary, and P. L. Grover, The metabolism of 7-ethylbenz[a]anthracene by rat liver microsomal preparations, *Drug Metab. Dispos.*, 1987, **15**, 682.

5-Methylchrysene, 6-Methylchrysene

Use/occurrence:	Model compounds
Key functional groups:	Chrysene, polycyclic aromatic hydrocarbon
Test system:	9000g liver supernatant (rat, mouse), mouse (topical, 0.07 μmol)

Structure and biotransformation pathway

This investigation was designed to determine the stereoselectivity of the metabolism of the strong carcinogen 5-methylchrysene (1) and of the weak carcinogen 6-methylchrysene (2) to dihydrodiols. It is believed that (1) is metabolically activated to the 1,2-dihydrodiol (3) and thence to the genotoxic 1,2-diol 3,4-epoxide. Compound (1) is also metabolized to the 7,8-dihydrodiol (4) although this is less active as a carcinogen than the 1,2-isomer. A major metabolite of (2) is the 1,2-dihydrodiol (5).

[³H]-(1) and -(2) were incubated with 9000 g liver supernatant (Arochlor 1254 pretreated rat, control, and 3-methylcholanthrene pretreated mouse), and the dihydrodiols produced were separated by reverse phase and chiral stationary phase HPLC. The assignment of configurations was made from CD spectra. The 1,2-diol (3) of (1), the 7,8-diol (4) of (1), and the 1,2-diol (5) of (2) were shown to be only of the (R,R)-configuration. This result was shown not to be due to rapid metabolism of any (S,S)-isomer that had been formed by studies of the metabolism of synthetic racemic (3) and (5).

Mice were treated with [³H]-(1) or -(2) by skin painting. Epidermal metabolites were isolated after 2 hours and were also shown to be of the

(R,R)-configuration, with less than 7% of (S,S)-diols being present. The (R,R)-diols were shown to be more tumorigenic than the (S,S)-enantiomers, the 1,2-diols of (1) being most active.

Reference

S. Amin, K. Huie, G. Balanikas, S. S. Hecht, J. Pataki, and R. G. Harvey, High stereoselectivity in mouse skin metabolic activation of methylchrysenes to tumorigenic dihydrodiols, *Cancer Res.*, 1987, **47**, 3613.

6-Nitro-5-methylchrysene, 5-Methyl-chrysene

Use/occurrence:	Experimental carcinogen (5-methylchrysene)
Key functional groups:	Arylmethyl, chrysene, nitrophenyl, polycyclic aromatic hydrocarbon
Test system:	Mouse (topical, 0.07 μmol), rat liver 9000g supernatant

Structure and biotransformation pathway

5-Methylchrysene (1) is a potent carcinogen which is thought to be activated by metabolism to a 1,2-diol 3,4-epoxide. 6-Nitro-5-methylchrysene (2) is, however, not carcinogenic; its metabolism has therefore now been studied to elucidate the reason.

Incubation of [³H]-(2) with rat liver supernatant (9000g) yielded three major radioactive peaks upon HPLC. These were 6-nitro-5-methylchrysene-1,2-diol (3), 6-nitro-5-methylchrysene-9,10-diol (4), and unchanged (2). A minor amount of 6-nitro-5-hydroxymethylchrysene (5) was also detected. Structures were confirmed by mass spectrometry, UV, and NMR.

The metabolism of [³H]-(1) and [³H]-(2) (0.07 μmol) after application to mouse skin was compared. Ethyl acetate extracts of epidermis following [³H]-(2) treatment yielded more than 92% of the radioactivity. These metabolites were identified by HPLC as (3), (4), and (5). Three more metabolites, of similar amounts, were unidentified. In comparison with the metabolites of [³H]-(1), the 1,2-diol (3) was much more abundant in the metabolite mixture from (2). DNA adduct formation from [³H]-(1) and [³H]-(2) was also compared in mouse epidermis, by Sephadex LH-20 and HPLC chromatography of DNA enzymatic hydrolysates. The binding of the diol epoxides from [^{3}H]-(1) was about 15 times higher than that from [^{3}H]-(2). The conclusion that the 3,4-epoxide from (3) must be a weak DNA-binding agent compared with 5-methylchrysene-1,2-diol 3,4-epoxide was confirmed by *in vitro* studies with the synthesized materials.

Reference

G. H. Shiue, K. El. Bayoumy, and S. S. Hecht, Comparative metabolism and DNA binding of 6-nitro-5-methylchrysene and 5-methylchrysene, *Carcinogenesis*, 1987, **8**, 1327.

Benzo[a]pyrene

Use/occurrence:	Environmental carcinogen
Key functional groups:	Polycyclic aromatic hydrocarbon
Test system:	Cell cultures from rodents, fish, and humans

Structure and biotransformation pathway

R = sulphate or glucuronic acid

An analysis system was established for the separation of three classes of polar conjugates of benzo[a]pyrene (1), the glucuronide and sulphate conjugates of 3-hydroxybenzo[a]pyrene (2), and the glutathione conjugates of the epoxides, 4,5-oxide, 7,8-oxide, and 7,8-diol 9,10-epoxide. Following incubation of [³H]benzo[a]pyrene with early passage Syrian Hamster, Wistar rat, Sencar mouse embryo, and bluegill fry cell lines the major water-soluble metabolite was 3-hydroxybenzo[a]pyrene glucuronide (48–62%). Some 9-hydroxyglucuronide was also detected after glucuronidase treatment of this metabolite. Human hepatoma cells produced 3-hydroxy- and 9-hydroxy-benzo[a]pyrene sulphates (37%), but no glucuronide conjugates and possibly only minor amounts of glutathione conjugates. The analyses were carried out by HPLC using a reverse phase ion pair column and step gradient of increasing amounts of tetrabutylammonium bromide/methanol with ammonium formate buffer.

Reference

I. Plakunov, T. A. Smolarek, D. L. Fischer, J. C. Wiley, and W. M. Baird, Separation by ion-pair high-performance liquid chromatography of the glucuronide, sulphate, and glutathione conjugates formed from benzo[a]pyrene in cell cultures from rodents, fish, and humans, *Carcinogenesis*, 1987, **8**, 59.

Benzo[*a*]pyrene

Use/occurrence:	Environmental carcinogen
Key functional groups:	Polycyclic aromatic hydrocarbon
Test system:	Rat liver phosphatidylcholine liposomes

Structure and biotransformation pathway

Liposomes were prepared from a rat liver phosphatidylcholine fraction and used to examine the quinone formation from benzo[*a*]pyrene (1), mediated by non-enzymatic lipid peroxidation. This oxidation was initiated by addition of 2,2′-azobis(2-amidinopropane) hydrochloride. Benzo[*a*]pyrene and its metabolites were extracted from the peroxidized phosphatidylcholine liposomes and analysed by HPLC using a reverse phase column and UV detection. After 24 hours of incubation benzo[*a*]pyrene 1,6-, 3,6-, and 6,12-quinones (2), (3), and (4) were detected at levels of 3.2 ± 0.2, 3.9 ± 0.5, and $4.0 \pm 0.6\%$ of the initial dose of (1). Structures were confirmed by mass spectrometry and by UV absorption. Benzo[*a*]pyrene was also metabolized under $^{18}O_2$ or in the presence of [^{18}O] methyl linoleate hydroperoxide, which showed that the origin of the two oxygens in the quinones was molecular oxygen, and that this oxygen had partly originated from the fatty acid hydroperoxide. It was suggested that the mechanism involves a direct attack on (1) by a lipid peroxy radical, followed by autoxidation. Other pentacyclic aromatic hydrocarbons that had lower carcinogenic activity showed little reactivity to lipid peroxy radicals.

Reference

J. Terao, B. P. Lin, H. Murakami, and S. Matsushita, Quinone formation from carcinogenic benzo[a]pyrene mediated by lipid peroxidation in phosphatidylcholine liposomes, *Arch. Biochem. Biophys.*, 1987, **254**, 472.

Benzo[a]pyrene

Use/occurrence:	Environmental carcinogen
Key functional groups:	Polycyclic aromatic hydrocarbon
Test system:	Perfused rat lung and liver

Structure and biotransformation pathway

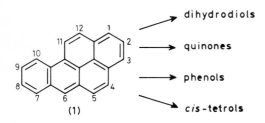

Benzo[α]pyrene (1) is known to be metabolically activated by microsomes. In the rat the lung is a target organ for the carcinogenic action of (1), although *in vitro* experiments have demonstrated that lung microsomes are much less effective than those from liver (a non-target organ) at metabolizing (1). Despite this, metabolites of (1) bind to a similar extent to nucleic acid and protein in isolated perfused rat lung and liver. The present study explores the mechanisms of this by investigating the metabolic profile of (1) in these organs.

[^{3}H]-(1) was perfused through isolated perfused liver and lung from 5,6-benzoflavone treated rats at a concentration of 1 μM. The tissue was homogenized and analysed by reverse phase HPLC for (1) and its unconjugated metabolites. Conjugated metabolites were separated on TLC and identified with reference to synthetic compounds. Amongst the unconjugated metabolites were *cis*-tetrols, 9,10-, 4,5-, and 7,8-dihydrodiols, 1,6-, 3,6-, and 6,12-quinones, and 9- and 3-phenols. The conjugates included the glucuronides and sulphates of the phenols and of the diols and tetrols.

Lung and liver produced similar total amounts of metabolites in 2 hour perfusion, although the lung produced more diols. The liver perfusate contained only low amounts of unconjugated diols and phenols but the bile contained large amounts of conjugates. Thus the liver shows an effective elimination of phenols and of the diol precursors for the carcinogenic metabolites (diol epoxides), resulting in a higher tissue concentration of diols and phenols in the lung tissue.

Reference

M. Mollière, H. Foth, R. Kahl, and G. F. Kahl, Comparison of benzo[a]pyrene metabolism in isolated perfused rat lung and liver, *Arch. Toxicol.*, 1987, **60**, 270.

Benzo[a]pyrene

Use/occurrence: Environmental carcinogen

Key functional groups: Polycyclic aromatic hydrocarbon

Test system: Human hepatocytes (0.1–100 μM)

Structure and biotransformation pathway

Although the metabolism of benzo[a]pyrene (1) has been well studied there has still been some doubt as to whether or not the pathways demonstrated *in vitro* are relevant to the *in vivo* carcinogenesis experiments, in which high concentrations of (1) are used. The currently reported study involves the analysis of metabolites of (1) by human hepatocytes over a four log concentration range.

[³H]-(1), at concentrations of 0.1, 1.0, 10, or 100 μM, was incubated with human hepatocytes for 24 hours at 37 °C. Unconjugated metabolites of (1) were extracted with acetone/ethyl acetate. Glucuronide and sulphate conjugates in the aqueous phase were enzymatically cleaved with β-glucuronidase and aryl sulphatase and the products extracted similarly. Analysis was with a reverse phase HPLC system, allowing the quantitation of the 9,10-dihydrodiol (2), 4,5-dihydrodiol (3), 7,8-dihydrodiol (4), and the 9-hydroxy- (5) and 3-hydroxy-derivative (6) of (1). At 100 μM (1) three samples of hepatocytes metabolized 74–89% (1) and at 0.1 μM 94–99% (1). The most abundant metabolites were unidentified 'polar materials', followed in abundance by (2), (3), (6), (4), and (5). A nearly linear dose–response relationship was observed, *i.e.* there was no saturation of metabolism at the highest doses of (1). However, the ratio of dihydrodiols to phenols increased slightly from 1.9 at 0.1 μM (1) to 3.1 at 100 μM (1). Conjugation ranged from 31 to 96%.

71

DNA binding of ^{3}H was also determined for cells after incubation with [^{3}H]-(1). A sharp increase was seen (64- to 844-fold) on increasing the concentration of (1) from 10 μM to 100 μM, and this appeared to be associated with an increase in the proportion of the dihydrodiol metabolites of (4) [which is a precursor of the proposed genotoxic metabolite 7,8-dihydroxy-9,10-epoxy-7,8,9,10-tetrahydro-(1)].

Reference

D. K. Monteith, A. Novotny, G. Michalopous, and S. C. Strom, Metabolism of benzo[a]pyrene in primary cultures of human hepatocytes: dose-response over a four-log range, *Carcinogenesis*, 1987, **8**, 983.

Benzo[a]pyrene

Use/occurrence:	Environmental carcinogen
Key functional groups:	Polycyclic aromatic hydrocarbon
Test system:	Mouse (topical, 200 nmol)

Structure and biotransformation pathway

Benzo[a]pyrene (1) is believed to be metabolically activated to *trans*-7β,8α-dihydroxy-9α,10α-7,8,9,10-tetrahydrobenzo[a]pyrene (BPDE), which is known to bind to the guanine residues in nucleic acids. The excretion of the deoxyguanosine adduct (2) of the epoxide in mouse faeces after topical administration of benzo[a]pyrene is now described.

Initially rats were injected (intraperitoneal or intravenous) with the reaction product formed between [³H]-BPDE and calf thymus DNA. Most of the radioactivity was excreted in the faeces, and was shown by HPLC to be associated with the deoxyguanosine adduct (2). Compound (2) was not formed in *in vitro* incubations of the [³H]-BPDE–DNA adduct at 37 °C for 24 hours.

[³H]Benzo[a]pyrene (1) was then painted on to the backs of mice (0.4 mCi, 200 nmol), and faeces and urine were collected. The urine contained less than 0.5% of the applied radioactivity, but 79% of the applied radioactivity could be recovered by extraction of the faeces. Sephadex LH-20 chromatography was used to purify this extract and HPLC confirmed that the adduct (2) was present. Acid hydrolysis released a product which co-chromatographed with the tetrol of (1).

Reference

B. Tierney, C. N. Martin, and R. C. Garner, Topical treatment of mice with benzo[a]pyrene or parenteral administration of benzo[a]pyrene diol epoxide–DNA to rats results in faecal excretion of a putative benzo[a]pyrene diol epoxide–deoxyguanosine adduct, *Carcinogenesis*, 1987, **8**, 1189.

Benzo[a]pyrene

Use/occurrence:	Carcinogen
Key functional groups:	Polycyclic aromatic hydrocarbon
Test system:	Purified pulmonary cytochrome P-450

Structure and biotransformation pathway

Pulmonary P-450$_{MC}$ was prepared from lung microsomes of 3-methylcholanthrene treated rats. The metabolism of (1) by pulmonary P-450$_{MC}$ was investigated in reconstituted systems in the presence and absence of epoxide hydrolase (EH). Metabolites were tentatively identified by comparison with authentic standards on HPLC. The metabolites produced by pulmonary P-450$_{MC}$ were compared to those formed by hepatic P-450$_{MC}$ from the same rats. The major metabolite produced by both enzymes was the 3-hydroxy derivative (2). In the absence of added EH, (2) represented a greater proportion of total metabolites formed by pulmonary P-450$_{MC}$ than by hepatic P-450$_{MC}$. Addition of EH decreased the ratio of (2) to total metabolites more with pulmonary P-450$_{MC}$ than with hepatic P-450$_{MC}$. 1,6-(3) and 3,6- (4) quinones were also major metabolites produced by both enzyme systems, and their proportion of the total metabolites was not affected by the addition of EH. The 9,10-dihydro-9,10-hydroxy metabolite (5) was the major dihydrodiol formed by pulmonary P-450$_{MC}$. Addition of EH increased the formation of dihydrodihydroxybenzo[a]pyrenes, particularly the 7,8-metabolite (6). The metabolism of (1) in pulmonary and

74

hepatic P-450$_{MC}$ did not reveal any mechanism that would explain the difference between lung and liver in hydrocarbon-induced tumorigenesis.

Reference

I. Sagami, T. Ohmachi, H. Fuji, and M. Watanabe, Benzo[a]pyrene metabolism by purified cytochrome P-450 from 3-methylcholanthrene treated rats, *Xenobiotica*, 1987, **17**, 189.

Benzo[*a*]pyrene-7,8-dihydrodiol

Use/occurrence:	Model compound
Key functional groups:	Benzo[*a*]pyrene, polycyclic aromatic hydrocarbon
Test system:	Rat intestinal mucosa microsomes

Structure and biotransformation pathway

[Analogous mechanism would apply for (+)-(1)]

Benzo[*a*]pyrene-7,8-dihydrodiol (1) is known to be metabolically activated to the 9,10-epoxides (2) and (3) by the cytochrome P-450 mixed function oxidase system. This study explored the potential of microsomal preparations of rat intestinal mucosa (which are known to be low in P-450) to carry out this oxidation.

Microsomes were incubated with (1) at a concentration of 20 μM, in the presence either of NADPH or of a lipid peroxide system (ascorbate, ferrous sulphate). Products were extracted into ethyl acetate and determined by reverse phase HPLC and fluorescence detection.

The major product from both systems was the tetrol (4) derived from the epoxide (2). The total proportion of tetrols from (2) was $53.3 \pm 14.5\%$ for microsomes + NADPH and $66.4 \pm 4.5\%$ for the peroxidation-catalysed system. Maximal rate for the latter system was found in the middle region of the intestine; the production of oxidized metabolites was also stimulated for microsomes from the intestine of rats fed a diet containing highly unsaturated cod liver oil. Rats fed a fat-free diet, a saturated lard diet, or a corn oil diet showed lower rates of lipid peroxidation and tetrol formation from (1).

Reference

J. D. Gower and E. D. Wills, The oxidation of benzo[*a*]pyrene-7,8-dihydrodiol mediated by lipid peroxidation in the rat intestine and the effect of dietary lipids, *Chem. Biol. Interact.*, 1987, **63**, 63.

Benzo[a]pyrene-7,8-dihydrodiol 9,10-oxide

Use/occurrence:	Metabolite of the carcinogen benzo[a]pyrene
Key functional groups:	Benzo[a]pyrene, epoxide, polycyclic aromatic hydrocarbon
Test system:	Rat hepatocytes

Structure and biotransformation pathway

(2)   (1)   (3) GS = glutathionyl

Benzo[a]pyrene-7,8-dihydrodiol 9,10-oxide (1) is an active mutagenic metabolite of benzo[a]pyrene. Hydrolysis of the epoxide occurs in aqueous solution giving the tetrol (2). However, in rat liver or in isolated hepatocytes conjugation with glutathione (3) occurs, and tetrol formation is a minor pathway. The half-life for the epoxide in Krebs buffer is 3.3 min, whereas the presence of hepatocytes (5×10^6 ml^{-1}, containing 50 nmol glutathione/10^6 cells) decreased this to 0.9 min. Depletion of glutathione to < 10 nmol/10^6 cells (by pretreatment of the rats from which the hepatocytes were derived with diethyl maleate) resulted in an increase in the half-life to 3.4 min. The tetrols and glutathione conjugates were separated and quantitated by reverse phase HPLC techniques.

Reference

L. Dock, M. Martinez, and B. Jernström, Increased stability of ($\pm$)-7β,8α-dihydroxy-9α,10α-epoxy-7,8,9,10-tetrahydrobenzo[a]pyrene through interaction with subcellular fractions of rat liver, *Chem. Biol. Interact.*, 1987, **61**, 31.

Benzo[*j*]fluoranthene

Use/occurrence:	Environmental carcinogen
Key functional groups:	Polycyclic aromatic hydrocarbon
Test system:	Rat liver 9000g supernatant

Structure and biotransformation pathway

(2) (*trans*)

(3) (*trans*)

(8)

(1)

(4)

(7)

(6)

(5)

Benzo[*j*]fluoranthene (1) was incubated at a concentration of 0.38 mM with 9000g supernatant from liver of Aroclor pretreated rats. Products were extracted with ethyl acetate and separated by reverse phase HPLC. The major peak was identified as *trans*-4,5-dihydro-4,5-dihydroxy-(1) (2) by comparison of UV spectra and chromatographic mobility with those of an authentic standard. The enantiomeric purity of (2) was determined by chiral phase HPLC to be 20% (2:3 mixture of enantiomers). It was tentatively assumed that the major isomer was 4*R*,5*R*. The second most abundant metabolite was identified by similar means to be *trans*-9,10-dihydro-9,10-dihydroxy-(1) (3) of enantiomeric purity 46% (73:27 diastereomer ratio). No 2,3-diol of (1) could be detected. Other metabolites that were tentatively identified were benzo[*j*]fluoranthene-4,5-dione (4) and 10-hydroxy-,

78

6-hydroxy-, 3-hydroxy-, and 4-hydroxy-benzo[*j*]fluoranthene, (5), (6), (7), and (8) respectively.

Reference

J. E. Rice, N. G. Geddie, and E. J. Lavoie, Identification of metabolites of benzo[*j*]-fluoranthene formed *in vitro* in rat liver homogenate, *Chem. Biol. Interact.*, 1987, **63**, 227.

Benzo[*b*]fluoranthene

Use/occurrence: Environmental pollutant

Key functional groups: Fluoranthene, polycyclic aromatic hydrocarbon

Test system: Mouse epidermis, rat liver 9000g supernatant

Structure and biotransformation pathway

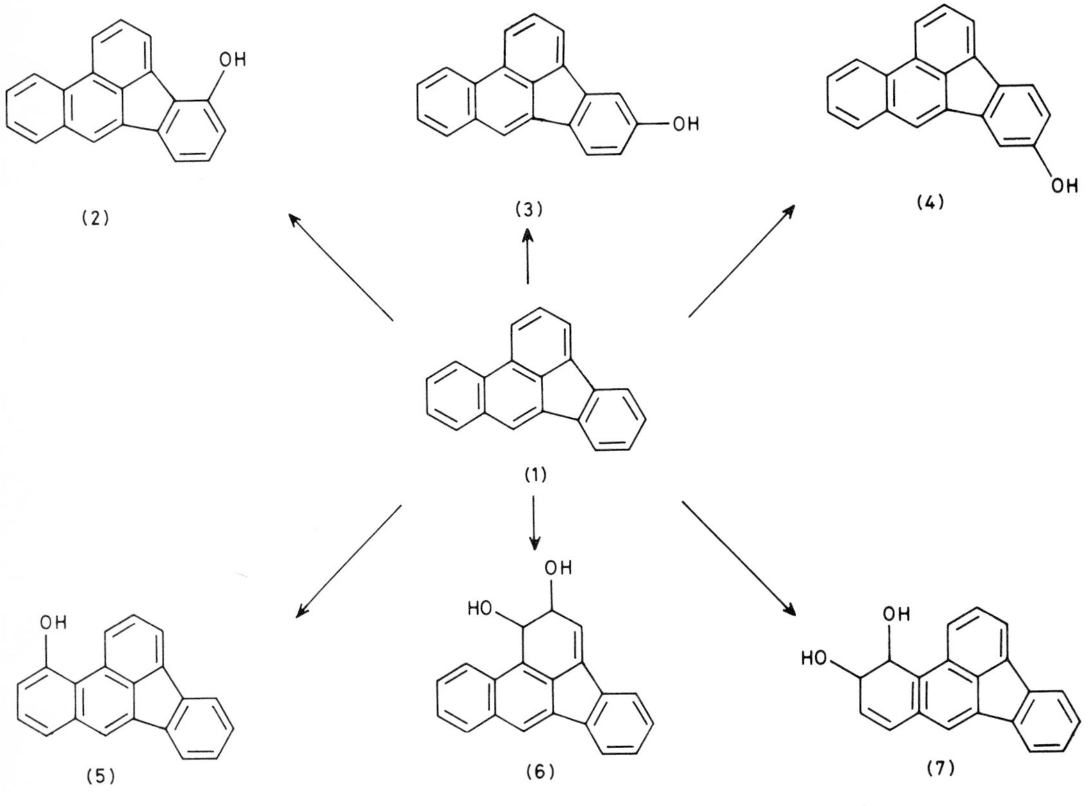

(2)

(3)

(4)

(1)

(5)

(6)

(7)

The mechanism of metabolic activation of benzo[*b*]fluoranthene (BbF) (1), a potent environmental carcinogen, is not known. BbF-9,10-diol was found to be tumorigenic in mouse skin. Since this metabolite can form a bay region dihydrodiol epoxide, it may be the proximate carcinogen. The present study was therefore carried out using HPLC to investigate the biotransformation of (1) in its target tissue, mouse epidermis. The major metabolites were 4-, 5-, and 6-hydroxy-BbF, (2), (3), and (4), along with their sulphate and glucuronide conjugates. 12-OH-BbF (5), BbF-1,2-diol (6), and BbF-11,12-diol (7) were minor metabolites, but BbF-9,10-diol was not detected. Further biotransformation of the latter by rat liver 9000g supernatant yielded 5- and 6-OH-BbF-9,10-diol, neither of which were detected in mouse epidermis.

This study did not support the hypothesis that (1) is metabolically activated through formation of the bay region diol epoxide.

Reference

J. E. Geddie, S. Amin, K. Huie, and S. S. Hecht, Formation and tumorigenicity of benzo[*b*]fluoranthene metabolites in mouse epidermis, *Carcinogenesis*, 1987, **8,** 1579.

Benzo[*f*]quinoline

Use/occurrence: Cigarette smoke, coal tar, petroleum distillates

Key functional groups: Polycyclic aromatic hydrocarbon, quinoline

Test system: Rat liver microsomes

Structure and biotransformation pathway

1,3-Benzo[*f*]quinoline (1) is widespread in environmental pollution and has proved positive in the Ames test in the presence of an activating system from rat liver homogenates. There are also preliminary data for carcinogenic activity. Because (1) requires metabolic activation, the present study was designed to examine the profile of metabolites produced by control, phenobarbital (PB)-induced, and 3-methylcholanthrene (3-MC)-induced rat liver microsomes. The respective rates of metabolism of (1) by these three systems were 0.5, 2.4, and 3.6 nmol min^{-1} mg^{-1}. The metabolic profile of (1) from PB-induced microsomes was BfQ-*N*-oxide (41%) (2), 9-hydroxy-BfQ

(20%) (3), BfQ-9,10-dihydrodiol (12%) (4), BfQ-7,8-dihydrodiol (3%) (5), 7-hydroxy-BfQ (13%) (6), and BfQ-5,6-dihydrodiol (0.5%) (7). A similar pattern was obtained from control microsomes. However, with 3-MC microsomes the most predominant metabolite was (5) (41%), a precursor of the bay region diol epoxide. Compounds (2) (23%), (6) (15%), (3) (9%), (4) (6%), and (7) (1.0%) were also obtained. Although both PB- and 3-MC-induced microsomes generate hydroxylated metabolites from (1), they exhibit positional selectivity, with 3-MC preferentially attacking at the 7,8-position to form the potentially mutagenic BfQ-7,8-dihydrodiol. This suggests a role for cytochrome P-450c, the major inducible form by 3-MC, in metabolic activation of (1).

Reference

C. Kandaswami, S. Kumar, S. K. Dubey, and H. C. Sikka, Metabolism of benzo[*f*]quinoline by rat liver microsomes, *Carcinogenesis*, 1987, **8**, 1861.

Dibenz[a,j]acridine

Use/occurrence: Tobacco smoke condensate

Key functional groups: Polycyclic aromatic hydrocarbon

Test system: Rat liver and lung microsomes

Structure and biotransformation pathway

According to the bay region theory, dibenz[a,j]acridine (1) should be activated through the *trans*-3,4-dihydroxy-3,4-dihydrodibenz[a,j]acridine (2), the predicted proximate carcinogen, to the *anti*-diol epoxide (3).

The distribution of previously identified metabolites of (1) resulting from 3-methylcholanthrene (MC)-induced microsomal incubations, and including (2), were examined in this study using liver and lung microsomes from control and pretreated male rats.

Both liver and lung incubations produced extensive attack at the 3,4- and 5,6-positions, with very little metabolism occurring at the ring N or the 1,2-positions. Attack at the 5,6-position was greatest for 3-MC-induced liver microsomes compared with control or phenobarbital (PB)-induced microsomes, but for all three liver systems activation at the 3,4-position was predominant (>61% for control and PB). With lung microsomes the proportion of 3,4-oxidative attack was 50–55% and 5,6-oxidation for control and PB microsomes was slightly greater than in comparable liver preparations. For all preparations (2) was the major metabolite (30–40%), with the 5,6-oxide (4) always the major component of the 5,6-oxidation products. Phenols were also quantitatively important.

When epoxide hydratase was inhibited by 3,3,3-trichloropropene 1,2-oxide, formation of the 3,4- and 5,6-dihydrodiols was substantially inhibited (85%). Conversely, pretreatment of liver microsomes with *trans*-stilbene oxide which induces epoxide hydratase, led to increased formation of dihydrodiols, particularly (5). The *N*-oxide (6) was also identified (*ca.* 1%) in incubations with control and PB-induced liver microsomes.

The stereochemistry of the metabolite (2) is unknown, but studies with the biosynthetic derivative suggest a low conversion into the *anti*-diol epoxide, and are consistent with a relatively low level of DNA binding and mutagenicity in the Ames test for (1).

Reference

J. H. Gill, C. C. Duke, A. J. Ryan, and G. M. Holder, Dibenz[*a,j*]acridine: distribution of metabolites formed by liver and lung microsomes from control and pretreated rats, *Carcinogenesis*, 1987, **8**, 425.

2-Acetylaminofluorene

Use/occurrence:	Environmental carcinogen
Key functional groups:	N-Acetyl aryl amine, polycyclic aromatic amine
Test system:	Human lymphocytes

Structure and biotransformation pathway

[³H]Acetylaminofluorene (1) was incubated with lymphocytes obtained from peripheral human blood samples, and formation of metabolites investigated by chromatographic comparison with authentic standards. The 7-hydroxy (2) and N-hydroxy (4) derivatives were the main metabolites detected in ether extracts while the 1-hydroxy compound, a major liver microsomal metabolite, was a minor metabolite and was not detectable with lymphocyte samples from five out of 23 subjects. The rate of formation of both major metabolites was approximately linear with time up to 10 hours and up to a cell concentration of 7.5×10^6 cells ml^{-1} of incubation mixture. There was a significantly greater 7-hydroxylation activity in lymphocytes from smokers compared with non-smokers, but no significant difference in N-hydroxylation.

Reference

M. E. McManus, K. J. Trainor, A. A. Morley, W. Burgess, I. Stupans, and D. J. Birkett, Metabolism of 2-acetylaminofluorene in cultured human lymphocytes, *Res. Commun. Chem. Pathol. Pharmacol.*, 1987, **55**, 438.

N-Sulphonoxy-2-acetylaminofluorene

Use/occurrence:	Metabolite of environmental carcinogen
Key functional groups:	Aryl amine, N-hydroxy, N-sulphonoxy
Test system:	Rat liver homogenate, hepatic cytosol

Structure and biotransformation pathway

The solvolysis and *in vitro* metabolism of N-sulphonoxy-2-acetyl-aminofluorene (1) was studied using a tritiated compound and products were identified by chromatographic comparison with authenticated reference compounds. The compound undergoes some rearrangement in aqueous media to give small amounts of the 1- and 3-sulphates (5) and (6) of 2-acetylaminofluorene. However, the extent of formation of these components was even less in the presence of a rat liver homogenate, possibly due to substrate binding to protein. Surprisingly there was almost quantitative conversion into the isomeric sulphate esters on incubation in the presence of bovine serum albumin. 2-Acetylaminofluorene (2) was a major metabolite formed by liver homogenate together with a smaller amount of 4-hydroxy-2-acetylaminofluorene (4). The latter was the major solvolysis product in the absence of liver cytosol. Experimental evidence was obtained that 2-acetylaminofluorene was formed directly from the N-sulphonoxy derivative in a one-step non-enzymatic reaction catalysed by liver cytosol. Subsequent

N-hydroxylation and enzymatic sulphation of the arylhydroxamic acid would lead to metabolic recycling of the ultimate carcinogen (1). A competing reaction in this recycling process would be reaction with cellular nucleophiles as shown by formation of 1-methylthio-2-acetylaminofluorene on incubation in the presence of the model nucleophile *N*-acetyl-L-methionine with corresponding reduction in the amount of acetylaminofluorene formed.

Reference

B. A. Smith, J. R. Springfield, and H. R. Gutman, Solvolysis and metabolic degradation, by rat liver, of the ultimate carcinogen, *N*-sulphonoxy-2-acetylaminofluorene, *Mol. Pharmacol.*, 1987, **31**, 438.

2-Nitrofluorene

Use/occurrence:	Environmental carcinogen
Key functional groups:	Fluorene, nitrophenyl, polycyclic aromatic hydrocarbon
Test system:	Rat (oral, 5 mg)

Structure and biotransformation pathway

2-Nitrofluorene (1) is a mutagenic, carcinogenic environmental contaminant. Its metabolism has now been studied after oral dosing of [9-^{14}C]-(1) to male rats. Metabolites were separated from urine by SepPak chromatography and resolved by reverse phase HPLC. Identification of some metabolites was obtained by HPLC–MS, using a moving belt interface, and mutagenicity tests were also carried out by the Ames test on some HPLC fractions. The urinary profile of metabolites from a naphthoflavone-induced rat contained about 43% of ^{14}C as unconjugated metabolites, which represented about 75% of the mutagenicity. Two major unconjugated metabolites were identified by MS as 5-hydroxy-2-acetylaminofluorene (2) and 7-hydroxy-2-acetylamino-fluorene (3). Also present were peaks that co-chromatographed with the 9-, 8-, 3-, 1-, and N-hydroxy-2-acetylaminofluorenes (4)–(8). A second group of

89

metabolites was thought to consist of hydroxylated isomers of (1). No 9-hydroxy-(1) was present. Faeces contained a similar range of metabolites with (2) and (3) being the major constituents.

The observation that (1) yields acetylaminofluorene metabolites suggests that the first metabolic step is reduction of the nitro-group to the amine. Acetylaminofluorene is a well known potent carcinogen.

Reference

L. Möller, J. Rafter, and J. A. Gustafsson, Metabolism of the carcinogenic air pollutant 2-nitrofluorene in the rat, *Carcinogenesis*, 1987, **8**, 637.

2-Nitrofluorene

Use/occurrence:	Environmental pollutant
Key functional groups:	Fluorene, nitrophenyl, polycyclic aromatic hydrocarbon
Test system:	Isolated perfused rat lung and liver

Structure and biotransformation pathway

(1) * = ^{14}C   (2)

2-Nitrofluorene (NF) (1), identified in diesel exhaust and urban air in particulate or gaseous form, is used as a model compound for nitrated polycyclic aromatic hydrocarbons (nitro-PAHs), a number of which are mutagenic and carcinogenic. After oral administration (1) is metabolized by the same pathway as 2-acetylaminofluorene to a range of OH-AAFs. It also undergoes direct cytochrome P-450-dependent ring hydroxylations which are inducible. Since humans are also exposed to (1) by inhalation, the studies have been extended to isolated perfused rat lung and to further characterization of metabolites in isolated perfused liver using LC–MS techniques. The lung was found to metabolize (1) to hydroxylated NFs, particularly 9-OH-NF (2), irrespective of route of administration. However, after intertracheal administration (1) was rapidly excreted unchanged into the perfusate. The liver cleared (1) as glucuronides in bile which were non-mutagenic until bile was treated with β-glucuronidase. The mutagenic metabolites identified in treated bile were the same as those in the perfusate of isolated lung. Thus (1) may be inhaled, passed through the lungs, hydroxylated and conjugated in liver, and excreted into the intestine where the conjugates may be hydrolysed by β-glucuronidase, releasing genotoxic metabolites.

Reference

L. Möller, S. Törnquist, B. Beije, J. Rafter, R. Toftgard, and J.-A. Gustafsson, Metabolism of the carcinogenic air pollutant 2-nitrofluorene in the isolated perfused rat lung and liver, *Carcinogenesis*, 1987, **8**, 1847.

1-Nitropyrene, 2-Nitrofluorene

Use/occurrence:	Environmental pollutants
Key functional groups:	Nitrophenyl, polycyclic aromatic hydrocarbon
Test system:	Rat and rabbit (oral, 100 mg kg^{-1})

Structure and biotransformation pathway

Both 1-nitropyrene (1) and 2-nitrofluorene (5) were dosed orally to rats and rabbits, and metabolites were extracted from urine and faeces with ethyl acetate. Metabolites were separated and purified by TLC and compared with authentic reference compounds. Further confirmation of structure was obtained by comparison of mass and UV spectra.

Aminopyrene (2) and both the *N*-formyl (3) and *N*-acetyl (4) derivatives were identified as metabolites of 1-nitropyrene in rabbit urine but were only detected in rat faeces. The corresponding metabolites of 2-nitrofluorene were also isolated from rabbit urine and rat faeces. No quantitative data on the amounts of each metabolite were reported. It was demonstrated that liver cytosol from rabbit, rat, mouse, guinea pig, and hamster catalysed formation of the novel *N*-formyl derivatives in the presence of *N*-formyl-L-kynurenine.

Reference

K. Tatsumi and H. Amano, Biotransformation of 1-nitropyrene and 2-nitrofluorene to novel metabolites, the corresponding formylamino compounds, in animal bodies, *Biochem. Biophys. Res. Commun.*, 1987, **142**, 376.

1-Nitropyrene

Use/occurrence: Environmental carcinogen

Key functional groups: Nitrophenyl, polycyclic aromatic hydrocarbon, pyrene

Test system: Rabbit tracheal epithelial cells

Structure and biotransformation pathway

The high activity of tracheal tissue in the metabolism of 1-nitropyrene (1) prompted the present study, in which $[4,5,9,10-^{14}C]-(1)$ was incubated (8.1 μM) with tracheal epithelial cells from the rabbit. The medium and the lysed cells were extracted with ethyl acetate/acetone (3:1) either before or after treatment with β-glucuronidase and sulphatase, and the metabolites analysed by reverse phase HPLC. Compounds were identified by comparison with authentic samples. In the medium the main metabolites were 1-nitro-4,5-dihydro-4,5-dihydroxypyrene (2), 6-hydroxy-1-nitropyrene (3), 8-hydroxy-

93

1-nitropyrene (4) 3-hydroxy-1-nitropyrene (5), 10-hydroxy-1-nitropyrene (6), and the reduction products *N*-acetyl-1-aminopyrene (7) and 1-aminopyrene (8). Intracellular metabolites were similar, although present at much lower concentrations. After 4 hours incubation there was considerable DNA-binding of metabolites of (1), reaching levels of 198 adducts/10^6 nucleotides, of particular significance in view of the known susceptibility of tracheal tissue to (1).

Reference

L. C. King, M. Jackson, L. M. Ball, and J. Lewtas, Metabolism and DNA binding of 1-nitro[^{14}C]pyrene by isolated rabbit tracheal epithelial cells, *Carcinogenesis*, 1987, **8**, 675.

1-Nitropyrene 4,5-oxide and 9,10-oxide

Use/occurrence:	Environmental pollutant
Key functional groups:	Epoxide, nitrophenyl, polycyclic aromatic hydrocarbon, pyrene
Test system:	Rat liver enzymes, rat (intravenous and intraperitoneal)

Structure and biotransformation pathway

1-Nitropyrene (1), the predominant nitrated polycyclic aromatic hydrocarbon in the environment, is both mutagenic and carcinogenic. In mammalian systems, DNA-binding appears to result from nitroreduction, K-region epoxidation, and nitro-reduction of oxidized metabolites. However, ring-oxidized products may also be substrates for glutathione transferases (GST) thus reducing the genotoxic potential of (1). The present study investigates the conjugation of 1-nitropyrene 4,5-oxide (2) and 1-nitropyrene 9,10-oxide (3) with GSH and compares the GSH conjugates excreted in bile with those produced *in vitro*. Both K-region epoxides reacted slowly with GSH, but this was greatly enhanced by addition of purified GST, in particular 3-3 and 4-4, with 1-1, 2-2, and 7-7 being less effective. The oxide (2) gave a 1:1 mixture of 4-(glutathion-S-yl)-5-hydroxy-4,5-dihydro-1-nitropyrene (4) and 5-(glutathion-S-yl)-4-hydroxy-4,5-dihydro-1-nitropyrene (5), while (3) gave a 2:1 mixture of 9-(glutathion-S-yl)-10-hydroxy-9,10-dihydro-1-nitropyrene (6) and 10-(glutathion-S-yl)-9-hydroxy-9,10-dihydro-1-

95

nitropyrene (7). Both K-region oxides were converted by microsomal epoxide hydratase into *trans*-dihydrodiols.

After administration of [4,5,9,10-^{3}H]-1-nitropyrene to Sprague–Dawley rats, analysis of biliary metabolites by HPLC showed four GSH conjugates identical to those described above for (2) and (3) *in vitro*. Glucuronide conjugates were also observed for 1-nitropyrene *trans*-4,5-dihydrodiol.

It therefore appears that (1) is oxidized preferentially to (2) which is eliminated via GSH conjugation or *O*-glucuronidation following dihydrodiol formation. The oxide (3) is eliminated via GSH conjugation.

Reference

Z. Djuric, B. Coles, E. K. Fifer, B. Ketterer, and F. A. Beland, *In vivo* and *in vitro* formation of glutathione conjugates from the K-region epoxides of 1-nitropyrene, *Carcinogenesis*, 1987, **8**, 1781.

6-Aminochrysene, *N*-Hydroxy-6-aminochrysene, 6-Nitrochrysene

Use/occurrence:	Environmental pollutants
Key functional groups:	Aryl amine, chrysene, *N*-hydroxy, nitrophenyl, polycyclic aromatic hydrocarbon
Test system:	Isolated rat hepatocytes

Structure and biotransformation pathway

dRib = 2-dexyribose

Members of the nitrated polycyclic aromatic hydrocarbons are mutagens and carcinogens, but information is lacking on their biotransformation in intact cells and particularly their possible interactions with DNA.

In this study *N*-hydroxy-6-aminochrysene (1), a possible proximate carcinogen for 6-aminochrysene (2) and 6-nitrochrysene (3) was synthesized and its binding to DNA *in vitro* examined. The rate of binding was 20–30 nmol mg^{-1} DNA/30 min, some 2–10 times greater than for other carcinogenic *N*-hydroxy-arylamines. Three major DNA-adducts were identified in enzymatic hydrolysates by mass and ^{1}H NMR spectroscopy as *N*-(deoxyinosin-8-yl)-6-aminochrysene (32%) (4) 5-(deoxyguanosin-*N*^{2}-yl)-6-aminochrysene (28%) (5) and *N*-(deoxyguanosin-8-yl)-6-aminochrysene (22%) (6). The authors suggest that the deoxyinosine adduct results from spontaneous oxidation of the corresponding deoxyadenosine adduct during sample preparation. In isolated hepatocytes from Sprague–Dawley rats treated with [^{3}H]-(2) or -(3) up to 12 pmol adducts mg^{-1} DNA were found. HPLC analysis showed that (4) and (6) accounted for 45% and 30% of the total adducts in these cells. Neither the preferential modification of deoxyadenosine nor the ease of oxidation of this adduct to a deoxyinosine derivative would be predicted on the basis of reactivity and this is so far unique among reactions of N-hydroxyarylamines with DNA.

Reference

K. B. Delclos, D. W. Miller, J. O. Lay, D. A. Casciano, R. P. Walker, P. P. Fu, and F. F. Kadlubar, Identification of C-8-modified deoxyinosine and N^2- and C-8-modified deoxyguanosine as major products of the *in vitro* reaction of N-hydroxy-6-aminochrysene with DNA and the formation of these adducts in isolated rat hepatocytes treated with 6-nitrochrysene and 6-aminochrysene, *Carcinogenesis*, 1987, **8**, 1703.

Aristolochic acid I and II

Use/occurrence:	Roots and leaves of *Aristolochia* species
Key functional groups:	Acetal, aryl carboxylic acid, methoxyphenyl, nitrophenyl, phenanthrene
Test system:	Rat (oral, 3 mg), guinea pig (oral, 3 mg), rabbit (oral, 10 mg), mouse (oral, 30–85 mg kg^{-1}), dog (oral, 10 mg), man (0.9 mg)

Structure and biotransformation pathway

Aristolochic acid I (1) and aristolochic acid II (2) were administered to several animal species and human volunteers. Metabolites, isolated by Sephadex LH-20 chromatography and TLC from urine and faecal extracts, were characterized by MS and NMR. HPLC was used for the quantification of metabolites. Both (1) and (2) underwent considerable metabolism in all species: most of the metabolites were reduction products. In the rat, the principal metabolite of (1), aristolactam Ia (3), was demonstrated by administration of the appropriate precursors, aristolactam I (4) and aristolochic acid Ia (5), as the stable metabolic end product of two biotransformation pathways.

In urine, most of (3) was present as a conjugate, believed to be a *N*-glucuronide. Compound (3) represented 46% of the dose in urine and 37% in the faeces. Other identified compounds were present as minor metabolites. The identified metabolites of (2) in total represented only 5% of the dose in urine and 10% in faeces and major pathways for this compound remain to be elucidated. Aristolactam II (6) was the major metabolite and small quantities of (3) were formed.

Reductive replacement of the nitro-group occurred forming aristolic acid I (7) from (1) and 3,4-methylenedioxy-1-phenanthrenecarboxylic acid (8) from (2). This unusual metabolic pathway is attributed to the intestinal flora. Further metabolism of (7) to yield the 8-hydroxy metabolite (9) was observed. Only the mouse had the same metabolic pathways for (1) and (2) as the rat. Not all the metabolites were found in urine from guinea pigs, rabbits, dogs, and man.

Reference

G. Krumbiegel, J. Hallensleber, W. H. Mennicke, N. Rittman, and H. J. Roth, Studies on the metabolism of aristolochic acids I and II, *Xenobiotica*, 1987, **17**, 981.

Alkenes, Halogenoalkanes, and Halogenoalkenes

Isoprene (2-methylbuta-1,3-diene)

Use/occurrence:	Chemical intermediate
Key functional groups:	Alkadiene, alkene
Test system:	Rat (inhalation exposure for 66 hours to atmospheres of 8, 266, 1480, and 8200 p.p.m.)

Structure and biotransformation pathway

Following exposure to various atmospheric concentrations of isoprene, rats were transferred to glass metabolism cages. Urine, faeces, and expired air were then collected for a further 66 hours. Independent dose, 70–80% of the metabolites were excreted via the urine. It was calculated that during exposure only a small proportion (3–25%) of the inhaled dose was metabolized. In separate experiments the blood and tissue concentrations of isoprene and its metabolites (1)–(5) were measured at intervals following exposure and recovery. The diepoxide (3), was found in all tissues analysed. Body fat appeared to be a reservoir for isoprene and its metabolites. Metabolites (1), (2), and (3) have been shown to be mutagenic and are close analogues of the mutagenic metabolites of buta-1,3-diene, a known animal carcinogen. Carbon dioxide collected from expired air after exposure accounted for 2–5% of the metabolized dose. Volatile metabolites were analysed by a vacuum line distillation technique.

Reference

A. R. Dahl, L. S. Birnbaum, J. A. Bond, P. G. Gervasi, and R. F. Henderson, The fate of isoprene inhaled by rats: comparison to butadiene, *Toxicol. Appl. Pharmacol.*, 1987, **89**, 237.

Styrene

Use/occurrence: Plastic monomer

Key functional groups: Arylethylene

Test system: Human occupational exposure

Structure and biotransformation pathway

Styrene (1) is known to be metabolized to styrene 7,8-oxide (2), which is a chiral compound. The oxide (2) is further transformed into other chiral metabolites such as phenylethylene glycol (3) and mandelic acid (4). The stereoselectivity of these metabolic pathways has now been investigated in humans occupationally exposed to styrene. Enantiomeric composition of urinary metabolites was determined by capillary GC using a chiral stationary phase. In the single example of human exposure presented the excretion of (3) was increased during exposure and showed an increasing L/D ratio (maximum *ca.* 3). Mandelic acid (4) was also excreted with a rather lower L/D ratio (maximum *ca.* 2). The L/D ratios returned to their original levels 40 hours after the end of the styrene exposure. It was concluded that epoxide hydrolase favours formation of the L-enantiomer of (3) and that D-(2) is likely to persist longer in the body.

Reference

M. Korn, R. Wodarz, K. Drysch, and F. W. Schmahl, Stereometabolism of styrene in man. Urinary excretion of chiral styrene metabolites, *Arch. Toxicol.*, 1987, **60**, 86.

Anethole, Estragole, Propylanisole

Use/occurrence:	Flavour chemicals
Key functional groups:	Arylalkene, arylalkyl, methoxyphenyl
Test system:	Human (male) (oral, 1 mg anethole, 0.1 mg estragole, 0.1 mg propylanisole)

Structure and biotransformation pathway

The excretion of radioactivity in urine and in expired air was measured for the three compounds in two male volunteers. The excretion of $^{14}CO_2$ was higher for p-propylanisole (41.8, 43.5%) than for *trans*-anethole (19.8, 22.7%) or for estragole (10.4, 12.9%) but the excretion in urine was lower (p-propylanisole: 23.1, 26.4%, *trans*-anethole: 62.8, 66.9%; estragole: 54.0, 62.1%). Urinary metabolites were separated by TLC and HPLC. p-Propylanisole was primarily metabolized to 4-methoxyhippuric acid (1) (12%) and to 1-(4-methoxyphenyl)propan-1-ol (2) (2%) and -2-ol (3) (8%). *trans*-Anethole was converted into nine metabolites of which 4-methoxyhippuric acid (56%) was the major. Estragole was converted into five metabolites which included 1'-hydroxyestragole (4). The overall metabolic fate of these three anisole food flavours in man is qualitatively similar to that reported previously in rats and mice. However, the extensive formation of $^{14}CO_2$ indicates that phenolic metabolites formed by O-demethylation would not have been identified.

Reference

S. A. Sangster, J. Caldwell, A. J. Hutt, A. Anthony, and R. L. Smith, The metabolic disposition of [*methoxy*-^{14}C]-labelled *trans*-anethole, estragole, and *p*-propylanisole in human volunteers, *Xenobiotica*, 1987, **17**, 1223.

1,2-Epoxy-3-phenoxypropane, 1,2-Epoxy-3-(4-methoxyphenoxy)propane, 1,2-Epoxy-3-(4-nitrophenoxy)propane

Use/occurrence:	Model compounds
Key functional groups:	Aryl ether, epoxide
Test system:	Rat liver microsomes, 9000g supernatant, 100 000g supernatant

Structure and biotransformation pathway

(1)   (4) R = OH
                    (7) R = Glutathione

(2)   (5) R = OH
                    (8) R = Glutathione

(3)   (6) R = OH
                    (9) R = Glutathione

The metabolism of epoxides to glutathione conjugates and to diols (by glutathione S-transferase and microsomal epoxide hydrolase respectively) is a detoxification process, resulting in the loss of mutagenicity of the compounds. Three model epoxides (phenyl glycidyl ethers) have now been investigated in an attempt to correlate the loss of mutagenicity as measured by the Ames test with these metabolic pathways.

1,2-Epoxy-3-phenoxypropane (phenylglycidyl ether) (1) and its 4-methoxyphenyl (2) and 4-nitrophenyl (3) analogues were incubated in the presence and absence of glutathione with 9000g supernatant, microsomes, and 100 000g supernatant from rat liver. Aliquots were analysed by reverse phase HPLC, using 4-hydroxypropiophenone as internal standard, allowing the determination of the diols (4)–(6) and glutathione conjugates (7)–(9). With inactivated rat liver fractions no diols were seen, although some conversion of (3) into (9) was observed. In the presence of the 9000g supernatant and glutathione there was a greater production of the glutathione conjugates compared with the diols. A ratio of 15:1 was measured for (9):(6) and 2:1 for

106

(7):(4). All of the glutathione conjugates were obtained also from the 100 000g liver fraction. The production of the glutathione conjugates (7)–(9) bore some correlation with decreases in mutagenicity measured by the Ames test, although some detoxification was also observed when (1) and (2) were incubated with microsomes in the absence of glutathione.

Reference

J. E. Sinsheimer, E. Van den Eeckhout, B. H. Hooberman, and V. G. Beylin, Detoxification of aliphatic epoxides by diol formation and glutathione conjugation, *Chem. Biol. Interact.*, 1987, **63**, 75.

Tridiphane

Use/occurrence:	Herbicide
Key functional groups:	Chloroalkyl, chlorophenyl, epoxide
Test system:	*In vitro* (mouse liver cell fractions)

Structure and biotransformation pathway

GS = glutathionyl

In mouse liver microsomes tridiphane was metabolized to the corresponding diol metabolite (1). Microsomal epoxide hydrolases were thought to be responsible for this metabolism, since the reaction was prevented in the presence of 1 mM cyclohexane oxide, a known inhibitor of these enzymes. Cytosolic epoxide hydrolase did not appear to metabolize tridiphane. The formation of the glutathione conjugates (2) and (3) was investigated in cytosol. Tridiphane was found to be metabolized to a product with the same TLC R_f as a synthetic glutathione conjugate. The glutathione conjugate was found to be an inhibitor of glutathione transferase activity.

Reference

J. Magdalou and B. D. Hammock, Metabolism of tridiphane [2-(3,5-dichlorophenyl)-2(2,2,2-trichloroethyl)oxirane] by hepatic epoxide hydrolases and glutathione S-transferases in mouse, *Toxicol. Appl. Pharmacol.*, 1987, **91**, 439.

(R)-(+)-Pulegone

Use/occurrence:	Naturally occurring plant oil, with flea repellent and abortifacient properties
Key functional groups:	Alkene, cyclohexanone, terpene
Test system:	Mouse (intraperitoneal, 300 mg kg^{-1}) and *in vitro* mouse liver microsomes

Structure and biotransformation pathway

(1) X = H or D   (2)

The hepatotoxic effects of (R)-(+)-pulegone (1) and its deuterated analogue were compared 24 hours after a single intraperitoneal dose. Three out of 10 mice dosed non-deuterated (R)-(+)-pulegone did not survive to 24 hours, whereas all of the 10 mice given [^{2}H$_6$]-(R)-(+)-pulegone survived to 24 hours. Furthermore the (R)-(+)-pulegone treatment dramatically increased alanine aminotransferase activity (ALT), an indication of hepatocellular toxicity. In contrast the [^{2}H$_6$]-(R)-(+)-pulegone treatment had little effect on ALT activity. In further experiments it was shown that pretreatment of mice with phenobarbital caused an even greater increase in ALT activity 24 hours after a dose of (R)-(+)-pulegone, whereas pretreatment with β-naphthoflavone, cobaltous chloride, or piperonyl butoxide all reduced the effect of (R)-(+)-pulegone on ALT activity. Thus it was concluded that P-450 isozymes inducible by phenobarbital were responsible for activating (R)-(+)-pulegone to the proximate hepatotoxin.

Incubation of (R)-(+)-pulegone with mouse liver microsomes led to the production of several metabolites. A major metabolite was identified by GC–MS as menthofuran (2). Studies on the formation of this metabolite using ^{18}O$_2$ and H$_2$^{18}O showed that the oxygen incorporated into the metabolite was derived from molecular oxygen. A deuterium isotope effect was seen when [^{2}H$_6$]-(R)-(+)-pulegone was included in the microsomal incubations. The proposed route of formation of menthofuran (2) involved oxidation of an allylic methyl group, intramolecular cyclization, and dehydration to form the furan. Menthofuran is believed to be the proximate hepatoxin of (R)-(+)-pulegone.

Reference

W. Perry Gordon, A. C. Huitric, C. L. Seth, R. H. McClanahan, and S. D. Nelson, The metabolism of the abortifacient terpene, (R)-(+)-pulegone, to a proximate toxin menthofuran, *Drug Metab. Dispos.*, 1987, **15**, 589.

Spydust

Use/occurrence:	Espionage
Key functional groups:	Alkyl aldehyde, nitrophenyl
Test system:	Rat (dermal, intravenous 0.8 mg kg^{-1}, oral 0.8–80 mg kg^{-1})

Structure and biotransformation pathway

In rats spydust is rapidly absorbed from the gastrointestinal tract whereas absorption from the skin is slower and dose dependent. Once absorbed spydust is rapidly metabolized and excreted. The metabolites identified result from the oxidative metabolism of the pentadienal side-chain, (1) and (2), and reduction of the nitro group followed by acetylation, (3) and (5). The hippuric acid analogue (4) was also found. Metabolites were identified by HPLC and comparison with authentic standards.

Reference

L. T. Burka, J. M. Sanders, Y. C. Kim, and H. B. Mathews, Absorption and metabolism of 5-(nitrophenyl)penta-2,4-dienal (Spydust), *Toxicol. Appl. Pharmacol.*, 1987, **87**, 121.

trans-4-Hydroxyhex-2-enal

Use/occurrence:	Pyrrolizidine metabolite and a lipid peroxidation product
Key functional groups:	Alkene, alkyl aldehyde, secondary alkyl alcohol
Test system:	Single doses of 15 mg administered to anaesthetized rats (210–372 g body weight) via the portal vein

Structure and biotransformation pathway

$$CH_3CH_2\overset{*}{C}HCH = CHCHO$$

(1)   * = ^3H

(2)

(3)

At 24 hours after administration of $[^3H]$-(1) less than 3% of the radiolabel dosed remained in the liver and the majority of the radioactivity had been eliminated in the urine (77–83% of the dose). A high proportion of the radioactivity in the urine was associated with an acidic fraction. The occurrence of the mercapturic acid metabolite (3) in this fraction was demonstrated by tandem mass spectrometry using negative ion fast atom bombardment in combination with mass analysed kinetic energy spectrometry and comparison with a synthetic standard. The proposed mechanism of formation of the mercapturic acid (3) is a Michael addition of glutathione to the alkene (1). Formation of a hemiacetal gave (2), and subsequent cleavage of glutamic acid and glycine moieties followed by acetylation gave the terminal metabolite (3). Thus the occurrence of this metabolite provides *in vivo* evidence for the occurrence of the Michael addition of glutathione.

Reference

C. K. Winter, H. J. Segall, and A. Jones, Distribution of *trans*-4-hydroxyhex-2-enal and tandem mass spectrometric detection of its urinary mercapturic acid in the rat, *Drug Metab. Dispos.*, 1987, **15**, 608.

Allylamine

Use/occurrence:	Cardiovascular toxin
Key functional groups:	Allyl, alkyl amine
Test system:	Rat (oral, 150 mg kg^{-1})

Structure and biotransformation pathway

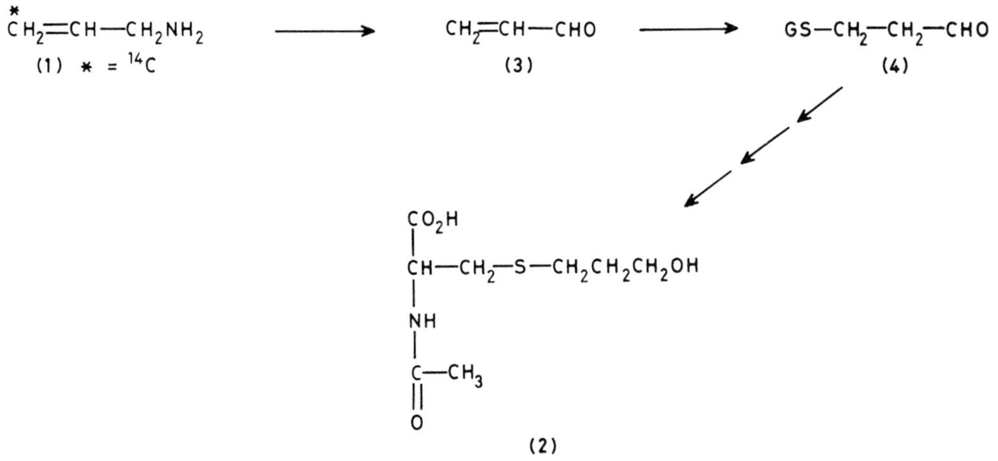

$$\overset{*}{C}H_2=CH-CH_2NH_2 \longrightarrow CH_2=CH-CHO \longrightarrow GS-CH_2-CH_2-CHO$$

(1) $* = {}^{14}C$ (3)   (4)

$$CO_2H$$
$$|$$
$$CH-CH_2-S-CH_2CH_2CH_2OH$$
$$|$$
$$NH$$
$$|$$
$$C-CH_3$$
$$||$$
$$O$$

(2)

After administration of (1) to rats, 62% of the radioactive dose was recovered in 0–24 hour urine. A single metabolite was found which, by HPLC, was identified as 3-hydroxypropylmercapturic acid (2) and confirmed by NMR and MS. It is known from previous studies *in vitro* that (1) is metabolized to the highly reactive aldehyde acrolein (3). It is proposed that *in vivo* the formation of (2) arises from the glutathione conjugate of acrolein (4).

Reference

P. J. Boor, R. Sanduja, T. J. Nelson, and G. A. S. Ansari, *In vivo* metabolism of the cardiovascular allylamine, *Biochem. Pharmacol.*, 1987, **36**, 4347.

2-Chlorobenzylidenemalononitrile

Use/occurrence: Sensory irritant

Key functional groups: Alkene, chlorophenyl, nitrile

Test system: Rat (intravenous or intragastric 0.08–159 μmol kg^{-1})

Structure and biotransformation pathway

The proposed metabolic scheme for 2-chorobenzylidenemalononitrile (1) was elucidated in the rat from intravenous or intragastric administration of ^{3}H-ring-labelled and two ^{14}C-labelled forms. At all doses after both routes of administration the majority (44–74%, intravenous; 67–100%, intragastric) of the radioactivity was excreted in urine. The urinary recovery of radioactivity did not depend on the administered dose and more than 85% of the total urinary radioactivity was eliminated in the first 24 hours. 2-Chlorohippuric acid (2) was the major metabolite and together with its immediate precursor,

114

2-chlorobenzoic acid (3), accounted for 72% of the urinary radioactivity. The acid (3) is a product of 2-chlorobenzaldehyde (4) oxidation, and the remainder of the aldehyde was reduced to 2-chlorobenzyl alcohol (5). This was further metabolized to (2-chlorobenzyl)acetylcysteine (6) and 2-chlorobenzylglucuronic acid (7). The remainder of the urinary radioactivity arose from the reduction of the olefinic side-chain of (1) to yield dehydro-CS (8) followed by hydrolysis via the amide (9) to (2-chlorophenyl)-2-cyanopropionic acid (10). 2-Chlorophenylacetylglycine (11) was also formed from (8).

The proposed pathway was confirmed by studies of the metabolism of (4) and (8). Each purified metabolite was identified as the methyl ester by GC–MS and the structures were confirmed by reference to mass spectra of synthetic standards. In addition, reverse isotope dilution analysis of the underivatized metabolites with standards was carried out.

Reference

K. Brewster, J. M. Harrison, L. Leadbetter, J. Newman, and D. G. Upshall, The fate of 2-chlorobenzylidenemalononitrile (CS) in rats, *Xenobiotica*, 1987, **17**, 911.

Acrylonitrile

Use/occurrence:	Manufacture of acrylic fibres and plastics, fumigant
Key functional groups:	Alkene, nitrile
Test system:	Rat (inhalation)

Structure and biotransformation pathway

The metabolism of acrylonitrile (1) has previously been well studied, with reports of the formation of glycidonitrile, cyanoacetaldehyde, glycolaldehyde, cyanoethanol, cyanoacetic acid, hydroxyethylmercapturic acid (2), carboxy-methylcysteine (3) thiodiglycolic acid (4), cyanoethylmercapturic acid (5), *N*-acetyl-3-carboxy-5-cyanotetrahydro-1,4-2[*H*]thiazine, and glucuronides. In the present study the urinary excretion of (2)–(5) was studied in the rat in an attempt to determine whether one of these metabolites was suitable as a biological monitor for acrylamide exposure in man.

Compounds (2), (3), and (5) were measured with a modified amino acid analysis. Compound (1) was determined by GC and (4) by GC–MS of its methyl ester. At an inhaled concentration of (1) of 5 p.p.m. 20% of the dose was excreted as (2)–(5), but at higher concentrations excretion fell to *ca.* 10%. The excretion of (5) was the most sensitive metabolite monitor of exposure to (1). Metabolite (4) was also concluded to be a possibly useful monitor, but (2) and (3) were not recommended on the basis of the low amounts of these compounds excreted. Acrylonitrile (1) was also excreted unchanged, although dose levels greater than 10 p.p.m. were necessary for significant detection of this compound.

Reference

G. Müller, C. Verkoyen, N. Soton, and K. Norpoth, Urinary excretion of acrylonitrile and its metabolites in rats, *Arch. Toxicol.*, 1987, **60**, 464.

116

Diethylstilbestrol

Use/occurrence:	Synthetic oestrogen
Key functional groups:	Alkene, aryl alkene, phenol
Test system:	Isolated hamster hepatocytes

Structure and biotransformation pathway

When freshly isolated hamster hepatocytes were incubated for 60 minutes with 50 nM diethylstilbestrol/mg cellular protein, yields of metabolites (1), (2), and (3) were 9.1, 14.2, and 0.3% respectively. Glucuronides (4%) and sulphates (2.8%) of diethylstilbestrol and its oxidative metabolites were also found. Metabolites were assayed by HPLC and the conjugates were hydrolysed with β-glucuronidase and sulphatase enzymes. Metabolites (1) and (2) were confirmed following GC–MS of their trimethylsilyl derivatives. Metabolite (3) was found in the sulphate/glucuronide fraction and the identity of the aglycone was confirmed by HPLC co-chromatography with a standard. The presence of reactive intermediates was postulated to account for non-extractable protein bound radioactivity.

Reference

G. Blaich, E. Pfaf, and M. Metzler, Metabolism of diethylstilbestrol in hamster hepatocytes, *Biochem. Pharmacol.*, 1987, **36**, 3135.

2-Bromo-octane, 2-Iodo-octane

Use/occurrence:	Chiral model substrates
Key functional groups:	Bromoalkyl, iodoalkyl, chiral carbon
Test system:	Rat liver partially purified glutathione S-transferase, rat liver cytosol

Structure and biotransformation pathway

$$CH_3 \overset{X}{\underset{H}{\overset{\dagger}{-}C-}}C_6H_{13} \longrightarrow CH_3 \overset{SG}{\underset{H}{-C-}}C_6H_{13}$$

X = Br or I
† = chiral carbon

GS = Glutathionyl

   The reactions of enantiomeric 2-bromo-octane and 2-iodo-octane with glutathione proceeded with inversion of configuration at the chiral carbon of the substrate. No evidence for substrate specificity was obtained. Incubation of racemic substrates resulted in approximately equal amounts of the corresponding conjugates. Incubation of either enantiomer resulted in formation of only one of the two diastereomers of the glutathione conjugate. Yields of glutathione conjugates were higher from 2-iodo-octane. 2-Chloro-octane was not a substrate for the enzyme systems used in this study.

Reference

R. E. Ridgewell and M. M. Abdel-Monem, Stereochemical aspects of the glutathione S-transferase-catalysed conjugations of alkyl halides. *Drug Metab. Dispos.*, 1987, **15**, 82.

α-Bromoisovalerylurea

Use/occurrence:	Model compound
Key functional groups:	Alkyl carboxamide, bromoacetyl, chiral carbon, ureide
Test system:	Isolated rat hepatocytes ($250–500\ \mu M$)

Structure and biotransformation pathway

(1) * = ^{14}C
† = chiral carbon

(2) GS = glutathionyl

(3)

(4)

(5)

α-Bromoisovalerylurea (1) had previously been used as a model substrate for the study of glutathione conjugation *in vivo* and in isolated perfused liver. In isolated hepatocytes a major proportion of this substrate was conjugated with glutathione to yield (2), but amidase-catalysed hydrolysis also occurred resulting in formation of α-bromoisovaleric acid (3) and urea (4). The hydrolysis product (3) was also efficiently conjugated with glutathione to yield (5). In glutathione-depleted hepatocytes, no conjugates, but only the hydrolysis products (3) and (4), were formed. A pronounced stereoselectivity in the metabolism of the enantiomers of (1) was observed; the R-enantiomer was conjugated with glutathione much faster than the S-enantiomer. The S-enantiomer was substantially hydrolysed, ultimately leading to the formation of the conjugate (5) of the hydrolysis product (3). Analysis and characterization of products in this study were by TLC (radiolabelled metabolites) and HPLC (non-radiolabelled metabolites and separation of diastereomers).

Reference

J. M. te Koppele, I. A. M. de Lannoy, K. S. Pang, and G. J. Mulder, Stereoselective glutathione conjugation and amidase-catalysed hydrolysis of α-bromoisovaleryl-urea enantiomers in isolated rat hepatocytes, *J. Pharmacol. Exp. Ther.*, 1987, **243**, 349.

1,2-Dibromo-1-phenylethane (DBPE)

Use/occurrence:	Model compound
Key functional groups:	Benzyl bromide, bromoalkyl
Test system:	Rat (oral, 66–330 mg kg^{-1})

Structure and biotransformation pathway

(3) GS = Glutathionyl (1) (2) RS = *N*-Acetylcysteinyl

The metabolism of DBPE was investigated to study the influence of aromatic substitution on the metabolism of a vicinal dibromo compound. The relationship between metabolism and the mutagenicity of 1,2-dibromoethane (DBE) has been well studied. DPBE is also mutagenic in the Ames test.

Following oral administration of DBPE, 41% of the dose was recovered in the 0–24 hour urine as the mercapturic acid (2). Pretreatment of rats with 1-phenylimidazole (a P-450 inhibitor) reduced the excretion of this metabolite as did pretreatment with diethyl maleate (which depletes glutathione). Thus two possible pathways were proposed, one with a primary oxidative step and one primarily involving glutathione. An episulphonium ion intermediate was proposed for the latter pathway. The intermediate glutathione conjugate (1) was found in the bile as an *R/S* enantiomeric mixture. The positional isomer (3) was also found in the bile as an *R/S* enantiomeric mixture. However, the corresponding mercapturic acid could not be found in the urine of non-cannulated rats dosed with DPBE. Identity of the mercapturic acid (2) was confirmed by GC–MS analysis and the glutathione conjugates (1) and (3) were determined by HPLC and comparison with synthetic standards.

Reference

C. E. M. Zoetemelk, W. van Hove, W. L. J. van der Laan, B. van Meeteren-Walchli, A. van der Gen, and D. D. Breimer, Glutathione conjugation of 1,2-dibromo-1-phenylethane in rats *in vivo*, *Drug Metab. Dispos.*, 1987, **15**, 418.

α-Chloro-DDT, Dicofol

Use/occurrence:	Acaricide
Key functional groups:	Benzhydrol, chloroalkane, chlorophenyl
Test system:	Male mice (intraperitoneal, 30 mg kg^{-1}), rat liver microsomal and cytosolic fractions

Structure and biotransformation pathway

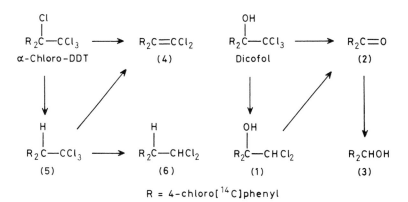

R = 4-chloro[^{14}C]phenyl

Dicofol was reductively dechlorinated to the dichloroethanol (1) *in vitro* and *in vivo* and both compounds were metabolized to dichlorobenzophenone (2) and dichlorobenzhydrol (3) *in vivo*. α-Chloro-DDT (an impurity in dicofol) was dechlorinated to the dichloroethylene (4) and to DDT (5) *in vivo* in the mouse. *In vitro* an anaerobic rat liver microsome system with NADPH also gave the reductive dechlorination products, but a liver cytosolic system with glutathione was less effective. One- and two-dimensional TLC with GC analyses were used to identify the metabolites.

Reference

M. A. Brown and J. E. Casida, Metabolism of a dicofol impurity α-chloro-DDT, but not dicofol or dechlorodicofol, to DDE in mice, and a liver microsomal system, *Xenobiotica*, 1987, **17**, 1169.

Mitometh, Lysodren

Use/occurrence:	Anticancer drugs
Key functional groups:	Chloroalkane, chlorophenyl
Test system:	Rat (oral, 400 mg kg^{-1}), guinea pig (oral, 900 mg kg^{-1})

Structure and biotransformation pathway

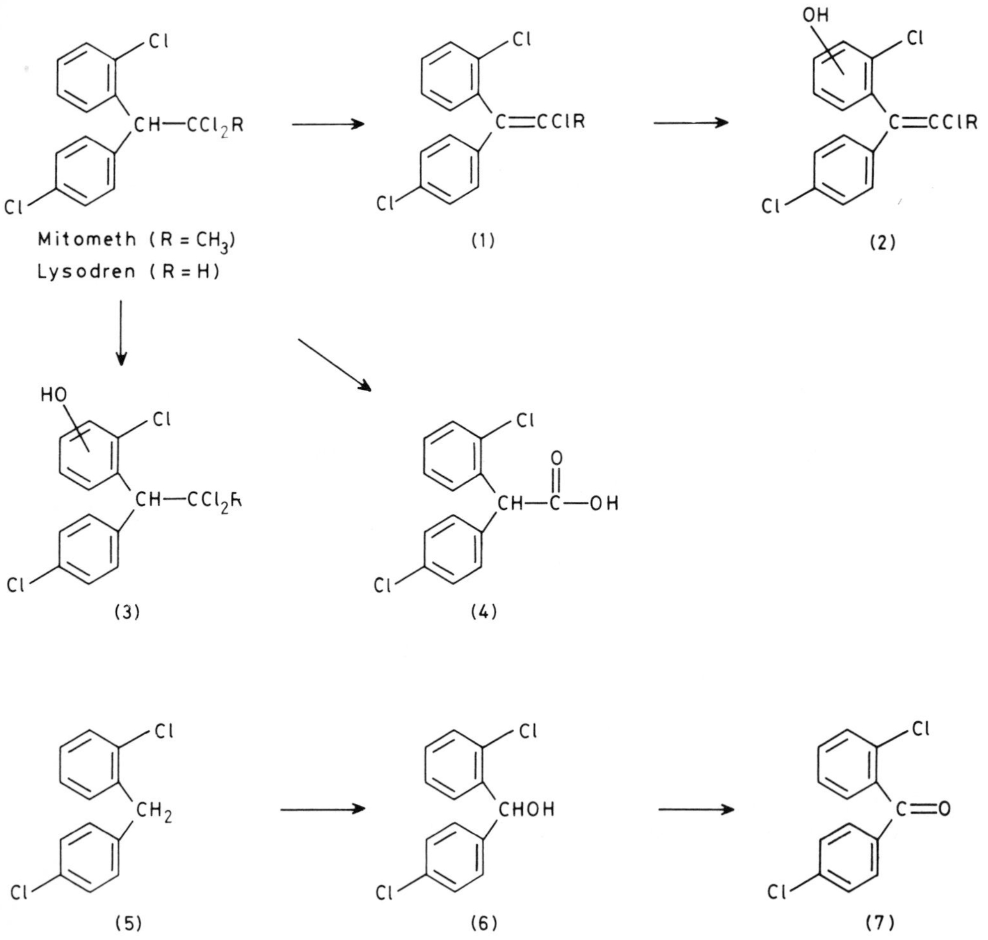

Mitometh (R = CH$_3$)
Lysodren (R = H)

(1)

(2)

(3)

(4)

(5)

(6)

(7)

The toxicity, metabolism, and therapeutic effect of lysodren and its methyl analogue mitometh were compared. Mitometh was found to have less severe side effects than lysodren yet still produce the desired therapeutic effect. It was suggested that there is a metabolic basis for this difference.

The unwanted side effects, including alopecia, diarrhoea, and weakness, were attributed to diphenylacetic acid (4). Diphenylacetic acid was the major metabolite of lysodren but could not be detected following administration of mitometh. In rats a large proportion of an oral dose of mitometh was excreted unchanged (40–60%). Only trace amounts of lysodren were found in the urine. The dehydrohalogenated metabolite (1) of mitometh accounted for 10–15% of the chlorine-containing metabolites in the rat and between 2 and 5% in the guinea pig. Metabolites were identified in extracts of urine by use of computerized MS interfaced to capillary GC. Reference standards were available for metabolites (1), (4), (5), (6), and (7). In addition several metabolites of both lysodren and mitometh were found giving M/Z values 16 greater than the compound administered and its corresponding dehydrohalogenated metabolite. Thus structures (2) and (3) were proposed. For both lysodren and mitometh, side-chain oxidation products (5)–(7) were also found.

Reference

B. L. Jensen, M. W. Caldwell, L. G. French, and D. G. Briggs, Toxicity, ultrastructural effects, and metabolic studies with 1-(o-chlorophenyl)-1-(p-chlorophenyl)-2,2-dichloroethane (o,p'-DDD) and its methyl analog in the guinea pig and rat, *Toxicol. Appl. Pharmacol.*, 1987, **87**, 1.

DDT, DDD, DDE, DDMU

Use/occurrence:	Insecticides
Key functional groups:	Chloroalkene, chloroalkyl, chlorophenyl
Test system:	Rat, Japanese quail (intraperitoneal, 200 mg kg^{-1})

Structure and biotransformation pathway

The excretion of radioactivity after doses of DDT (1), DDD (2), DDE (3), and DDMU (4) was determined. Metabolites in excreta were extracted into either hexane or methanol, separated by TLC, and examined by GC and GC–MS . Radioactivity was more rapidly excreted after doses of (4) than of

124

the other compounds. With the exception of (4) excretion was more rapid in rat than in quail. For (1), (2), and (3) the radioactivity in extracts of faeces was higher in quail (64%, 80%, and 63% dose) than in rat (53%, 38%, and 54% dose), indicating that the rat excretes significant amounts of these compounds as polar (non-extractable) metabolites. However, with (4) the extractable radioactivity in faeces was similar in rat (78%) and quail (70%). After administration of (1) to rats the acetic acid derivative (5) was the major metabolite. The alcohol (6), ketone (7), (2), (3), and unchanged (1) were also present. In quail, unchanged (1) was the major radiolabelled component and (5) was again the major metabolite. Small quantities of (2) and (3) were also detected. The major metabolite of (2) in both rat and quail was (5). In addition to unchanged (2), (3) and (7) were also present. The metabolism of (3) and (4) showed species-dependent differences. In the rat, the major metabolites of (3) were ring-hydroxylated derivatives (8): there was no evidence for this pathway in the quail. After administration of (4), (5) and (6) were the major metabolites in the rat although (7), (6), and the corresponding acetate (9) together with unchanged (4) were also present. In quail, (9) was the major metabolite identified and a ring-hydroxylated derivative (10) was significant. The conversion of (1) and (4) into (5) is postulated to proceed via the aldehyde (11).

This study has shown that (1), (2), and (3) are excreted more slowly by the Japanese quail than by the rat. This difference may be explained by the poorer ability of the quail to produce hydrophilic metabolites such as (5). However, although (4) was rapidly excreted by the quail it was not metabolized significantly to (5). From a comparison of the metabolism of (4) and other compounds it was concluded that (4) is not an intermediate in the metabolism of (1).

Reference

S. C. Fawcett, L. J. King, P. J. Bunyan, and P. I. Stanley, The metabolism of [14C]-DDT, [14C]-DDD, [14C]-DDE, and [14C]-DDMU in rats and Japanese quail, *Xenobiotica*, 1987, **17**, 525.

3-Chloro-2-methylpropene

Use/occurrence:	Insecticide, fumigant, synthetic intermediate
Key functional groups:	Alkene, chloroalkyl
Test system:	Rat (oral, 150 mg kg^{-1})

Structure and biotransformation pathway

In rats, about 7% of the dose of 3-chloro-2-methylpropene (1) was exhaled unchanged during 24 hours after dosing while about 10% dose was exhaled as $^{14}CO_2$. About 58% of the dose was excreted in the 0–24 hour urine. Eight metabolites in urine were separated and detected by radio-HPLC. The mercapturic acid conjugate (2) was the most important (45% urinary radioactivity) and was identified by ^{1}H NMR, FAB-MS, and by comparison with a synthetic standard. The other metabolites were not investigated.

Reference

B. I. Ghanayem and L. T. Burka, Comparative metabolism and disposition of 1-chloro- and 3-chloro-2-methylpropene in rats and mice, *Drug Metab. Dispos.*, 1987, **15**, 91.

1-Chloro-2-methylpropene (dimethylvinyl chloride)

Use/occurrence: Chemical intermediate

Key functional groups: Chloroalkene

Test system: Rat, mouse (oral, 150 mg kg^{-1})

Structure and biotransformation pathway

Rats and mice each exhaled 25% dose as $^{14}CO_2$ during 24 hours after administration of [2-^{14}C]dimethylvinyl chloride (1), while 30% and 5% dose were exhaled unchanged by rats and mice respectively. The 24 hour urine of rats and mice contained 35% and 47% dose respectively. Patterns of metabolites in urine from the two species were quantitatively similar. Up to nine metabolites were separated and detected by radio-HPLC. Metabolites (2) and (3) were identified by ^{1}H NMR, ^{13}C NMR [not (3)], FAB-MS, and comparison with synthetic standards which confirmed the *trans* stereochemistry of the metabolites. The cysteine conjugate (2) was the major urinary metabolite and accounted for 23% (rat) or 35% (mouse) urinary radioactivity. The mercapturic acid (3) accounted for 9% (rats) or 12% (mice) urinary radioactivity.

Reference

B. I. Ghanayem and L. T. Burka, Comparative metabolism and disposition of 1-chloro- and 3-chloro-2-methylpropene in rats and mice, *Drug Metab. Dispos.*, 1987, **15**, 91.

Tetrachloroethylene, *S*-(1,2,2-Trichlorvinyl)-L-cysteine

Use/occurrence:	Solvent and its metabolite
Key functional groups:	Amino acid, alkyl thioether, chloroalkane
Test system:	*In vitro* rat subcellular fractions and bacterial β-lyase enzyme

Structure and biotransformation pathway

In the presence of NADPH and rat hepatic microsomal fractions tetrachloroethylene was metabolized to the water-soluble metabolites trichloroacetic acid (1) and oxalic acid (2). In addition a metabolite was found which was largely bound to microsomal macromolecules; the majority of the alkylated macromolecules were identified as *N*-trichloroacetylated phospholipids by HPLC.

When tetrachloroethylene was incubated with microsomes (in the absence of NADPH) or with cytosol in the presence of 10 mM glutathione the glutathione conjugate (3) was formed.

In further experiments the hydrolysis product of (3), *S*-(1,2,2-trichlorovinyl)-L-cysteine (4), was cleaved by a bacterial β-lyase to dichloroacetic acid (5) and pyruvate. Metabolites were identified following derivatization by GC–MS. It was postulated that the oxidative pathway was a detoxification pathway and that the glutathione pathway could contribute to the observed nephrocarcinogenic effect of tetrachloroethylene in the rat. A mutagenic thiol intermediate was proposed in the pathway leading to dichloroacetic acid.

Reference

W. Dekant, G. Martens, S. Vamvakas, M. Metzler, and D. Henschler, Role of glutathione-transferase-catalysed conjugation *versus* cytochrome P-450-dependent phospholipid alkylation, *Drug Metab. Dispos.*, 1987, **15**, 702.

Clomiphene

Use/occurrence:	Antioestrogen, anti-cancer agent
Key functional groups:	Alkyl aryl ether, chloroalkene, dialkylaminoalkyl
Test system:	Rat (intraperitoneal, 0.15 mg), rat liver microsomes

Structure and biotransformation pathway

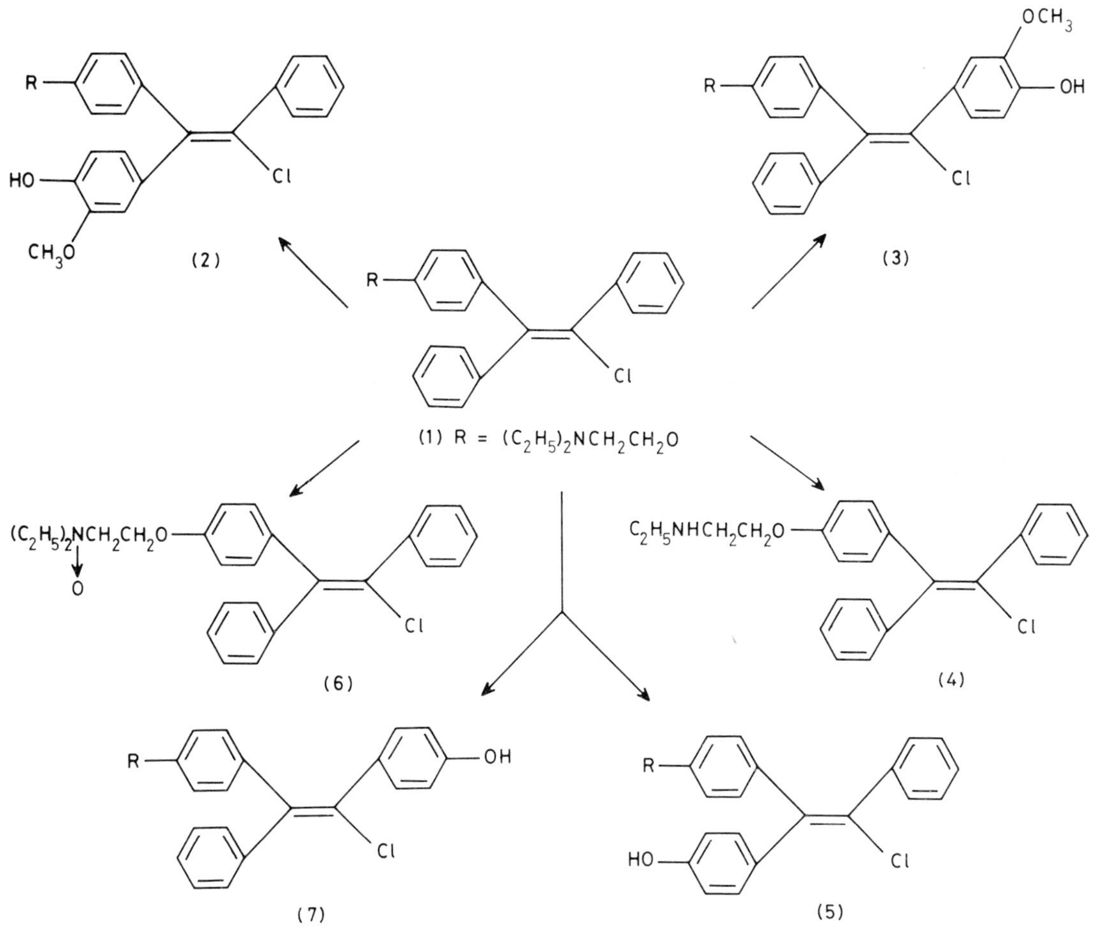

R
HO
CH3O
Cl
(2)
OCH3
OH
(3)
(1) R = (C2H5)2NCH2CH2O
(C2H5)2NCH2CH2O
O
(6)
C2H5NHCH2CH2O
(4)
(7)
(5)

Intraperitoneal administration of [^{3}H,^{14}C]clomiphene (1) to female rats resulted in a total of 85% of the dose being excreted in the faeces. TLC examination of faecal extracts showed the presence of (1) together with 3-methoxy-4-hydroxyclomiphene (2), 3′-methoxy-4′-hydroxyclomiphene (3), desethylclomiphene (4), 4-hydroxyclomiphene (5), and clomiphene N-oxide (6).

Metabolism of (1) by liver microsomes yielded (6) as the major product (TLC, HPLC) followed by (5), 4'-hydroxyclomiphene (7), and (4). Pretreatment of the rats with phenobarbital caused (4) to be the major product followed by (5), (7), and (6). 4'-Hydroxyclomiphene (7) reacted readily with alcohols; mass spectrometry of metabolite (7) showed that the chlorine had been replaced with methoxy or ethoxy, resulting from contact with methanol or ethanol in the chromatography systems. (7) Also reacted with γ-(p-nitrobenzyl)pyridine (used as a test for electrophiles). Metabolites (5) and (7) could be converted into (2) and (3) respectively by incubation with liver microsomes in the presence of S-adenosyl-L-methionine and NADPH. The guaiacol (3) also showed reactivity with γ-(p-nitrobenzyl)pyridine, possibly due to formation of a quinonoid structure.

Reference

P. C. Ruenitz, R. F. Arrendale, G. D. George, C. B. Thompson, C. M. Mokler, and N. T. Nanavati, Biotransformation of the antiestrogen clomiphene to chemically reactive metabolites in the immature female rat, *Cancer Res.*, 1987, **47**, 4015.

Aldrin

Use/occurrence:	Model compound
Key functional groups:	Alkene, chlorocycloalkene
Test system:	Reconstituted monooxygenase systems and microsomes from rat liver

Structure and biotransformation pathway

The ability of ten cytochrome P-450 isozymes isolated from untreated and inducer treated rat livers to catalyse aldrin epoxidation was investigated. Seven of these isozymes were found to catalyse aldrin epoxidation; a novel aldrin metabolite *endo*-dieldrin (3) was produced in a six-fold excess over *exo*-dieldrin (1) (the usual form of the metabolite) by one isozyme form. Further studies of aldrin epoxidation were performed with liver microsomes prepared from rats with varying physiological status. The results of these studies suggested that aldrin epoxidation is a reaction involving male-specific and phenobarbital inducible P-450 isozymes in rat liver. Metabolites were determined by GC assay. The structure of the *endo*-dieldrin (3) was confirmed by GC–MS comparison with a standard.

Reference

T. Woolf and F. P. Guenerich, Rat liver cytochrome P-450 isozymes as catalysts of aldrin epoxidation in reconstituted monooxygenase systems and microsomes, *Biochem. Pharmacol.*, 1987, **36**, 2581.

Acyclic Functional Compounds

Sodium monofluoroacetate

Use/occurrence:	Rodenticide
Key functional groups:	Fluoroacetyl
Test system:	Mouse (intravenous, 0.5 mg kg^{-1})

Structure and biotransformation pathway

$$FCH_2CO_2Na \longrightarrow F^-$$

For this study the test compound was labelled with the short-lived ($t_{1/2}$ = 110 min), positron-emitting ^{18}F radionuclide, and the distribution and excretion of the radiolabel was studied up to 4 hours after dosing. The presence of $^{18}F^-$ in plasma was demonstrated by paper chromatography. Comparison of data obtained after administration of Na^{18}F led to the suggestion that following administration of [^{18}F]fluoroacetate about 25% of the radiolabel was present in osseous tissue as F$^-$ at 2 hours after dosing. It was also suggested that urinary excretion of ^{18}F following administration of [^{18}F]fluoroacetate was due to renal elimination of $^{18}F^-$ generated from *in vivo* degradation of [^{18}F]fluoroacetate, but no experimental evidence was presented to support this hypothesis.

Reference

T. R. Sykes, J. H. Quastel, M. J. Adam, T. J. Ruth, and A. A. Noujaim, The disposition and metabolism of [^{18}F]fluoroacetate in mice, *Biochem. Arch.*, 1987, **3**, 317.

Acetonitrile

Use/occurrence: Solvent

Key functional groups: Nitrile

Test system: Rat hepatocytes

Structure and biotransformation pathway

$$CH_3CN \longrightarrow HCN + \ ?$$

It is known that saturated nitriles may be metabolized to cyanide although the mechanism of this transformation is not completely understood. Two possibilities are (a) interaction with glutathione and glutathione transferase (nucleophilic substitution) or (b) cytochrome P-450 oxidation to a cyanohydrin intermediate which would decompose to hydrogen cyanide and formaldehyde. These possibilities have now been explored using hepatocytes from rats (female Sprague–Dawley) that had been pretreated with buthionine sulfoximine (BSO) (to reduce the glutathione content) or with cobalt heme (to reduce cytochrome P-450 content) or acetone (to induce cytochrome P-450).

BSO treatment reduced the glutathione content of hepatocytes to 18% of control levels, but did not significantly affect the cyanide production from acetonitrile. Cobalt heme reduced cytochrome P-450 to 41% of control levels and significantly decreased cyanide production (13% of control at an acetonitrile concentration of 20 mM). Pretreatment with acetone doubled the yield of cyanide from acetonitrile in hepatocytes (even though *in vitro* experiments had demonstrated a direct inhibiting effect on acetonitrile metabolism by acetone).

It thus appears that acetonitrile is not metabolized by nucleophilic substitution by glutathione, but by an acetone-inducible isoenzyme of cytochrome P-450. The possible cyanohydrin intermediate has not been identified, however, and attempts to show production of formaldehyde have been unsuccessful.

Reference

J. J. Freeman and E. P. Hayes, The metabolism of acetonitrile to cyanide by isolated rat hepatocytes, *Fund. Appl. Toxicol.*, 1987, **8**, 263.

Ethylene glycol ethyl ether, ethylene glycol acetate ethyl ether

Use/occurrence: Solvent

Key functional groups: Alkyl alcohol, dialkyl ether

Test system: Occupational human exposure (urinary excretion)

Structure and biotransformation pathway

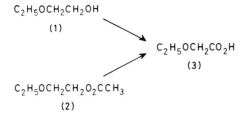

Urine was collected from five women with occupational exposure to the ethyl ether of ethylene glycol (1) and the ethyl ether of ethylene glycol acetate (2). The air concentrations of (1) and (2) were determined by GC. The ethers (1) and (2) were metabolized to ethoxyacetic acid (3) which was determined in the urine by GC, as its methyl ester derivative. Excretion of (3) increased over the working week, reaching more than 100 mg g^{-1} creatine on Friday. Clearance was slow, with 1.2–2.6 mg g^{-1} creatine still being detectable 12 days later. The concentration of (3) at the end of a week's exposure correlated with the exposure dose of (1) or (2), confirming the validity of measurements of (3) as a biological monitor for (1) or (2).

Reference

H. Veulemans, D. Groeseneken, R. Masschelein, and E. van Vlem, Field study of the urinary excretion of ethoxyacetic acid during repeated daily exposure to the ethyl ether of ethylene glycol and the ethyl ether of ethylene glycol acetate, *Scan. J. Work Environ. Health*, 1987, **13**, 239.

n-Butoxyethanol

Use/occurrence:	Solvent
Key functional groups:	Alkyl alcohol, dialkyl ether
Test system:	Rat (subcutaneous, dermal $118-200$ mg kg^{-1})

Structure and biotransformation pathway

$$n-C_4H_9O\overset{*}{C}H_2\overset{*}{C}H_2OH \longrightarrow n-C_4H_9OCH_2CO_2H$$

(1) $* = {}^{14}C$ (2)

Following subcutaneous application of $[1,2-{}^{14}C]$-n-butoxyethanol (1) to rats (118 mg kg^{-1}) 10% of the dose was eliminated as ${}^{14}CO_2$ in exhaled air. 78% of the radioactivity was eliminated in the urine in 72 hours. Percutaneous absorption of (1) was also studied, in view of the widespread use of the compound in cleaning products that are in contact with the human skin. 20–23% of the radioactivity from the dose (200 mg kg^{-1}) was excreted in the urine in 48 hours. The highest blood and plasma levels of radioactivity were seen after 2 hours. The major metabolite found in the plasma ultrafiltrate was n-butoxyacetic acid (2), which was determined by HPLC.

Reference

F. G. Bartnik, A. K. Reddy, G. Klecak, V. Zimmermann, J. J. Hostynek, and K. Kunstler, Percutaneous absorption, metabolism, and hemolytic activity of n-butoxyethanol, *Fund. Appl. Toxicol.*, 1987, **8**, 59.

n-Butoxyethanol (ethylene glycol monobutyl ether)

Use/occurrence:	Solvent, chemical intermediate
Key functional groups:	Alkyl alcohol, dialkyl ether
Test system:	Rat (oral, 125 and 500 mg kg^{-1})

Structure and biotransformation pathway

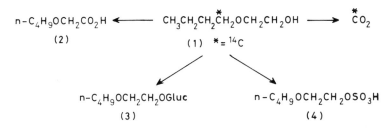

Approximately 18% and 10% dose respectively were exhaled as $^{14}CO_2$ during 48 hours following single 125 or 500 mg kg^{-1} doses. No more than 2% dose was exhaled as other volatiles and less than 3% dose was excreted in faeces. Urinary excretion during 48 hours amounted to 70% dose (125 mg kg^{-1}) or 45% dose (500 mg kg^{-1}). Butoxyacetic acid (2) was the major urinary metabolite at both dose levels and was identified by HPLC and ^1H NMR comparison with a synthetic standard. The glucuronide conjugate (3) of the parent alcohol was also an important urinary metabolite at both dose levels whereas the sulphate conjugate (4) was detected as a minor metabolite only at the lower dose. Conjugates were identified by FAB-MS, ^1H NMR, and enzymic deconjugation. No significant amounts of unchanged (1) were excreted in urine.

Reference

B. I. Ghanayem, L. T. Burke, J. M. Sanders, and H. B. Matthews, Metabolism and disposition of ethylene glycol monobutyl ether (2-butoxyethanol) in rats, *Drug Metab. Dispos.*, 1987, **15**, 478.

Polyethylene glycols (200, 400, 1000, and 6000)

Use:	Solvents
Key functional groups:	Alkyl alcohol, dialkyl ether
Test system:	Perfused rat and guinea pig livers

Structure and biotransformation pathway

$$H-[OCH_2CH_2]_nOCH_2CH_2OH$$

Sulphate conjugate

$$H-[OCH_2CH_2]_nOCH_2CH_2OSO_3H \ (?)$$

The conjugation of polyethylene glycol (PEG) with [^{35}S]sulphate was investigated in isolated perfused rat and guinea pig livers. The addition of PEG 200 (*ca.* 40 mM) to the perfusate of a rat liver containing [^{35}S]sulphate caused a small increase in the concentrations of ester sulphate in the bile and plasma. TLC analysis of bile showed that this was caused by the appearance of two sulphated metabolites one of which was present in only trace amounts. The addition of PEG 200 to the perfusate of a guinea pig liver caused a rapid increase in the concentration of ester [^{35}S]sulphate in the bile. Three metabolites were found, the major one having TLC characteristics identical to the major metabolite found in the rat experiment. The sulphation of other PEGs was also studied in the isolated perfused guinea pig liver. PEG 400 and PEG 1000 formed smaller amounts of sulphate conjugates than PEG 200 but these were chromatographically identical to the minor metabolites formed from PEG 200. It was concluded that the latter were therefore formed from high molecular weight impurities in the PEG 200. In contrast to the other PEGs, PEG 6000 was not sulphated by the perfused guinea pig liver and neither was ethylene glycol.

The perfused guinea pig liver can sulphate PEG 200 at a higher rate than the perfused rat liver but even so the amount sulphated is less than 1% of the PEG 200 added to the perfusate. Although the concentration of [^{35}S]sulphate ester formed is similar in bile and plasma for both species, the proportion of the sulphate metabolite eliminated in plasma is much greater than in bile because of the volume of perfusate used.

Reference

A. B. Roy, C. G. Curtis, and G. M. Powell, The metabolic sulphation of polyethylene glycols by isolated perfused rat and guinea pig livers, *Xenobiotica*, 1987, **17**, 725.

Mannose

Use/occurrence:	Carbohydrate
Key functional groups:	Hexose
Test system:	Hamster (intraperitoneal, $31\ \mu g\ kg^{-1}$)

Structure and biotransformation pathway

(1) $*$ = ^3H (2)

In order to study the possible formation of retinyl phosphate mannose, hamsters were treated intraperitoneally with $[2\text{-}^3H]$mannose (4 mCi kg^{-1}, 31 $\mu g\ kg^{-1}$) (1). In fact no evidence was found for retinyl phosphate mannose but a novel radiolabelled metabolite was detected by chromatography of extracts of intestinal epithelial cells. The compound co-chromatographed with lactic acid (2) and was a substrate for lactate oxygenase and L-lactate-2-monooxygenase (leading to compounds that co-chromatographed with pyruvate and acetate respectively). Radiolabelled lactic acid was also formed following intraperitoneal injection of $[15\text{-}^3H]$retinol which was thought to result from transfer of the label to NADH during the NAD$^+$ oxidation of retinol to retinaldehyde.

Reference

K. E. Creek, S. Shankar, and L. M. DeLuca, *In vivo* formation of tritium-labelled lactic acid from $[2\text{-}^3H]$mannose or $[15\text{-}^3H]$retinal by hamster intestinal epithelial cells, *Arch. Biochem. Biophys.*, 1987, **254**, 482.

Potassium nonane-5-sulphate

Use:	Surfactant
Key functional groups:	Alkyl sulphate
Test system:	Rat (oral, intravenous, intraperitoneal, 5 mg kg^{-1})

Structure and biotransformation pathway

$$CH_3CH_2CH_2\overset{*}{C}H_2CH\overset{*}{C}H_2CH_2CH_2CH_3 \longrightarrow CH_3CH_2CH_2CH_2CHCH_2CH_2CH_2CH_2OH$$

$$\underset{OSO_3H}{|} \qquad\qquad\qquad \underset{OSO_3H}{|}$$

$$* = {}^{14}C \qquad\qquad\qquad\qquad (1)$$

$$CH_3CH_2CH_2CH_2CHCH_2CH_2CH_2CO_2H$$

$$\underset{OSO_3H}{|}$$

$$(2)$$

Following administration of either ^{14}C- or ^{35}S-labelled potassium nonane-5-sulphate to rats by a variety of routes *ca.* 90% of the radioactivity was eliminated via urine within 6 hours. Three major components were present in urine and these were identified as the unchanged parent, nonan-1-ol-5-sulphate (1), and nonanoic acid 5-sulphate (2). Small amounts of inorganic sulphate were also detected. The structures of these compounds were deduced using a combination of GC–MS (after acid hydrolysis of purified compound), electrophoresis, and TLC. There was no evidence for β-oxidation as has been found for some other alkyl sulphates.

Reference

S. K. Bains, A. H. Olavesen, J. G. Black, D. Howes, C. G. Curtis, and G. M. Powell, Metabolism in the rat of potassium nonane-5-sulphate, a symmetrical anionic surfactant, *Xenobiotica*, 1987, **17**, 709.

Valproic acid

Use/occurrence:	Antiepileptic, anticonvulsant
Key functional groups:	Isoalkyl carboxylic acid, fatty acid (saturated)
Test system:	Rat liver microsomes, reconstituted cytochrome P-450

Structure and biotransformation pathway

Microsomes from phenobarbital treated rats catalysed the production of Δ^4-VPA (1) in addition to the previously identified metabolites 3-OH-VPA (4), 4-OH-VPA (3), and 5-OH-VPA (2). The relative rates of production of these four metabolites (pmol per nmol cytochrome P-450 in 20 minutes) were 66:1502:4318:497. The formation of Δ^4-VPA (1) by microsomes was inhibited by the addition of cytochrome P-450 inhibitors (*e.g.* CO, metyrapone) and enhanced by the P-450 inducer phenobarbital. A second unsaturated metabolite was tentatively identified as Δ^3-VPA. Reconstituted preparations of cytochrome P-450 also catalysed the formation of Δ^4-VPA, together with 4-OH-VPA and 5-OH-VPA, but not 3-OH-VPA. Δ^4-VPA was not produced by metabolism of 4-OH-VPA or 5-OH-VPA with microsomes from phenobarbital treated rats. Thus the formation of Δ^4-VPA represents a novel metabolic route (desaturation of a non-activated alkyl substituent) for cytochrome P-450. The metabolites were identified by GC–MS as their trimethylsilyl derivatives: Δ^4-VPA was confirmed to be present by selected ion recording and comparison with the authentic compound. Δ^4-VPA is a potent hepatotoxin and may be responsible for the liver damage that is occasionally seen with valproic acid therapy.

Reference

A. E. Rettie, A. W. Rettenmeier, W. N. Howald, and T. A. Baillie, Cytochrome P-450-catalysed formation of Δ^4-VPA, a toxic metabolite of valproic acid, *Science*, 1987, **235**, 890.

Valproic acid

Use/occurrence:	Antiepileptic, anticonvulsant
Key functional groups:	Isoalkyl carboxylic acid, fatty acid (saturated)
Test system:	Male albino rats (intraperitoneal, 100 mg)

Structure and biotransformation pathway

Labelled with ^2H at C-2 or $^2H_2$ at C-3

The biotransformation of valproic acid (1) in rats has been re-investigated using two new forms of the compound labelled with deuterium in different positions (C-2 or C-3). Eleven metabolites of each form were detected in urine by GC–MS procedures. The comparative retention of deuterium was used to determine the role of β-oxidation and/or direct hydroxylation in the metabolism of valproic acid: *e.g.* 3-oxo-VPA (3) appeared to be formed by

144

oxidation of Δ^2-VPA (2) rather than by oxidation of 3-hydroxy-VPA (4). A revised scheme for the biotransformation of valproic acid was presented.

Reference

A. W. Rettenmeier, W. P. Gordon, H. Barnes, and T. A. Baillie, Studies on the metabolic fate of valproic acid in the rat using stable isotope techniques, *Xenobiotica*, 1987, **17**, 1147.

Busulfan

Use/occurrence:	Antineoplastic agent
Key functional groups:	Alkyl sulphonate, methanesulphonate
Test system:	Rat (intraperitoneal, 15 mg kg^{-1})

Structure and biotransformation pathway

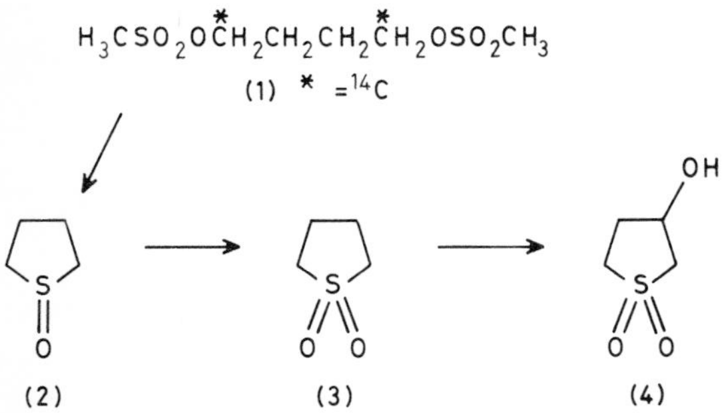

$$H_3CSO_2O\overset{*}{C}H_2CH_2CH_2\overset{*}{C}H_2OSO_2CH_3$$

(1)   * = ^{14}C

(2)   (3)   (4)

Excretion in urine accounted for 69% dose during 72 hours after dosing. HPLC analysis of urine separated at least eight radioactive fractions of which unchanged busulfan (6%), (2) (20%), (3) (13%), and (4) (39%) were identified by GC–MS and NMR (values as % urine radioactivity). A minor radioactive fraction (2%) co-eluted with reference tetrahydrofuran. A mechanism for formation of the observed metabolites via a glutathione conjugation pathway was discussed.

Reference

M. Hassan and H. Ehrsson, Urinary metabolites of busulfan in the rat, *Drug Metab. Dispos.*, 1987, **15**, 399.

Busulfan

Use/occurrence:	Antineoplastic agent
Key functional groups:	Alkyl sulphonate, methanesulphonate
Test system:	Isolated perfused rat liver

Structure and biotransformation pathway

$$H_3CSO_2O\overset{*}{C}H_2CH_2CH_2\overset{*}{C}H_2OSO_2CH_3 \longrightarrow$$

(1) $* = {}^{14}C$

structure (2):

H₂C——CH₂ / H₂C CH₂ ring with S⁺ center, CH₂ substituent connecting to

$$H_2NCHCH_2CH_2CONHCHCONHCH_2CO_2H$$

with CO₂H below

(2)

[^{14}C]Busulfan (1) yielded a single major metabolite during 4 hour cyclic perfusion. This metabolite was identified as γ-glutamyl-β-(S-tetrahydro-thiophenium)alanylglycine (2) by ^{252}Cf plasma desorption time-of-flight mass spectrometry and by comparison with a synthetic standard.

A mean ($n = 6$) of $38 \pm 11\%$ of the total administered radioactivity was excreted as the metabolite in bile whereas only about 1% was excreted as unchanged (1). In the perfusate, a mean of $32.5 \pm 13.5\%$ of the radioactivity present was busulfan and $43 \pm 10\%$ corresponded to the metabolite at the end of the perfusion. Formation of (2) was inhibited by the presence of the glutathione-S-transferase inhibitor ethacrynic acid, indicating that the reaction of (1) with glutathione was enzymatic in nature. At physiological pH in bile, (2) decomposed to tetrahydrothiophene, the probable intermediate in the formation of urinary sulpholane metabolites of (1) described in a separate publication.

Reference

M. Hassan and H. Ehrsson, Metabolism of [^{14}C]busulfan in isolated perfused rat liver, *Eur. J. Drug Metab. Pharmacokinet.*, 1987, **12**, 71.

Hexamethylene bisacetamide

Use/occurrence:	Anti-cancer agent
Key functional groups:	Acetamide, alkyl amide
Test system:	Human (infusion, 4.8–43.2 $g\ m^{-2}\ day^{-1}$)

Structure and biotransformation pathway

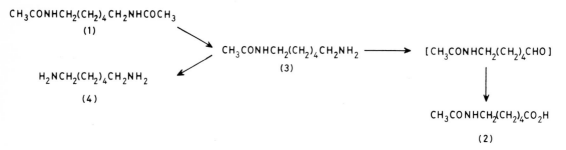

Hexamethylene bisacetamide (1) is a tumour cell differentiating agent which is undergoing Phase I clinical trial as an anti-cancer agent. It was infused through a free-flowing peripheral or central venous catheter to cancer patients. Blood and urine were collected and analysed by GC for (1) and its metabolites. The procedure involved addition of 1,2-diphenylethylamine or cadaverine as internal standard, followed by derivatization with trifluoroacetic anhydride and 2,2,2-trifluoroethanol. Detection was with a nitrogen–phosphorus detector.

The major plasma metabolite was 6-acetamidohexanoic acid (2). As the dose of (1) was increased the ratio of the steady-state concentration of (2) to that of (1) decreased. At doses of (1) above 16 $g\ m^{-2}\ day^{-1}$, N-acetyl-1,6-diaminohexane (3) was also detected and quantified in plasma. 1,6-Diaminohexane (4) and 6-aminohexanoic acid could not be detected in plasma.

In urine (2) was again the major metabolite (9.6–18.7% of the daily dose) followed by (3) (4.2–14.2%) and (4) (0–1.4%). Unchanged (1) was also present (18.3–46.7% of the dose). Increasing the dose of (1) decreased the proportion of (3) in urine but had no effect on the proportion of (2).

Reference

M. J. Egorin, E. G. Zuhowski, A. S. Cohen, L. A. Geelhaar, P. S. Callery, and D. A. VanEcho, Plasma pharmacokinetics and urinary excretion of hexamethylene bisacetamide metabolites, *Cancer Res.*, 1987, **47**, 6142.

Tiadenol disulphoxide

Use/occurrence:	Hypolipidaemic agent
Key functional groups:	Alkyl alcohol, alkyl sulphoxide
Test system:	Rat (oral, 200 mg kg^{-1})

Structure and biotransformation pathway

$$HOCH_2CH_2-\underset{\underset{O}{\|}}{S}-(CH_2)_{10}-\underset{\underset{O}{\|}}{S}-CH_2CH_2OH \longrightarrow \text{Glucuronide}$$

(1)   (2)

$$HOCH_2CH_2-\underset{\underset{\|}{O}}{\overset{\overset{O}{\|}}{S}}-(CH_2)_{10}-\underset{\underset{O}{\|}}{S}-CH_2CH_2OH$$

(3)

$$HOCH_2CH_2-\underset{\underset{O}{\|}}{S}-(CH_2)_{10}-\underset{\underset{O}{\|}}{S}-CH_2CO_2H$$

(5)

$$HOCH_2CH_2-\underset{\underset{O}{\|}}{S}-(CH_2)_{10}-\underset{\underset{\|}{O}}{\overset{\overset{O}{\|}}{S}}-CH_2CO_2H$$

(7)

$$HOCH_2CH_2-\underset{\underset{\|}{O}}{\overset{\overset{O}{\|}}{S}}-(CH_2)_{10}-\underset{\underset{\|}{O}}{\overset{\overset{O}{\|}}{S}}-CH_2CH_2OH$$

(4)

$$HO_2CCH_2-\underset{\underset{O}{\|}}{S}-(CH_2)_{10}-\underset{\underset{O}{\|}}{S}-CH_2CO_2H$$

(6)

This study was conducted using non-radioactive methodology. About 45% of an oral dose of tiadenol disulphoxide (1) was excreted unchanged in urine during 48 hours after administration. A further 10% dose was associated with the glucuronide conjugate (2) of the parent compound as evidenced by enzymic deconjugation experiments. The neutral metabolites (3) and (4), resulting from oxidation of the sulphoxide groups, accounted together for about 15% dose in the 48 hour urine. The acidic metabolites (5)–(7), resulting from oxidation of the terminal hydroxy-groups, together accounted for a further 15% dose and urinary excretion of these metabolites was complete by 12 hours after dosing. No conjugates of metabolites (3)–(7) were detected in urine and the remainder of the dose was excreted unchanged in faeces. The neutral metabolites were identified using EI and FD-MS and the acid metabolites using positive and negative FAB-MS.

Reference

R. Maffei Facino, M. Carini, O. Tofanetti, I. Casciarri, and E. Longoni, Experimental studies on pharmacology, metabolism, and toxicology with tiadenol disulphoxide, *Arzneim. Forsch.*, 1987, **37** (I), 682.

Geranylgeranylacetone

Use/occurrence: Anti-ulcer drug

Key functional groups: Alkyl ketone, isoprenoid

Test system: Rat (oral, 125–500 mg), rat liver microsomes

Structure and biotransformation pathway

Information on the initial biotransformation of (1) in rat was obtained from experiments with liver microsomes and subsequent metabolic routes by analysis of the products excreted in urine. Metabolites of (1) were isolated by solvent extraction of acidified urine or microsomes and were characterized by GC–MS and NMR. Three metabolites, all of which were more hydrophilic than (1), were identified in microsomal extracts. Reduction of the ketone group produced (2) and ω-oxidation of (1) and (2) produced the corresponding primary alcohols (3) and (4). Other, unidentified, metabolites were also produced. *In vivo*, 17% of the radioactive dose was excreted within 24 hours. Neither unchanged (1) nor the metabolites formed *in vitro* were detected. The urinary metabolites were dicarboxylic acids of different carbon chain length which, it is suggested, arise from sequential β-oxidation of (3)

150

and (4). The major metabolite, accounting for 32% of the urinary radioactivity, was the end product with a chain length of C_7 (5); C_9 (6) and C_{11} (7) diacids represented 15% and 5% respectively of the urinary radioactivity.

Reference

Y. Nishizawa, S. Abe, K. Yamada, T. Nakamura, I. Yamatsu, and K. Kinoshita, Identification of urinary and microsomal metabolites of geranylgeranylacetone in rats, *Xenobiotica*, 1987, **17**, 575.

Enisoprost

Use/occurrence: Cytoprotective (anti-ulcer)

Key functional groups: Alkyl ester, prostanoid

Test system: Human (oral, 0.45 mg)

Structure and biotransformation pathway

The metabolism of [³H]enisoprost (1), an analogue of misoprostol, was studied in five male volunteers. 59.0 ± 2.98% of the dose was excreted in urine and 17.4 ± 1.51% in the faeces over nine days. Enisoprost was rapidly de-esterified to the acid (2) which was converted into (3) (3.6% dose in urine),

152

(4) (4.8%), the dicarboxylic acids (5) (22%) and (6) (8.5%), and a γ-lactone (7) (2.6%). The acid (2), the five urinary metabolites and two other minor metabolites (8) and (9) were also detected in plasma. Metabolite separation and isolation was carried out using C18-Bond Elut columns and HPLC with characterization by GC–MS. About 5–10% of the dose was excreted as ^3H_2O in urine indicating that oxidation of the 11α-hydroxy position also occurred.

Reference

L. M. Allan, A. J. Hawkins, C. W. Vose, J. Firth, R. D. Brownsill, and J. A. Steiner, Disposition of [^{3}H]enisoprost, a gastric antisecretory prostaglandin, in healthy humans, *Xenobiotica*, 1987, **17**, 1233.

Arachidonic acid hydroperoxide

Use/occurrence:	Fatty acid metabolite
Key functional groups:	Fatty acid (unsaturated) hydroperoxide
Test system:	Rat liver microsomes

Structure and biotransformation pathway

(1) * = ^{14}C

(2)

(3)

Incubation of [^{14}C]arachidonic acid 15-hydroperoxide (1) with liver microsomes from phenobarbital treated rats yielded two major metabolites. These were separated by reverse phase HPLC, methylated with diazomethane, and repurified. The metabolites were identified by the mass spectra of the trimethylsilyl derivatives of the purified products (either directly or after catalytic hydrogenation or acidic hydrolysis), as 11,14,15-trihydroxyeicosa-5,8,12-trienoic acid (2) and 13-hydroxy-14,15-epoxy-eicosa-5,8,11-trienoic acid (3). The metabolism was confirmed to be cytochrome P-450-dependent by inhibition experiments with clotrimazole, although the absence of inhibition by carbon monoxide and the lack of effect on addition of NADPH suggested that the metabolism was NADPH-independent.

Reference

R. H. Weiss, J. L. Arnold, and R. W. Estabrook, Transformation of an arachidonic acid hydroperoxide into epoxyhydroxy and trihydroxy fatty acids by liver microsomal cytochrome P-450, *Arch. Biochem. Biophys.*, 1987, **252**, 334.

5-Fluoroarachidonic acid

Use/occurrence:	Model compound
Key functional groups:	Unsaturated fatty acid
Test system:	Human platelets

Structure and biotransformation pathway

(1)   OH   (2)

Human platelets were incubated with 5-fluoroarachidonic acid (1) or arachidonic acid followed by HPLC analysis of the products. The major product formed from the fluorinated analogue was isolated, purified as the methyl ester and analysed, by mass spectrometry after trimethylsilylation. The structure was confirmed as the 12-hydroxyeicosatetraenoic acid (2).

Reference

T. Taguchi, T. Takigawa, A. Igarashi, Y. Kobayashi, Y. Tanaka, W. Jubiz, and R. C. Briggs, Synthesis of 5-fluoroarachidonic acid and its biotransformation to 5-fluoro-12-hydroxyeicosatetraenoic acid, *Chem. Pharm. Bull.*, 1987, **35**, 1666.

Substituted Monocyclic Aromatic Compounds

2,5,4′-Trichlorobiphenyl (TCB)

Use/occurrence:	Industrial chemical
Key functional groups:	Biphenyl, chlorophenyl
Test system:	Rat (oral)

Structure and biotransformation pathway

TCB ✳ = ^{14}C

(1)

(2)

(3)

(4)

(5)

(6) R = OCH$_3$ or OH

40% and 55% of an oral dose was excreted by male and female rats respectively during 0–24 hours after administration (dose not specified). Unchanged TCB accounted for 12% of the radioactivity extracted from faeces and 3% of that in urine. Ten metabolites were present in extracts of faeces; these were either free or conjugated derivatives of the phenols (1)–(4). About 28% of the extractable radioactive metabolites were monohydroxy-, 30% dihydroxy-, and 30% trihydroxy-TCBs. Seven metabolites were identified in urine including (3) and the dechlorinated metabolites (5) and (6), together representing 23% of the extracted radioactivity. The remainder were free or conjugated derivatives of (1), (4), and possibly (2). Metabolites were identified partly after hydrolysis and/or methylation (no further details given).

Reference

M. Kamal, J. P. Lay, and I. Scheunert, Metabolism of trichlorobiphenyl in different test species, *Chemosphere*, 1987, **16**, 599.

2,2',4,4',5,5'-Hexachlorobiphenyl

Use/occurrence:	Environmental pollutant
Key functional groups:	Biphenyl, chlorophenyl
Test system:	Rat and dog liver microsomes, reconstituted cytochrome P-450

Structure and biotransformation pathway

(1)

Microsomes from dog liver metabolize 2,2',4,4',5,5'-hexachloro-biphenyl (1) (2,4,5-HCB) at much higher rates (15-fold) than those from rat liver, and consequently dogs have an enhanced ability to eliminate this polychlorinated biphenyl (PCB). Phenobarbital pretreatment was found to induce at least two cytochrome P-450 isozymes in the dog and one of these (PBD-2) was purified to 95% homogeneity. Antibodies to PBD-2 inhibited more than 90% of (1) metabolism in control and phenobarbital pretreated dog microsomes. Rat hepatic microsomes were also found to contain a homologous phenobarbital-inducible cytochrome P-450 isozyme (PB-B) although in a reconstituted system this only metabolized (1) at one eleventh the rate seen with purified PBD-2. Antibody inhibition studies showed that PB-B accounted for only about one-half of the metabolism of (1) in phenobarbital-induced rats. The postulate that PBD-2 is largely responsible for the microsomal metabolism of (1) in the dog was further supported by the observation that phenobarbital pretreatment increased by nearly six-fold the level of this isozyme (measured by immunoblot analysis), which was in parallel with the overall enhancement caused by phenobarbital hepatic metabolism (1).

Reference

D. B. Duignan, I. G. Sipes, T. B. Leonard, and J. R. Halpert, Purification and characterization of the dog hepatic cytochrome P-450 isozyme responsible for the metabolism of 2,2',4,4',5,5'-hexachlorobiphenyl, *Arch. Biochem. Biophys.*, 1987, **255**, 290.

4-Chlorodiphenyl ether (4-CDE)

Use/occurrence:	Dielectric fluid component
Key functional groups:	Aryl ether, chlorophenyl
Test system:	Rat (intravenous, 850 nmol kg^{-1}), rat liver microsomes (3.2 nmol mg^{-1} protein)

Structure and biotransformation pathway

About 41% and 33% dose were excreted in urine and faeces respectively during one week after intravenous administration of [^{14}C]-CDE. The phenol (2) accounted for about 90% of the radioactivity in acid-hydrolysed urine. Unchanged 4-CDE accounted for less than 1% urinary radioactivity. No quantitative data were given for other metabolites. More than 70% of 4-CDE was converted into (2) after 30 minutes microsomal incubation. About 7% of the ^{14}C label became irreversibly bound to microsomal protein after one hour, which was taken to indicate formation of reactive arene oxide intermediates. Identification of (2) was based on GC–MS comparison with a synthetic standard. Identification of other metabolites was based on MS fragmentation patterns alone and was only tentative.

Reference

Y. C. Chui, R. F. Addison, and F. C. P. Law, Studies on the pharmacokinetics and metabolism of 4-chlorodiphenyl ether in rats, *Drug Metab. Dispos.*, 1987, **15**, 44.

161

Mitotane

Use/occurrence:	Adrenolytic agent, adrenocortical carcinoma treatment agent
Key functional groups:	Chloroalkyl, chlorophenyl
Test system:	Isolated perfused dog adrenal glands

Structure and biotransformation pathway

The objective of this study was to identify metabolites produced at the site of action of the parent drug (1). Metabolic pathways of (1) in the dog adrenal gland involved oxidation of the chloroalkyl function to yield the carboxylic acid (2) which was then hydroxylated either in the *para*-position of the aromatic ring bearing the *ortho*-chloro substituent, or on the α-carbon. Following three hours perfusion the aromatic hydroxy compound (4) and its sulphate were the major metabolites, accounting for 8.0% and 5.1% respectively of the original substrate radioactivity. The intermediate acid (2) accounted for a further 4.1%. The α-hydroxy compound was associated with only 1.3% of the starting radioactivity, but it was of interest that the adrenal cortex was able to effect both aromatic and aliphatic hydroxylations as well as the sulphate conjugation reaction. Identification of metabolites in this study rested on TLC and HPLC comparison with synthetic standards (2)–(4) and in the case of the sulphate conjugate on chemical and selective enzymic hydrolysis to yield (4). The position of the ^{14}C label was not stated.

Reference

J. E. Sinsheimer and C. J. Freeman, Short communication: Mitotane (1-(o-chloro-phenyl)-1-(p-chlorophenyl)-2,2-dichloroethane) metabolism in perfusion studies with dog adrenal glands, *Drug Metab. Dispos.*, 1987, **15**, 267.

Bromobenzene

Use/occurrence: Model compound

Key functional groups: Bromophenyl

Test system: Rat (intraperitoneal, 65, 130, and 1200 mg kg^{-1}), guinea pig (intraperitoneal, 650, 130, and 314 mg kg^{-1})

Structure and biotransformation pathway

GS = Glutathionyl
CS = Cysteinyl

Proportions of urinary metabolites in 0–24 hour urine were determined. In the guinea pig 3-bromophenol (8) was the major urinary metabolite, representing up to 37% of the administered dose. In the rat 4-bromophenol (9) and 3-bromophenol (8) were excreted in the urine in approximately equal proportions, representing up to 8% of the dose. 4-Bromocatechol (5) was a

163

major urinary metabolite in both guinea pig and rat, representing up to 10% of the dosed radioactivity. 2-Bromophenol (10) was a minor metabolite in both species, representing less than 3% of the dose. A number of minor metabolites were identified, including the 3,4-dihydrodiol (2). Methylthio analogues of dihydrodiols were found in the guinea pig but not in the rat. Two di(methylthio)dihydroxytetrahydrobromobenzenes were formed by the rat but not by the guinea pig. In the guinea pig the acidic urinary metabolites, presumably derived from (3), (4), (6), and (7), were a mercaptoacetate, a mercaptolactate, and a mercapturate. In the rat acidic metabolites were a mercapturic acid and a premercapturic acid. This species difference reflected the different acetylation/deacetylation processes for cysteine conjugates. Metabolites were identified as trimethylsilyl derivatives by GC–MS and acidic metabolites were methylated prior to silylation.

Reference

K. Lertratanangkoon and M. G. Horning, Bromobenzene metabolism in the rat and the guinea pig, *Drug Metab. Dispos.*, 1987, **15**, 1.

1-Naphthol, Phenol

Use/occurrence:	Model compounds
Key functional groups:	Phenol
Test system:	Human neutrophils

Structure and biotransformation pathway

The metabolism of 1-naphthol (1) and phenol (2) by a superoxide anion generating system (hypoxanthine/xanthine oxidase) and by phorbol myristate acetate stimulated human neutrophils has now been demonstrated. Incubation of (1) with hypoxanthine/xanthine oxidase yielded 1,4-naphtho-quinone (3), as detected by HPLC with electrochemical detection. 33% stimulation of the reaction was achieved by addition of catalase, and 96% inhibition by addition of superoxide dismutase, confirming the involvement of O_2^-. A similar incubation of (2) did not yield 1,4-benzoquinone, although hydroquinone (4), catechol (5), and an unidentified product (metabolite of catechol) were detected by HPLC. O_2^- was again shown to be involved in the formation of these products, although inhibition experiments with catalase and mannitol revealed the participation of OH'. Consistent results were obtained with the superoxide generating system of phorbol myristate acetate stimulated neutrophils. 1-Naphthol interfered with superoxide-dependent cytochrome c reduction and with luminol-dependent chemiluminescence; [14C]-(1) also bound to leukocyte protein. Phenol did not affect superoxide-dependent cytochrome c reduction (at 10 or 30 μM) but did inhibit luminol-dependent chemiluminescence and bound to protein. This binding was inhibited by catalase and azide.

It was concluded that the oxidative metabolism differs for the two phenols; metabolism of (2) appears to be peroxidase-mediated whereas metabolism of

165

(1) is partially peroxidase-mediated and partially directly superoxide-dependent.

Reference

D. A. Eastmond, R. C. French, D. Ross, and M. T. Smith, Metabolic activation of 1-naphthol and phenol by a simple superoxide-generating system and human leukocytes, *Chem. Biol. Interact.*, 1987, **63**, 47.

Etoposide

Use/occurrence:	Anti-tumour agent
Key functional groups:	Glycoside, *o*-methoxy-phenol
Test system:	Mouse liver microsomes

Structure and biotransformation pathway

Etoposide (1) is metabolized by mouse liver microsomes to form a 3′,4′-dihydroxy derivative which was identified by HPLC and MS. The *o*-demethylation is mediated by cytochrome P-450. ESR studies showed that a semiquinone radical (3) could be formed enzymatically from (2) or the *ortho*-quinone derivative (4). The *ortho*-quinone and semiquinone radical are implicated as the bioalkylating species which bind covalently to mouse liver microsomal proteins to give (6).

Reference

N. Haim, J. Nemec, J. Roman, and B. K. Sinha, *In vitro* metabolism of etoposide (VP-16-213) by liver microsomes and irreversible binding of reactive intermediates to microsomal proteins, *Biochem. Pharmacol.*, 1987, **36**, 527.

Etoposide

Use/occurrence:	Anti-cancer agent
Key functional groups:	Glycoside, *o*-methoxy-phenol
Test system:	Rat liver microsomes, purified rat liver microsomal cytochrome P-450

Structure and biotransformation pathway

(1)   (2)

Cytochrome P-450 is believed to be involved in the production from etoposide (VP-16-213) (1) of a reactive metabolite which binds covalently to rat liver and HeLa cell microsomal proteins. In the present study it is suggested that this reactive metabolite is the catechol derivative (2), produced by *O*-demethylation of (1). This transformation was demonstrated in incubation systems containing rat liver microsomes or purified rat liver microsomal cytochrome P-450. For microsomes the K_m (μM) and V_{max} (nmol formaldehyde/min mg protein) values were 130, 8.5 (uninduced); 600, 11.8 (phenobarbital induced); and 160, 15.6 (3-methylcholanthrene induced) respectively, *i.e.* the affinity of etoposide for cytochrome P-450 is lower for phenobarbital induced microsomes and the demethylation process is induced with 3-methylcholanthrene induced microsomes. The product was identified as (2) by mass spectrometry (high resolution) and by its chromatographic identity (HPLC) with the synthetic catechol. The metabolite is toxic to Chinese hamster ovary cells, binds to calf thymus DNA (10 times more than

168

etoposide) and inactives single and double stranded DNA of the baterio-phage $\phi \times 174$.

Reference

J. M. S. van Maanen, J. de Vries, D. Pappie, E. van den Akker, M. V. M. Lafleur, J. Retel, J. van der Greef, and H. M. Pinedo, Cytochrome P-450-mediated *O*-demethylation: a route in the metabolic activation of etoposide (VP-16-213), *Cancer Res.*, 1987, **47**, 4658.

Butylated hydroxytoluene (2,6-di-t-butyl-4-methylphenol)

Use/occurrence:	Antioxidant in food
Key functional groups:	Aryl tertiary-alkyl, arylmethyl, phenol
Test system:	Rat and mouse, liver and lung microsomes (phenobarbital pretreated)

Structure and biotransformation pathway

Following microsomal incubation of [^{14}C]butylated hydroxytoluene (1) (position of radiolabel not stated), metabolites were quantified by HPLC and characterized by comparing chromatographic and UV spectral properties with authentic standards. UV spectral data were obtained using a diode array detector linked to the HPLC. In this way the formation of eight previously reported metabolites was confirmed. Structures of a further four previously

170

unknown metabolites (5), (6), (7), and (10) were additionally derived from MS data and ^{1}H NMR data [(5) and (6) only]. It was shown that (5) was the sole precursor of the other three novel metabolites.

Two main metabolic processes occurred, hydroxylation of the methyl and t-butyl substituents [metabolites (2)–(7)] and oxidation of the aromatic ring system [metabolites (8)–(13)]. Metabolites from the former process, containing the 4-hydroxymethyl grouping [(2) and (6)], were further oxidized to yield the aldehydes (3) and (7) and the acid (4). Oxidation of the π-electron system led to the quinone (9), quinol (12), and hydroxyquinol (13), formed *via* the peroxyquinol (8), and the quinone methide (11). All of these products were assumed to arise from the intermediate radical shown, either *via* a second one-electron oxidation to give (11) or combination with oxygen to give (8). The novel metabolite (10) resulted from a combination of aliphatic and aromatic oxidation processes. Quantitatively (limited data reported), (2) was the principal metabolite in rat liver and lung microsomes, whereas mice produced large quantities of both (2) and (5) in these tissues. Aromatic ring-oxidized products appeared to be minor metabolites. The metabolite profile was similar in rat liver and lung, but mouse lung produced more quinone (9) relative to other metabolites than mouse liver. Some quantitative differences in metabolite profiles were noted in incubations of microsomes from different strains of mice.

Reference

J. A. Thompson, A. M. Malkinson, M. D. Wand, S. L. Mastovich, E. W. Mead, K. M. Schullek, and W. G. Laudenschlager, Oxidative metabolism of butylated hydroxytoluene by hepatic and pulmonary microsomes from rats and mice, *Drug Metab. Dispos.*, 1987, **15**, 833.

Dihydrosafrole [1-(3,4-methylenedioxy-phenyl)propane]

Use/occurrence:	Herbs and spices, insecticide synergists in aerosol sprays
Key functional groups:	Acetal, alkylphenyl
Test system:	Dog, cynomolgus monkey (nasal instillation)

Structure and biotransformation pathway

Methylenedioxyphenyls, of which dihydrosafrole (1) is a representative compound, occur in many herbs and spices and are used as insecticide synergists. Compound (1) is hepatotoxic and carcinogenic in oesophagus and forestomach of animals. Inhalation is a major route of exposure in man.

In this study [^{3}H]-(1) was given by nasal instillation to beagles and cynomolgus monkeys, and nasapharyngeal mucus, blood, urine, and faeces were examined for clearance of ^{3}H-(1). Mucus was analysed by HPLC for metabolites. Some radioactivity was recovered in the mucus, but most was found in urine and blood over the first 24 hours. After instillation into the turbinate region of either species (1) was metabolized to 2-methoxy-4-propylphenol (2), 2-methoxy-4-propenylphenol (3), and 1-(3,4-methylene-dioxyphenyl)propan-1-ol (4). Instillation in the ethmoid region of one dog produced 1-(3,4-methylenedioxyphenyl)propene (isosafrole) (5). Results suggested that interspecies, interindividual, and inter-regional differences occur in metabolism of nasally deposited (1).

Reference

J. Petridou-Fischer, S. L. Whaley, and A. R. Dahl, *In vivo* metabolism of nasally instilled dihydrosafrole [1-(3,4-methylenedioxyphenyl)propane] in dogs and monkeys, *Chem. Biol. Interact.*, 1987, **64**, 1.

Phenol, 4-Nitrophenol, 2-Cresol, 4-Chlorophenol, Pentachlorophenol

Use/occurrence: Model compounds

Key functional groups: Arylmethyl, phenol

Test system: Zebra fish (in water)

Structure and biotransformation pathway

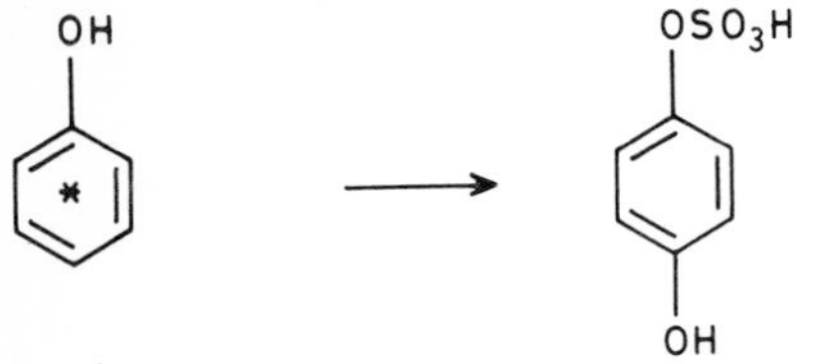

$* = {}^{14}C$ (1)

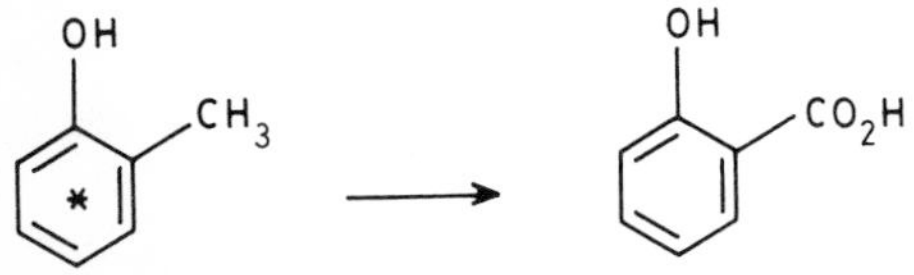

(2)

The biotransformation of five phenols was studied in zebra fish following administration of each compound to the aqueous medium. The medium was analysed by TLC and HPLC for parent compound and metabolites. All of the five phenols were converted into the glucuronide and sulphate conjugates. Phenol was also metabolized to quinol sulphate (1) and 2-cresol to 2-hydroxybenzoic acid (2). The metabolism of the phenols is not different from that previously reported for other fish species.

Reference

T. Kasokat, R. Nagel, and K. Urich, The metabolism of phenol and substituted phenols in zebra fish, *Xenobiotica*, 1987, **17**, 1215.

Phenol, 4-Nitrophenol, 2-Methylphenol

Use/occurrence:	Model compounds
Key functional groups:	Arylmethyl, nitrophenyl, phenol
Test system:	*Rana temporaria, Xenopus laevis* (injection, 3.5 mg kg^{-1})

Structure and biotransformation pathway

175

The excretion and metabolism of three phenols (^{14}C-labelled) were compared in two amphibians (*Rana, Xenopus*). Each excreted 90–95% of the dose within 24 hours and 50–65% was found to be metabolized. Metabolism was mostly by conjugation by sulphation or glucuronidation of the injected phenol. However, the *Xenopus* strain was unable to glucuronidate phenols but excreted an increased amount of other metabolites. Phenol (1) was also metabolized to quinol (2), which was excreted largely as the sulphate conjugate. Small amounts of catechol (3) and resorcinol (4) were excreted by *Rana*, but both compounds were largely excreted as the sulphate conjugates by *Xenopus*. 4-Nitrophenol (5) was found to be hydroxylated to 4-nitro-catechol (6) and reduced to give 4-acetamidophenol (7). Hydroxylation of 2-methylphenol (8) to 2-methylquinol (9) or 2-methylresorcinol (10) and oxidation to 2-hydroxybenzoic acid (11) was observed. All metabolites were identified by HPLC. Other metabolites were detected but not identified.

Reference

G. Gorge, J. Beyer, and K. Urich, Excretion and metabolism of phenol, 4-nitro-phenol, and 2-methylphenol by the frogs *Rana temporaria* and *Xenopus laevis*, *Xenobiotica*, 1987, **17**, 1293.

Dinitrobenzene (DNB; *o*-, *m*-, and *p*-isomers)

Use/occurrence:	Industrial chemicals
Key functional groups:	Nitrophenyl
Test system:	Erythrocytes from rat, rhesus monkey, and man

Structure and biotransformation pathway

o - DNB   *m* - DNB   *p* - DNB

ArNO$_2$ $\longrightarrow$ ArSG

Ar = o -, m-or p-nitrophenyl ; GS = glutathionyl

Erythrocytes from all three species metabolized *o*-DNB and *p*-DNB to *S*-(nitrophenyl)glutathione conjugates, although there were species differences in the rate and extent of conjugate formation. No metabolites of *m*-DNB were detected in the erythrocytes of any of the three species. The test compounds were incubated for 2 hours at 37°C at a concentration of 100 μM. Analysis of samples and characterization of metabolites, which were by HPLC, were not described in the paper under review. Rat and monkey erythrocytes metabolized *o*-DNB to the glutathione conjugate faster than human erythrocytes. Similarly, *p*-DNB was also conjugated more rapidly in monkey than in human erythrocytes. In the rat, the glutathione conjugate was not the only metabolite of *p*-DNB; another major metabolite, which co-eluted with *p*-nitrophenylhydroxylamine (results not shown in paper under review) was also detected. Some radioactivity from each DNB isomer became irreversibly bound to erythrocyte macromolecules. Species variations in this binding were small but binding of the isomers increased in the order *m*-DNB < *o*-DNB < *p*-DNB. Neither formation of the glutathione conjugate nor the extent of irreversible binding correlated wtih methaemoglobin production, suggesting that glutathione conjugation does not necessarily protect against methaemoglobinaemia and that the same metabolites may not be responsible for irreversible binding and methaemoglobin production.

Reference

P. A. Cossum and D. E. Rickert, Metabolism and toxicity of dinitrobenzene isomers in erythrocytes from Fischer-344 rats, rhesus monkeys, and humans, *Toxicol. Lett.*, 1987, **37**, 157.

Dinitrobenzene (DNB; *o*-, *m*-, and *p*-isomers)

Use/occurrence:	Industrial chemicals
Key functional groups:	Nitrophenyl
Test system:	Rat (oral, 0.15 mmol kg^{-1})

Structure and biotransformation pathway

Gluc = glucuronyl
SCysAc = *N*-acetylcysteinyl

Following separate administration of ^{14}C-labelled DNB isomers, the urine was the primary route of excretion, accounting for 82%, 63%, and 75% of the total dose of 1,2-, 1,3-, and 1,4-DNB respectively after 48 hours. Faecal excretion accounted for 8% and 9% of doses of 1,2- and 1,4-DNB respectively and 18% dose in the case of 1,3-DNB. The results showed that DNB isomers were initially metabolized *via* nitro-group reduction or glutathione displacement of a nitro-group. It was notable that 1,3-DNB (the only isomer producing toxicity in rats) was apparently metabolized

178

exclusively by reduction. The major urinary metabolites of 1,2-DNB were the N-acetylcysteine conjugate (2; 42% dose), which was presumably the terminal metabolite resulting from glutathione displacement of a nitro-group, and 4-amino-3-nitrophenyl sulphate (8; 17% dose). Minor metabolites were the N-glucuronide (5; 4%), the sulphate conjugate (3; 1.5%), and the hydroxylamine N-glucuronide (1; 0.6–2.0%). Two further minor metabolites were unidentified.

The major urinary metabolites of 1,3-DNB were reported as 3-amino-acetanilide (7; 21% dose), 1,3-diacetamidobenzene (9; 7% dose), 4-acet-amidophenyl sulphate (6; 6% dose) (a surprising metabolite of 1,3-DNB), and 3-nitroaniline N-glucuronide (5; 4% dose). Four metabolites, accounting for 2–6% dose each, were unidentified. None of these unknown metabolites corresponded to the glutathione or N-acetylcysteine conjugates of 1,3-DNB.

The major urinary metabolites of 1,4-DNB were 2-amino-5-nitrophenyl-sulphate (3; 33% dose), the N-acetylcysteine conjugate (2; 11% dose), and 1,4-diacetamidobenzene (9; 7% dose). Three metabolites were unidentified (2.3–4.7% dose each). Urinary metabolites were identified by HPLC and MS comparison with authentic standards. Conjugate assignations were based on identification of the aglycone after enzymic deconjugation.

Reference

D. D. Nystrom and D. E. Rickert, Metabolism and excretion of dinitrobenzenes by male Fischer-344 rats, *Drug Metab. Dispos.*, 1987, **15**, 821.

1,3-Dinitrobenzene

Use/occurrence:	Industrial chemical
Key functional groups:	Nitrophenyl
Test system:	Rat testicular cell culture

Structure and biotransformation pathway

Dinitrobenzene underwent limited metabolism in rat Sertoli cell and Sertoli-germ cell co-cultures. The major metabolites were *m*-nitroaniline (1) and *m*-nitroacetanilide (2), each accounting for 3% of the dinitrobenzene added to the culture. The metabolites were identified by either GC–MS or TLC comparisons with reference compounds.

Reference

P. M. D. Foster, S. C. Lloyd, and M. S. Prout, Toxicity and metabolism of 1,3-dinitro-benzene in rat testicular cell culture, *Toxicol. in vitro*, 1987, **1**, 31.

4-Nitroanisole, 4-Nitro-*m*-cresol, Methylparathion

Use/occurrence:	Pesticides
Key functional groups:	Nitrophenyl, phenol, *O*-methyl phosphorothioate
Test system:	Crustaceans $(0.01 \text{ mg l}^{-1}$, 0.1 mg l^{-1} in water)

Structure and biotransformation pathway

181

The biotransformation of the [*ring-U-*^{14}C] nitroaromatic compounds nitroanisole (1), 4-nitro-*m*-cresol (2), and methylparathion (3) was examined in Malaysian prawns, ridgeback prawns, and crayfish. The radioactivity eliminated and remaining in soft crustacean tissues was extracted, concentrated, and analysed by TLC and HPLC by comparison with authentic reference compounds. *O*-Demethylation of (1) to produce *p*-nitrophenol (4) occurred in each species although the extent of this reaction was greater in crayfish (98%) than in Malaysian prawns (29%) and ridgeback prawns (11%). All three species readily hydrolysed (3) to form (4). 5-Hydroxy-2-nitrobenzaldehyde (5) was the major metabolite of (2) in the crayfish. However, oxidation of (2) was not detected in the other two species.

Nitroreduction was not observed in the metabolism of the three substrates by Malaysian prawns or crayfish. However, 4-nitroso-*m*-cresol (6) arising from reduction of (2) was observed in ridgeback prawns. Conjugation was overall the dominant detoxication pathway. Crayfish conjugated the phenolic substrates (2) and (4) to form exclusively the corresponding *β*-D-glucosides (7) and (8). Malaysian prawns conjugated (2) and (4) to the *β*-D-glucosides (7) and (8) and sulphate monoesters (9) and (10). Ridgeback prawns formed (7) and (8) in small quantities but preferentially formed the sulphates (9) and (10) and unidentified conjugates.

Reference

G. D. Foster and D. G. Crosby, Comparative metabolism of nitroaromatic compounds in freshwater, brackish water, and marine water decapod crustaceans, *Xenobiotica*, 1987, **17**, 1393.

2,4-Dinitrotoluene, 2,4-Dinitrobenzyl alcohol, 2,4-Dinitrobenzaldehyde

Use:	Industrial chemicals
Key functional groups:	Aryl aldehyde, arylmethyl, benzyl alcohol, nitrophenyl
Test system:	Rat liver microsomes and cytosol

Structure and biotransformation pathway

2,4-Dinitrotoluene (1) was converted into 2,4-dinitrobenzyl alcohol (2) by rat liver microsomes *in vitro*, which could be further oxidized to 2,4-dinitrobenzaldehyde (3) by either microsomal or cytosolic fractions. Incubation of 2,4-dinitrobenzaldehyde with either cytosolic or microsomal preparations containing reduced and oxidized pyridine nucleotides resulted in the formation of 2,4-dinitrobenzyl alcohol (2) and 2,4-dinitrobenzoic acid (4) together with two unidentified metabolites. All metabolites were identified by HPLC comparison with reference compounds. Experiments with different pyridine nucleotides and enzyme inhibitors showed that these biotransformations were mediated by cytochrome P-450, alcohol, and aldehyde reductase and aldehyde dehydrogenase. 2,4-Dinitrobenzaldehyde is known to be a direct acting mutagen. Although it has not been detected in rat urine, its formation from 2,4-dinitrotoluene *in vitro* may explain why 2,4-dinitrotoluene is carcinogenic in the rat.

Reference

M. Shoji, M. Mori, T. Kawajiri, M. Sayama, Y. Mori, T. Miyahara, T. Honda, and H. Kozuka, Metabolism of 2,4-dinitrotoluene, 2,4-dinitrobenzyl alcohol, and 2,4-dinitrobenzyaldehyde by rat liver microsomal and cytosol fractions, *Chem. Pharm. Bull.*, 1987, **35**, 1579.

Phenol, 4-Nitrophenol, and *N*-Hydroxy-2-acetylaminofluorene

Use/occurrence:	Model compounds
Key functional groups:	*N*-Acetyl aryl amine, *N*-hydroxy, phenol
Test system:	Male and female rat (intravenous, 60 or 133 μmol kg^{-1})

Structure and biotransformation pathway

* = ^{14}C

[ring-^{3}H]

X = Sulphate or glucuronide
Y = Glucuronide

The sex difference in sulphation by rats of three model compounds has been studied. For phenol there was no major difference in the extent of conjugation; both male and female rats excreted approximately half of the dose as the sulphate conjugate and the remainder as the *O*-glucuronide. Conjugation of 4-nitrophenol followed a similar pattern, with male and female rats excreting approximately 30% of the dose as the sulphate and 40% as the *O*-glucuronide. With *N*-hydroxy-2-acetylaminofluorene male rats excreted 26% of the dose as the *N*-*O*-glucuronide whereas female rats excreted 43% of the dose as this metabolite. The sulphation of *N*-hydroxy-2-acetylaminofluorene could not be determined directly *in vivo* owing to the instability of the *N*-*O*-sulphate. It was suggested that the degree of sulphation could be measured indirectly by measurement of the competing glucuronide pathway. Twice as much covalent binding of *N*-hydroxy-2-acetylaminofluorene was seen in the liver of male rats compared with that found in the liver of female rats. Metabolites were determined by HPLC or TLC.

Reference

J. H. N. Meerman, C. Nijland, and G. J. Mulder, Sex differences in sulfation and glucuronidation of phenol, 4-nitrophenol, and *N*-hydroxy-2-acetylaminofluorene in the rat *in vivo*, *Biochem. Pharmacol.*, 1987, **36**, 2605.

Benzoic acid

Use/occurrence:	Model compound
Key functional groups:	Aryl carboxylic acid
Test system:	Southern flounder (intramuscular, 0.25 mmol kg^{-1})

Structure and biotransformation pathway

(1) * = ^{14}C (2)

Excretion of injected radioactivity was slow (10% dose/day) and followed zero-order kinetics. More than 95% of the excreted radiolabel was associated with a single metabolite, which was shown by hydrolysis and TLC to be the taurine conjugate (2) of the parent compound (1). Two minor metabolites were also detected but not further characterized. Isolated renal tubules and both hepatic and renal mitochondria produced (2) *in vitro*. The slow excretion of the radiolabel following injection of (1) was shown to be due to a combination of factors, including slow metabolism to (2), saturation of the renal tubular transport of (2), and low affinity of secretory transport for (1) and (2) relative to hippuric acid, the major mammalian metabolite of (1).

Reference

M. O. James and J. B. Pritchard, *In vivo* and *in vitro* renal metabolism and excretion of benzoic acid by a marine teleost, the southern flounder, *Drug Metab. Dispos.*, 1987, **15**, 665.

Benzoic acid

Use/occurrence:	Model compound
Key functional groups:	Aryl carboxylic acid
Test system:	Rainbow trout (oral, 10 mg kg^{-1})

Structure and biotransformation pathway

(1) * = ^{14}C → (2)

[^{14}C]Benzoic acid (1) was absorbed on to diet contained in gelatine capsules. Capsules smeared with cod liver oil were consumed within 30 seconds of being placed in the fish tank. No leakage of radioactivity from the capsules occurred. The administered radioactivity was almost quantitatively (99%) recovered in the tank water during 48 hours after dosing and was almost exclusively (>98%) associated with a single metabolite, which was identified as *N*-benzoyltaurine (2) by FAB-MS, IR, and ^1H NMR. Acid hydrolysis of the metabolite produced a ninhydrin-positive product with TLC characteristics very similar to those of taurine. Hippuric acid (benzoylglycine) was not detected as a metabolite.

Reference

A. B. Burke, P. Millburn, K. R. Huckle, and D. H. Hutson, Formation of the taurine conjugate of benzoic acid in rainbow trout, *Salmo gairdneri*, *Drug Metab. Dispos.*, 1987, **15**, 581.

5-Aminosalicylate

Use/occurrence:	Anti-inflammatory drug
Key functional groups:	Aryl amine, aryl carboxylic acid, phenol
Test system:	*In vitro*, human mononuclear cells and granulocytes

Structure and biotransformation pathway

When control human mononuclear cells were incubated with 5-aminosalicylate no reaction products were detected. However, following pretreatment of the cells with cytocholasin B several 'metabolites' of 5-aminosalicylate were produced, but only two were identified. The major metabolite was gentisate (1), and the next most abundant metabolite was salicylic acid (2). The biotransformation was thought to be mediated by hydroxyl radicals generated by the activated cells. It was suggested that the ability of 5-aminosalicylate to trap such active species could explain its mode of action in the control of ulcerative colitis. Similar results were obtained with the granulocytes. Metabolites were determined by HPLC assay.

Reference

B. J. Dull, K. Salata, A. Van Langehove, and P. Goldman, 5-Aminosalicylate: oxidation by activated leukocytes and protection of cultured cells from oxidative damage, *Biochem. Pharmacol.*, 1987, **36**, 2467.

N,N-Disubstituted 2-hydroxyacetamide esters

Use/occurrence:	Prodrugs
Key functional groups:	Aryl carboxylate, glycolamide
Test system:	50% human plasma

Structure and biotransformation pathway

(1)

A series of benzoate esters of various *N*-substituted glycolamides were found to be hydrolysed in 50% human plasma solutions at 37°C quite rapidly but with vastly different rates. The *N,N*-disubstituted glycolamide esters were hydrolysed particularly rapidly with *N,N*-diethyl- and *N,N*-di-isopropyl-substituted glycolamide esters showing the highest reactivity (half-life of hydrolysis only 2–5 s). The simple methyl and ethyl esters of benzoic acid were hydrolysed much more slowly, with half-lives of 1.8 and 4 hours respectively. On the basis of total inhibition by 10^{-3} M physostigmine the plasma-catalysed hydrolysis was attributed to cholinesterase. The esters of *N,N*-disubstituted glycolamides were stable in aqueous solution and may represent potentially useful biolabile prodrugs for carboxylic acid agents.

Reference

H. Bundgaard and N. M. Nielsen, Esters of *N,N*-disubstituted 2-hydroxyacetamides as a novel highly biolabile prodrug type for carboxylic acid agents, *J. Med. Chem.*, 1987, **30**, 451.

2-Naphthylacetic acid

Use:	Model compound
Key functional groups:	Arylacetic acid
Test system:	Guinea pig, mouse, hamster (intraperitoneal, 5–200 mg kg^{-1})

Structure and biotransformation pathway

(1) — CH$_2$CONHCH$_2$CO$_2$H

CH$_2$CO$_2$H

(2) — CH$_2$CONHCH$_2$CH$_2$SO$_3$H

(3) — CH$_2$CONHCHCO$_2$H, CH$_2$CH$_2$CO$_2$H

(4) Gluc = glucuronyl — CH$_2$COGluc

Guinea pigs, mice, and hamsters were given a single intraperitoneal dose of [^{14}C]-2-naphthylacetic acid at four dose levels, *viz.* 5, 50, 100, and 200 mg kg^{-1}, and the urines were analysed for the presence of the glycine (1), taurine (2), glutamine (3), and glucuronide (4) conjugates. Guinea pigs eliminated 70–81% of the dose in urine and the major metabolite was the glycine conjugate (1). The proportion of this decreased from 72 to 32% as the dose increased from 5 to 200 mg kg^{-1}, but there was no evidence for saturation of this pathway. The ester glucuronide (4) and the unconjugated acid were present as minor metabolites at the lowest dose (2–3%) but their proportions increased with dose to 18% and 40% respectively at the highest dose. In mice the urinary elimination of radioactivity was 50% at the two lowest doses and 28% at the highest dose but rose to 70% at the 100 mg kg^{-1} dose.

The taurine conjugate was the major metabolite at all dose levels, accounting for 31–56% of the dose at doses up to and including 100 mg kg^{-1} but only 28% at 200 mg kg^{-1} because of metabolic saturation. The glycine and glucuronic acid conjugates were also eliminated in the urine as minor metabolites (< 10% of the dose), and there was evidence for saturation of the former at the highest dose level. The free acid was also found in urine but only in small amounts (< 5% of the dose). The urinary elimination of radioactivity by hamsters decreased from 86 to 50% over the dose range. The glycine and glutamate conjugates were the major metabolites at all dose levels, with small amounts of the taurine and glucuronide metabolites also being found. There was no evidence for dose dependent metabolism, the glycine and glutamine metabolites accounting for 14–33% and 19–27% of the dose respectively and the taurine and glucuronide conjugates for 5–14%

and 4–10% of the dose respectively. The formation of several amino acid conjugates observed with 2-naphthylacetic acid is not due to saturation of some metabolic pathways at low doses with the resulting formation of unusual conjugates at higher doses.

Reference

T. S. Emudianughe, J. Caldwell, and R. L. Smith, Studies on the metabolism of arylacetic acids 7. The influence of varying dose size upon the conjugation pattern of 2-naphthylacetic acid in the guinea pig, mouse, and hamster, *Xenobiotica*, 1987, **17**, 823.

1- and 2-Naphthylacetic acid

Use:	Model compounds
Key functional groups:	Arylacetic acid
Test system:	Guinea pig, mouse, hamster, gerbil (intraperitoneal, 100 mg kg^{-1})

Structure and biotransformation pathway

$$RCH_2CONHCH_2CO_2H \longleftarrow RCH_2\overset{*}{C}O_2H \longrightarrow RCH_2CONHCH_2CH_2SO_3H$$

$$(1) \qquad\qquad * = {}^{14}C \qquad\qquad (2)$$

$$RCH_2CONHCHCO_2H$$
$$| $$
$$CH_2CH_2CO_2H$$
$$(3)$$

R = 1-naphthylacetic acid   or   2-naphthylacetic acid

When $[^{14}C]$-1-naphthylacetic acid was administered to guinea pigs, mice, hamsters, and gerbils by the intraperitoneal route a large proportion (76–93%) of the radioactivity was eliminated in urine over three days and mostly within 24 hours. Three radiolabelled components were detected in the urine by comparative TLC. These were the parent acid (guinea pigs 9%, mice 13%, gerbils 9%, and hamsters 19% of the dose), ad 1-naphthylacetyl-glucuronide (guinea pigs 33%, mice 52%, gerbils 28%, and hamsters 64%), and 1-naphthylacetylglycine (1) (guinea pigs 20%, mice 12%, gerbils 38%, and hamsters 10%). $[^{14}C]$-2-Naphthylacetic acid was also rapidly excreted by all four species, the recovery of radioactivity in 0–24 hour urines being 68–94% of the dose. Chromatographic examination of the urines showed that in all four species a proportion of the parent acid was eliminated unchanged (guinea pigs 20%, mice and gerbils 2%, and hamsters 9% of the dose), and the glucuronide conjugate was also formed (guinea pigs 32%, mice 33%, gerbils 3%, and hamsters 4%). Unlike the situation with the 1-isomer there were species differences in amino acid conjugation of the 2-isomer. Thus in guinea pigs only the glycine conjugate (32% of the dose) was found in urine, whereas in mice and gerbils both the glycine conjugate (mice 3%, gerbils 51%) and the taurine conjugate (2) (mice 44%, gerbils 4%) were present. In the hamster the glutamine conjugate (3) (18%) was formed in addition to the glycine (17%) and taurine conjugates (5%).

Reference

T. S. Emudianughe, J. Caldwell, and R. L. Smith, Studies on the metabolism of arylacetic acids 6. Comparative metabolic conjugation of 1- and 2-naphthylacetic acids in the guinea pig, mouse, hamster, and gerbil, *Xenobiotica*, 1987, **17**, 815.

Ketoprofen

Use/occurrence:	Anti-inflammatory drug
Key functional groups:	Aryl propionic acid, chiral carbon
Test system:	Rabbit (intravenous, 3 mg kg^{-1})

Structure and biotransformation pathway

† chiral carbon

Rabbits were administered bolus doses of *R*- or *S*- ketoprofen and concentrations of the separate enantiomers in urine and plasma determined. Ketoprofen acylglucuronides were also quantified in urine. A mean of 9% of a dose of *R*-ketoprofen was converted to the *S*-enantiomer but no *S*- to *R*-conversion was detected. The disposition of ketoprofen showed marked enantioselectivity, the mean clearance of the *R*-enantiomer being approximately eight times that of the *S*-enantiomer. This was mostly not accounted for by the degree of *R*- to *S*- inversion or by enantioselective glucuronide formation and was therefore due mainly to unknown processes. In rabbits with uranyl nitrate induced renal dysfunction, there was an approximate three-fold increase in the degree of *R*- to *S*-inversion.

Reference

A. Abas and P. J. Meffin, Enantioselective disposition of 2-arylpropionic acid non-steroidal anti-flammatory drugs. IV, Ketoprofen disposition, *J. Pharm. Exp. Ther.*, 1987, **240**, 637.

Fenoprofen

Use/occurrence:	Anti-inflammatory drug
Key functional groups:	Aryl propionic acid, chiral carbon
Test system:	Rabbit (intravenous, 5 mg kg^{-1})

Structure and biotransformation pathway

† chiral carbon

Rabbits were administered bolus doses of *R*-, *S*-, or racemic fenoprofen and concentrations of the separate enantiomers in urine and plasma determined. A mean of 73% of a dose of *R*-fenoprofen was converted into *S*-fenoprofen, although there was a large inter-animal variation. No *R*-fenoprofen was found in samples from animals administered *S*-fenoprofen. *R*- to *S*- inversion was not affected by phenobarbital pretreatment.

Reference

J. Hayball and P. J. Meffin, Enantioselective disposition of 2-arylpropionic acid non-steroidal anti-inflammatory drugs. III, Fenoprofen disposition, *J. Pharm. Exp. Ther.*, 1987, **240**, 631.

Benzyl acetate

Use/occurrence:	Flavouring agent, fragrance
Key functional groups:	Benzyl alcohol, methyl carboxylate
Test system:	Rat (dermal, 100–500 mg kg^{-1})

Structure and biotransformation pathway

[*Methylene*-^{14}C]benzyl acetate (1) was applied to rat skin, either neat or as a solution in 50% ethanol, at doses of 100, 250, and 500 mg kg^{-1}. After 6 hours 28–48% of radioactivity was still at the site of application. The radioactivity from the absorbed material was excreted largely in the urine, with less than 3% in the faeces. Retention of radioactivity in the carcass was less than 2% dose, 72 hours after dosing.

Urinary metabolites were separated and quantitated by radio-TLC and HPLC. Their identity was confirmed by chromatographic behaviour and colour reactions, in comparison with authentic standards. Approximately 95% of the absorbed dose was excreted as hippuric acid (2); about 1–2% each of benzoyl glucuronide (3), benzoic acid (4), and benzylmercapturic acid (5) were also detected. Increasing doses caused a larger proportion of metabolism to the glucuronide (3), but the use of ethanol in the dosing solution had no effect on the absorption of (1). The absorption was not affected by increases in the concentration of (1) applied to the skin.

Reference

M. A. J. Chidgey, J. F. Kennedy, and J. Caldwell, Studies on benzyl acetate. III, The percutaneous absorption and disposition of [*methylene*-^{14}C]benzyl acetate in the rat, *Food Chem. Toxicol.*, 1987, **25**, 521.

3,4-Dichlorobenzyloxyacetic acid

Use/occurrence:	Antisickling agent
Key functional groups:	Alkyl carboxylic acid, benzyl ether, chlorophenyl
Test system:	Rat (intraperitoneal, 100 mg kg^{-1})

Structure and biotransformation pathway

Within five days of administration of (1) to rats $77.4 \pm 4.6\%$ dose was excreted in urine with only $3.2 \pm 0.5\%$ being recovered in faeces. Urinary metabolites were separated by HPLC. The extent of metabolism to the taurine conjugate (3) (60% dose) was unusual. This conjugate was identified by EI–MS, LC–MS, ^{1}H NMR, and comparison with a synthetic standard. The glycine conjugate (2) accounted for only 2% dose and was identified by the same techniques but excluding NMR. The hippuric acid (5) which did not contain the radiolabel, and which was identified by EI–MS and comparison with a synthetic standard accounted for 12% dose. Its precursor acid (4) was detected in trace amounts. Unchanged (1) accounted for 13% dose. A highly polar unknown (2% dose) was thought to be a small molecule arising from side-chain cleavage. Enzyme incubation experiments yielded no evidence for the formation of glucuronide or sulphate conjugates. Excretion of radioactivity in expired air was not apparently monitored.

Reference

R. C. Peffer, D. J. Abraham, M. A. Zemaitis, L. K. Wong, and J. D. Alvin, 3,4-Dichlorobenzyloxyacetic acid is extensively metabolized to a taurine conjugate, *Drug Metab. Dispos.*, 1987, **15**, 305.

Use/occurrence:	Local anaesthetic
Key functional groups:	Alkoxyphenyl, alkyl ester, aryl amine, aryl carboxylate, dialkylamino-alkyl
Test system:	Man (oral, 100 mg)

Structure and biotransformation pathway

During 9 hours after administration 92.1% dose was excreted in urine. Metabolites (2)–(5) were quantified by GC–MS with selected ion monitoring. Metabolites (6)–(9) were quantified by HPLC [(7)–(9) after enzymic hydrolysis]. Benoxinate and its N-oxide (1) were quantified together by GC. The structure of (1) was proposed on the basis of its facile conversion into benoxinate, either thermally or by titanium trichloride treatment. The remaining metabolites were identified by GC–MS and TLC comparison with synthetic standards, the conjugates after enzymic hydrolysis. No differences in the extent of deconjugation were observed after hydrolysis with β-glucuronidase alone or a β-glucuronidase/sulphatase mixture, from which it was inferred that sulphate conjugates were absent. In urine, the substituted benzoic acid glucuronide (7) was by far the most important metabolite (83% dose). The free acid (2) accounted for 3.5% dose and its glycine conjugate (6) for only 0.4% dose. Proportions of unchanged benoxinate and its other metabolites were in the range 0.1–1.6% dose. Metabolites in which the O-butyl group was absent together amounted to $ca.$ 5% dose.

Reference

F. Kasuya, K. Igarashi, and M. Fukui, Metabolism of benoxinate in humans, *J. Pharm. Sci.*, 1987, **76**, 303.

Nafimidone

Use/occurrence:	Anticonvulsant
Key functional groups:	Aryl ketone, imidazole, naphthalene
Test system:	Rat (oral, 10 and 100 mg kg^{-1}; intravenous, 10 mg kg^{-1})

Structure and biotransformation pathway

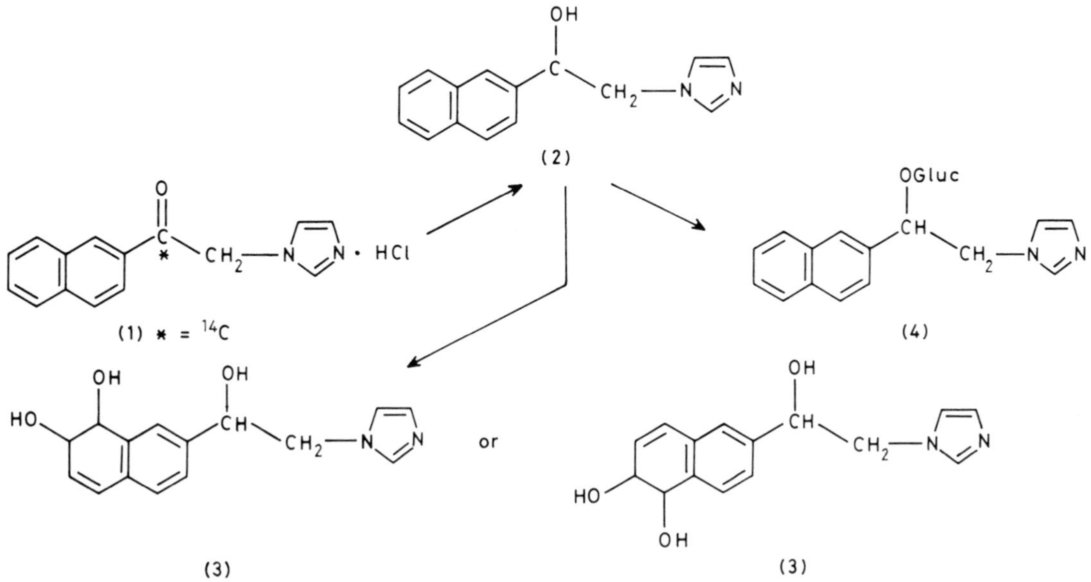

Following oral or intravenous administration, nafimidone (1) was rapidly eliminated from plasma ($t_{1/2}$ *ca.* 5 min) with concomitant formation of the alcohol (2) which was also pharmacologically active. About two-thirds of oral and intravenous doses were excreted *via* the urine during four days. Two major (no quantitative data shown) urinary metabolites were isolated from rats receiving oral 100 mg kg^{-1} doses and characterized by ^{1}H NMR. These metabolites were formed by further biotransformation of the alcohol (2) to yield a dihydrodiol (3), presumably *via* intermediate epoxidation of the naphthalene ring system, and the glucuronide conjugate (4). From the spectral data obtained it was not possible to distinguish between the two possible isomeric structures shown for (3).

Reference

D. J. M. Graham, K. M. Hama, S. A. Smith, L. Kurz, M. D. Chaplin, and D. J. Hall, Disposition of nafimidone in rats, *Drug Metab. Dispos.*, 1987, **15**, 565.

Ketorolac tromethamine

Use/occurrence:	Analgesic
Key functional groups:	Alkyl carboxylic acid, aryl ketone
Test system:	Man, monkey, rat, rabbit, mouse (oral, intramuscular, and intravenous doses in the range 0.25–16 mg kg^{-1})

Structure and biotransformation pathway

* = ^{14}C Ketorolac tromomethamine

(1)   (2)

Ketorolac was absorbed rapidly in all species following either oral or intramuscular administration. The plasma half-life of ketorolac (1) ranged from 1.1 hours (rabbits) to 6.0 hours (man). In the mouse linear pharmacokinetics occurred in the dose range 0.25–16 mg kg^{-1}. Most of the dose was excreted in the urine ranging from 78.9% (mouse) to 100% (monkey) following intravenous administration. The dose was excreted in urine primarily as conjugates of ketorolac (1), unconjugated ketorolac, and the *p*-hydroxy metabolite (2) in humans. The most comparable species to humans metabolically was the mouse. Metabolites were identified by enzyme hydrolysis, TLC, HPLC, NMR, and MS. There was no evidence of ring cleavage products in human urine.

Reference

E. J. Mroszczac, F. W. Lee, D. Combs, F. H. Sarnquist, B. Huang, A. T. Wu, L. G. Tokes, M. L. Maddox, and D. K. Cho, Ketorolac tromethamine absorption, distribution, metabolism, excretion, and pharmacokinetics in animals and humans, *Drug Metab. Dispos.*, 1987, **15**, 618.

450191-S (benzodiazepine precursor)

Use/occurrence:	Tranquilizer
Key functional groups:	Benzodiazepine (ring-opened), dimethylamide, triazole
Test system:	Dog (oral, 50 mg kg^{-1})

Structure and biotransformation pathway

(1)–(6)

Metabolite	CONR^{1}R^2
(1)	CON(CH$_3$)$_2$
(2)	CONHCH$_3$
(3)	CONH$_2$
(4)	CO$_2$H
(5)	CON(CH$_3$)CH$_2$OH
(6)	CONHCH$_2$OH

Metabolites (1)–(4), resulting from ring closure, *N*-demethylation, and deamination had been identified in a previous study. In the study under review, the *N*-hydroxymethyl intermediate metabolites (5) and (6) were isolated from plasma. The more important metabolite (6) was identified by EI- and FD-MS, IR, and comparison with a synthetic standard. The minor metabolite (5) was converted into (2) during isolation and was identified by EI-MS of its acetyl derivative and by comparison with a synthetic standard.

Reference

M. Koike, R. Norikura, S. Futaguchi, T. Yamaguchi, K. Sugeno, K. Iwatani, Y. Ikenishi, and Y. Nakagawa, *N*-Hydroxymethyl metabolites of 450191-S, a 1*H*-1,2,4-triazoylbenzophenone derivative, in dog plasma, *Drug Metab. Dispos.*, 1987, **15**, 426.

Melphalan

Use/occurrence:	Antineoplastic agent
Key functional groups:	Amino acid, *N*-chloro-ethyl, dialkyl aryl amine
Test system:	Immobilized microsomal glutathione-*S*-transferases from cynomolgus monkey liver

Structure and biotransformation pathway

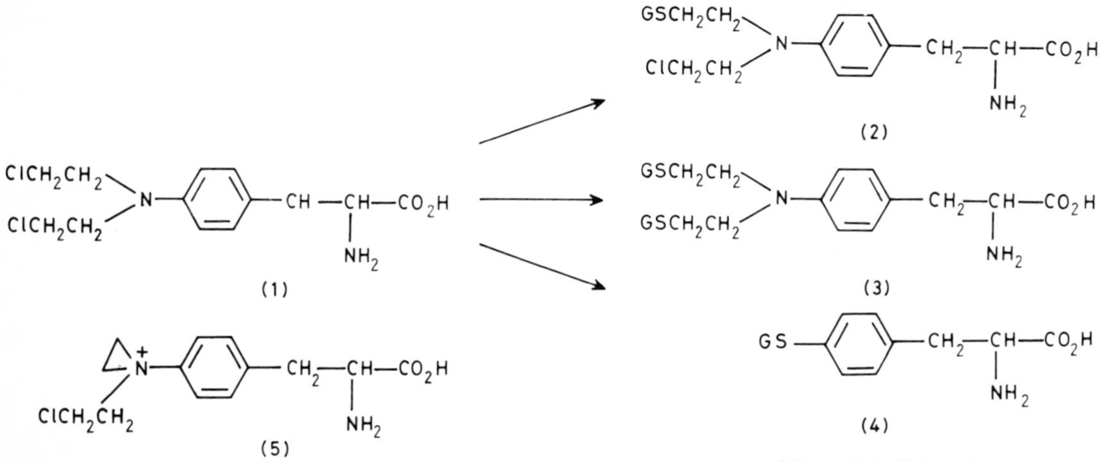

GS = glutathionyl

The conjugates (2) and (3), formed by displacement of one or both chlorines on the β-chloroethyl side-chains, had been previously identified. In this study the conjugate (4), formed by displacement of the entire mustard moiety, was characterized by FAB-MS and by TLC and HPLC comparison with the reaction product of the corresponding 4-halogenophenylalanines with glutathione under identical reaction conditions. It was proposed that the formation of (4) proceeded *via* an aziridinium ion intermediate (5).

Reference

D. M. Dulik and C. Fenselau, Conversion of melphalan to 4-(glutathionyl)phenyl-alanine: a novel mechanism for conjugation by glutathione *S*-transferases, *Drug Metab. Dispos.*, 1987, **15**, 195.

Phencyclidine

Use/occurrence:	Anaesthetic
Key functional groups:	Arylalkyl
Test system:	Rat liver microsomes

Structure and biotransformation pathway

(1)   (2)

The objective of this study was to obtain evidence for the metabolic formation of phenolic metabolites of phencyclidine (1), which could significantly contribute to its hallucinatory effects *in vivo*. For the microsomal incubation a 1:1 mixture of unlabelled (1) and [*phenyl*-^{2}H]-labelled (1) was used. Incubations were terminated after 20 minutes and extracts were methylated and analysed by GC–MS using selected ion monitoring. Reference standards of the *o*-, *m*-, and *p*-methoxy-substituted derivatives of (1) were available. The results showed that the *meta*-substituted phenol (2) was formed at rates of 0.15 and 0.2 nmol mg^{-1} protein in microsomes from non-treated and phenobarbital pretreated rats respectively. The reaction was shown to be NADPH-cytochrome P-450 dependent.

Reference

S. Ohta, H. Masumoto, K. Takeuchi, and M. Hirobe, A phenolic metabolite of phencyclidine: the formation of a pharmacologically active metabolite by rat liver microsomes, *Drug Metab. Dispos.*, 1987, **15**, 583.

Febantel

Use/occurrence:	Anti-nematode, anti-cestode
Key functional groups:	*N*-Arylimine, diaryl thioether, methoxyacetylamino, methyl carbamate
Test system:	Liver subcellular fractions from rat, horse, pig, cattle, sheep, chicken, and trout

Structure and biotransformation pathway

The capacity of the hepatic cytosol and microsomal fractions to metabolize (1) and its metabolites [febantel sulphoxide (2), fenbendasole (3), and oxfendazole (4)] were investigated using liver samples from rats, horses, pigs, cattle, sheep, chickens, and trout. Metabolism of (1) by the hepatic cytosol was negligible for all species except sheep and trout. The metabolites were determined by HPLC. There was evidence for significant species differences in the activities of the hepatic microsomes: *e.g.* the cyclization of (1) to (3) predominated in the pig and trout and sulphoxidation of (1) to (2) occurred readily in sheep and chickens. *In vitro* studies with ruminal fluids from cattle and sheep showed that (1) and (2) were only chemically modified to a slight extent but that (3) was oxidized to (4) and (4) was reduced to (3).

Reference

C. Beretta, L. Fadini, J. M. Stracciari, and C. Montesissa, *In vitro* febantel transformation by sheep and cattle ruminal fluids and metabolism by hepatic subcellular fractions from different animal species *Biochem. Pharmacol.*, 1987, **36**, 3107.

Propachlor

Use/occurrence:	Herbicide
Key functional groups:	Isoalkyl aryl amine, chloroacetyl
Test system:	Swine, rats $(5-200 \text{ mg kg}^{-1}$ dietary or oral)

Structure and biotransformation pathway

2-Chloro-N-isopropyl-[1-^{14}C]acetanilide (propachlor) (1) was administered orally to pigs and rats to determine the role of the intestinal microflora in producing methylsulphonyl-containing metabolites. Germ-free pigs did not excrete these metabolites in urine, and the addition of high levels of antibiotic to food diminished but did not abolish the excretion of these metabolites in

205

urine. In the bile duct-cannulated pig low levels of (7) and (8) as glucuronides and (9) were detected in the urine and bile. It was concluded that non-biliary excretion of propachlor metabolites into the intestine was more important in swine than in rats, and that the gut microflora produce the methylsulphonyl metabolites. The metabolites were isolated by means of Sephadex columns and HPLC and their identities confirmed by GC–MS. The formation of the terminal metabolites is believed to involve conjugation of propachlor with glutathione (2) followed by the action of peptidase and β-lyase to give the thiol (4). Subsequent S-methylation (5) and oxidation (6) introduces the methylsulphonyl function.

Reference

P. W. Aschbacher and C. B. Struble, Evidence for involvement of non-biliary excretion into the intestines in the formation of methylsulphonyl-containing metabolites of 2-chloro-N-isopropylacetanilide (propachlor) by swine and rats, *Xenobiotica*, 1987, **17**, 1047.

Bamipine

Use/occurrence:	Antihistamine
Key functional groups:	Aryl amine, dialkyl aryl amine, N-methylpiperidine
Test system:	Rat (oral, 60 mg kg^{-1})

Structure and biotransformation pathway

The major metabolites of (1) in rat urine were the ether glucuronides (4) and (5) of the phase 1 metabolites (2) and (3) resulting from aromatic hydroxylation and N-demethylation. Together (4) and (5) accounted for more than 50% of urinary radioactivity. The intact conjugates were characterized by FAB-MS and ^{1}H NMR.

Reference

R. Neidelein and M. Kleiser, Biotransformation und Pharmacokinetik von Bamipin an Ratten, *Arzneim. Forsch.*, 1987, **37** (**I**), 32.

Pyrilamine

Use/occurrence:	Antihistamine
Key functional groups:	Dialkyl aryl amine, dimethylaminoalkyl, methoxyphenyl, pyridine
Test system:	Rat (intraperitoneal, 20 mg kg^{-1})

Structure and biotransformation pathway

This study was conducted using non-radioactive methodology. Compounds (2)–(7) were reported as metabolites of pyrilamine (1) following GC–MS analysis (with and without prior trimethylsilylation) of extracts of enzymatically hydrolysed urine collected during 24 hours after dosing. Relative proportions of metabolites were estimated from measurements of areas under peaks in the ion current chromatogram, and should be regarded as approximate only. With the exception of compound (2), for which a synthetic standard was available, identifications were based on interpretation of MS fragmentation patterns of the components and/or their TMS derivatives, and were described as tentative. The major phase 1 metabolite, accounting for an estimated 62% of urinary metabolites, was the O-demethylated compound (4). The N-debenzylated metabolite (2) and the N-dealkylated metabolite (7) each accounted for an estimated 15% of urinary metabolites. The O-methylcatechol compounds (5) and (6) were estimated to represent 1% and 6% respectively of urinary metabolites. The structures of these compounds, which gave similar mass spectra, were assigned on the basis of comparison of the GC-retention times of their TMS derivatives with retention times of the analogous derivatives of 4-hydroxy-3-methoxy-amphetamine and 3-hydroxy-4-methoxyamphetamine. Compound (3), resulting from dealkylation of the benzylpyridyl-substituted nitrogen, was a minor metabolite (0.7% urinary metabolites), while only a trace of unchanged (1) was detected in urine.

Reference

S. Y. Yeh, N-Debenzylation of pyrilamine and tripelennamine in the rat, *Drug Metab. Dispos.*, 1987, **15**, 466.

Pyrilamine maleate

Use/occurrence: Antihistamine

Key functional groups: Dialkyl aryl amine, dimethylaminoalkyl, methoxyphenyl, pyridine

Test system: Rat (intravenous, 0.7, 7 mg kg^{-1})

Structure and biotransformation pathway

(1) * = ^{14}C

(2)

(3)

(4) Gluc = glucuronyl

Approximately 29% or 38% of high or low doses respectively of pyrilamine (1) were excreted in urine during 24 hours after intravenous administration. 24 hour faecal excretion accounted for 27% (high) or 30% (low) dose. One major (73–80% urinary ^{14}C) and four minor (*ca.* 10% urinary ^{14}C in total) components were separated by HPLC of the 24 hour urine from both dose groups. The major metabolite was identified as the glucuronic acid conjugate (4) by FAB-MS of the intact conjugate and by HPLC comparison of the aglycone (3) released by enzymic hydrolysis, with a reference standard. The free aglycone (3), resulting from O-demethylation of the parent compound (1), was also tentatively identified in urine by HPLC comparison with the reference standard. A further minor metabolite co-eluted (HPLC) with a reference standard of the N-oxide (2) of the parent compound and this was confirmed by CI-MS of the isolated metabolite and reference compound The parent compound (1) co-eluted with one of the other minor radiolabelled components in urine and one component was not identified. HPLC analysis of faeces extracts indicated that most of the radioactivity present co-eluted with the O-desmethyl compound (3) and that the N-oxide (2) was present, but no quantitative data were given. Mono-N-desmethyl-(1)

was not detected in this study, although it had previously been reported as a metabolite of (1) in the rat after oral administration.

Reference

D. W. Kelly and W. Slikker Jr., The metabolism and elimination of pyrilamine in the rat, *Drug Metab. Dispos.*, 1987, **15**, 460.

Tripelennamine

Use/occurrence:	Antihistamine
Key functional groups:	Dialkyl aryl amine, dimethylaminoalkyl, pyridine
Test system:	Rat (intraperitoneal, 20 mg kg^{-1})

Structure and biotransformation pathway

This study was conducted using non-radioactive methodology. Components (2)–(6) were reported as metabolites of tripelennamine (1) following GC–MS analysis (with and without prior trimethylsilylation) of extracts of enzymatically hydrolysed urine collected during 24 hours after dosing. The O-methylcatechol derivatives (6) and (7) were very minor metabolites and were tentatively assigned their structures on the basis of GC-retention time comparison with metabolites of the related compound pyrilamine, for which some mass spectral characterization had been obtained. The remaining metabolites (2)–(5) were assigned their structures on the basis of MS fragmentations. Additionally an authentic standard of (3) was available. Relative proportions of metabolites were estimated from measurements of areas under peaks in the ion current chromatogram, and should be regarded as approximate only. The major phase 1 metabolite, accounting for an estimated 43% of urinary metabolites, was the α-hydroxylated compound (2). This metabolite was considered to be the intermediate leading to formation of the N-debenzylated metabolite (3) (16% urinary metabolites). Less important metabolites were formed *via* N-dealkylation [(4), 10% urinary metabolites] and aromatic hydroxylation [(5), 7% urinary metabolites]. It is unusual that the intermediates (2) and (4) were sufficiently stable to be found as urine metabolites.

Reference

S. Y. Yeh, N-Debenzylation of pyrilamine and tripelennamine in the rat, *Drug Metab. Dispos.*, 1987, **15,** 466.

Methapyrilene

Use/occurrence:	H1 antagonist
Key functional groups:	Aminopyridine, dimethylaminoalkyl, thienylalkyl
Test system:	Rat, guinea pig, rabbit liver microsomes

Structure and biotransformation pathway

Methapyrilene (1) was converted into normethapyrilene (2), methapyrilene *N*-oxide (3), 2-thiophenyl methanol (4), thiophene-2-carboxylic acid (5), *N*-(2-pyridyl)-*N'*,*N'*-dimethylethylenediamine (6), and methapyrilene amide (7) by liver microsomes from rat, guinea pig, and rabbit. In the rat and rabbit liver preparations *N*-(2-thienylmethyl)-2-aminopyridine (8), 2-aminopyridine (9), and (5-hydroxypyridyl)methyapyrilene (10) were also detected. Novel metabolites, methapyrilenemethanol (11), *N*-hydroxynormethapyrilene (12), and *N*-hydroxy(hydroxypyridyl)normethapyrilene (13), were also identified. Metabolite structures were characterized by comparison of retention times on GC with those of authentic standards and confirmed by GC–MS. The formation of both (2) and (3) was quantitatively important in all three species but the formation of both (4) and (5) was only a minor pathway. The guinea pig did not form (8), (9), or (10), all three of which were formed in rat and rabbit.

Reference

R. C. Kammerer and D. A. Schmitz, Species differences in the *in vitro* metabolism of methapyrilene, *Xenobiotica*, 1987, **17**, 1121.

Methapyrilene

Use/occurrence:	H1 antagonist
Key functional groups:	Aminopyridine, dimethylaminoalkyl, thienylalkyl
Test system:	Rat (oral, 50 and 80 mg kg^{-1}), liver microsomes

Structure and biotransformation pathway

In vivo studies were carried out on the metabolism of methapyrilene (1) in F-344 and Sprague–Dawley rats and various new metabolites were separated and identified by capillary column GC–MS techniques or by reverse phase HPLC and direct insertion MS. New metabolites identified include N-(N',N'-dimethylaminoethyl)-2-aminopyridine (2) and the corresponding N-oxide (3), 2-(N',N'-dimethylamino)-N-2$'$-pyridylacetamide (4), N-(2-pyridyl)-N-(2-thienylmethyl)aminoacetaldehyde (5), 2-hydroxymethylthiophene (6), and a derivative hydroxylated on the 5-position of the pyridine ring (7). The amide (4) was identified only as an *in vitro* metabolite; otherwise there were no major differences between the *in vitro* and *in vivo* studies, nor between the strains of rat dosed *in vivo*. Metabolites were quantified by GC–MS SIMS analysis.

216

Reference

S. S. Singer, W. Ligensky, L. E. Kratz, N. Castagnoli, and J. E. Rose, A comparison of the *in vivo* and *in vitro* metabolites of the H1-antagonist *N,N*-dimethyl-*N'*-2-pyridyl-*N'*-(2-thienylmethyl)-1,2-ethanediamine (methapyrilene) in the rat, *Xenobiotica*, 1987, **17**, 1279.

Terodiline

Use/occurrence:	Anticholinergic/calcium antagonist
Key functional groups:	t-Alkyl amine, benzhydryl, chiral carbon, prochiral carbon
Test system:	Rat liver microsomes

Structure and biotransformation pathway

(1) † indicates chiral carbon

(2)

(3)

The biotransformation of racemic and *R*- and *S*- forms of terodiline (1) was investigated using liver microsomes from rats, with and without pretreatment with phenobarbitone. The overal metabolism of the racemate and the individual isomers was slow, with the *S*-isomer metabolized more rapidly than the *R*-isomer. The metabolism of terodiline, especially the *S*-isomer was enhanced by pretreatment with phenobarbitone. Metabolism occurred by both aromatic *para*-hydroxylation (2) and benzylic oxidation (3) producing a diastereomeric pair of metabolites (2). Metabolite (2) was the major metabolite (53%) of *R*-terodiline whereas metabolite (3) represented only 17% of the consumed substrate. With *S*-terodiline aromatic *para*-hydroxylation to give (2) represented only 5% as opposed to 14% of (3) produced by benzylic hydroxylation. The rate of aromatic hydroxylation of racemic terodiline followed that of the *R*-isomer, whereas benzylic hydroxylation followed that of the *S*-isomer. Phenobarbitone pretreatment enhanced the latter. Metabolites were quantified by GC–MS.

Reference

B. Lindeke, O. Ericsson, A. Jansson, B. Noren, S. Stromberg, and B. Vangbo, Biotransformation of terodiline III. Opposed stereoselectivity in benzylic and aromatic hydroxylation in rat liver microsomes, *Xenobiotica*, 1987, **17**, 1269.

Chlorpheniramine

Use/occurrence:	Antihistamine
Key functional groups:	Chlorophenyl, dimethylaminoalkyl, pyridine
Test system:	Rabbit liver microsomes

Structure and biotransformation pathway

Six metabolites of chlorpheniramine (1) were detected by GC and GC–MS following incubation of (1) with rabbit liver microsomes. The monodesmethyl metabolite (2) and the didesmethyl metabolite (3), 3-(*p*-chlorophenyl)-3-(2-pyridyl)propionic acid (4), and 3-(*p*-chlorophenyl)-3-(2-pyridyl)propyl

219

alcohol (5) were identified. These four metabolites were previously characterized as metabolites of (1) in *in vivo* studies in animals. In addition the amide (6), an intramolecular cyclization adduct (7), and a hydroxylated cyclization adduct (8) were identified by MS. The alcohol (5), the acid (4), and the cyclization adduct (7) were formed from the aldehyde (9) that results from deamination. The presence of this aldehyde was confirmed by trapping experiments with methoxylamine. Evidence was presented that metabolite (8) was an artefact of the derivatization of the hydroxylated amine (10), but it was argued that the indolisine (7) is the spontaneous cyclization adduct of the aldehyde (9).

Reference

R. C. Kammerer and M. A. Lampe, *In vitro* metabolism of chlorpheniramine in the rabbit, *Biochem. Pharmacol.*, 1987, **36**, 3445.

Procyclidine

Use:	Anticholinergic drug
Key functional groups:	Chiral carbon, cyclohexyl, benzyl alcohol, pyrrolidine
Test system:	Rat hepatocytes

Structure and biotransformation pathway

The mechanism of formation of the eight metabolites of procyclidine, (1)–(3), (5), and (8)–(11), previously identified from rat urine, has been investigated using hepatocytes from untreated and phenobarbitone treated rats. The rates of formation of metabolites and disappearance of procyclidine from the incubation medium were determined by GC. The major metabolites formed from procyclidine were the ketone (3) and the 4-hydroxy metabolites (1) and (2). Hydroxylation also occurred at the 3-position to give (6) and (7) but these were only minor metabolites. The 3,4-diols (11) and (5) were also identified as minor metabolites. In a further series of experiments possible intermediates in the formation of the 3,4-diols were incubated with rat hepatocytes. The monohydroxylated metabolite (7) was converted into (9) (10.6%) and (11) (3.2%), (6) was converted into (7) (17.6%) and (10) (8.6%),

221

(12) was converted into (13) (18.4%), and (4) into (5) (22.7%). The 4-hydroxy metabolites (1) and (2) and the corresponding 4-oxo-compound (3) were interconvertible in hepatocytes, with the *trans*-hydroxylated metabolite (2) always being the major compound found in the incubation medium. It was concluded that the 3,4-diols are possibly formed by two consecutive hydroxylations, the 3-hydroxy metabolites being intermediates. More of the *cis*-3,*trans*-4-diols (8) or (9) than the *cis*-3,*cis*-4-diols (10) or (11) are formed from the mono-hydroxylated metabolites (6) or (8) *in vitro* than would be expected from the *in vivo* data. This suggests that there may be alternative mechanisms for the formation of the *cis*-3,*cis*-4-diols.

Reference

V. Rogiers, G. Paeme, W. Sonck, and A. Vercrysse, Metabolism of procyclidine in isolated rat hepatocytes, *Xenobiotica*, 1987, **17**, 849.

Flumecinol

Use/occurrence:	Agent for the treatment of icterus neonatorum
Key functional groups:	Benzhydrol, tertiary alcohol, trifluoromethylphenyl
Test system:	Human (oral, 100 mg)

Structure and biotransformation pathway

Following oral administration of [^{14}C]flumecinol (1) to six male volunteers, $77.8 \pm 6.0\%$ of the dose was excreted in urine and $12.0 \pm 5.3\%$ in faeces over 120 hours. Urine and faeces were subject to solvent extraction procedures to isolate metabolites which were analysed by TLC, MS, GC–MS, and GC–IR. Flumecinol was not excreted unchanged in urine but was found in faeces both non-conjugated (1.2% dose) and as glucuronide and sulphate conjugates (10.8%). The urinary metabolites were all glucuronide and/or sulphate conjugates of the products arising from hydroxylation at 4′ (2), at 3′,4′ [dihydrodiol (3)], or in the alkyl side-chain (4) and (5). Two other metabolites (6) and (7) were hydroxylated in both the phenyl ring and the alkyl side-chain. The secondary hydroxy-flumecinol (4) was the major metabolite, amounting to 25% of the dose in human urine.

Reference

I. Klebovich, L. Vereskey, E. Toth, J. Tamas, M. Mak, G. Jalsovsky, and S. Holly, Metabolism of flumecinol in humans, *Xenobiotica*, 1987, **17**, 1247.

Inabenfide

Use/occurrence:	Plant growth regulator
Key functional groups:	Aryl amide, aryl carboxamide, benzhydrol, chlorophenyl, pyridine
Test system:	Rat (intraperitoneal, 400 mg, twice daily)

Structure and biotransformation pathway

After intraperitoneal administration of (1) to rats, urine and faecal excretion of radioactivity represented 21% and 32% of the dose respectively in the first 24 hours. The urinary metabolites were initially separated by TLC and identified by GC–MS. A [^{2}H$_5$]benzene labelled form of (1) was used to

225

assist in the detection of metabolites. Unchanged (1) was excreted in faeces but 4-hydroxy-inabenfide (2) was the major metabolite in urine. Hydroxylation is postulated to proceed via an unstable epoxide intermediate (3) from the fact that a dihydrodiol (4) was formed. The 3-hydroxy (5) and 4-hydroxy metabolites are probably formed non-enzymically. Isolation of the hydroxy-methoxy metabolite (6) implies the formation of a catechol (7) although this compound was not detected. Formation of the 4'-hydroxy metabolite (8) suggests an NIH shift in which the chloro substituent originally at this position is lost. Other metabolites identified include the product of amide hydrolysis (9), the *N*-oxide (10), and the ketone (11). This latter metabolite is of interest as it involves the relatively unusual oxidation of a carbinol between two benzene rings.

Reference

H. Kinoshita, Y. Tohira, H. Sugiyama, and M. Sawada, Studies on the metabolism of inabenfide (a new plant growth regulator) in rats, I. Characterization of metabolites in rats, *Xenobiotica*, 1987, **17**, 925.

Chlorphenoxamine

Use/occurrence:	Antihistamine
Key functional groups:	Alkyl ether, benzhydrol, chlorophenyl, dimethylaminoalkyl
Test system:	Human volunteer (oral, 120 mg day^{-1})

Structure and biotransformation pathway

This study used non-radiolabelled chlorphenoxamine (1), and metabolites isolated from urine were located during work-up by comparison with similar extracts of pre-dose urine. No quantitative data were presented. Metabolites were identified by MS, ^{1}H NMR, IR, UV, and TLC comparison with synthetic standards. Conjugates were assigned on the basis of enzymic deconjugation experiments. Detected pathways of metabolism involved N-demethylation, N-oxidation, ether cleavage, and aromatic hydroxylation, followed by methylation of dihydroxylated metabolites. Alkenes resulting from loss of water from the tertiary alcohols (6) and (8) were also detected, but were considered to be artefacts.

Reference

S. Goenechea, G. Rücker, H. Brzezinka, G. Hoffman, M. Neugebauer, and T. Pyzik, Isolierung von Chlorphenoxamin-Metaboliten aus menschlichem Harn und ihre Identifizierung, *Arzneim.-Forsch.*, 1987, **37** (**II**), 854.

Chlorphenoxamine

Use/occurrence:	Antihistamine
Key functional groups:	Alkyl ether, benzhydrol, chorophenyl, dimethylaminoalkyl
Test system:	Human volunteers (oral, 40 mg)

Structure and biotransformation pathway

Urine collected during 24 hours after a single dose was incubated with glucuronidase/sulphatase and extracted with diethyl ether (pH 3 or 10). Extracts were analysed by GC–MS before and after methylation or acetylation. The study used non-radioactive methodology and no quantitative data were presented. In addition to metabolites shown, numerous artefacts were detected, resulting from decomposition of metabolites during analysis. Structural assignments were based mainly on interpretation of MS fragmentation patterns of the metabolites and their derivatives or of decomposition products. The metabolic pathways found involved: (1) *N*-demethylation [(2) and (3), described as trace metabolites]; (2) oxidative

229

deamination (4) and (5); (3) ether cleavage (6); (4) aromatic ring hydroxylation (7).

Reference

C. Koppel, J. Tenczer, I. Arndt, and K. Ibe, Urinary metabolism of chlorphenoxamine in man, *Arzneim. Forsch.*, 1987, **37** (**II**), 1062.

Doxylamine

Use/occurrence:	Anti-nauseant
Key functional groups:	Alkyl ether, dimethylaminoalkyl pyridine
Test system:	Man (oral, 10–50 mg)

Structure and biotransformation pathway

Urine was collected from two volunteers who received oral doses of doxylamine (1). Urinary metabolites were extracted into organic solvent, separated by HPLC, and analysed by GC–MS as their trifluoroacetyl derivatives. Both *N*-desmethyl (2) and *N,N*-didesmethyldoxylamine (3) were identified in urine but neither phenolic nor *N*-oxide metabolites, previously reported for rats and monkeys, were detected. However, the *N*-acetyl conjugates of the *N*-desmethyl (4), and *N,N*-didesmethyl (5) metabolites were identified in the urine of both volunteers. The authors believe this to be the first report of acetylation *in vivo* of primary and secondary aliphatic amines in man.

Reference

D. A. Ganes and K. K. Midha, Identification in *in vivo* acetylation pathway for *N*-dealkylated metabolites of doxylamine in humans, *Xenobiotica*, 1987, **17**, 993.

p-Tyramine

Use/occurrence:	Adrenergic
Key functional groups:	Alkyl amine, phenol
Test system:	Human (oral, 125 mg)

Structure and biotransformation pathway

HO—⟨benzene⟩—$C^2H_2CH_2NH_2$ (1) ⟶ HO—⟨benzene⟩—$C^2HCH_2NH_2$ (OH) (3) ⟶ HO—⟨benzene⟩—C^2HCO_2H (OH) (4)

(1) ⟶ HO—⟨benzene⟩—C^2H_2CHO ⟶ HO—⟨benzene⟩—$C^2H_2CO_2H$ + HO—⟨benzene⟩—C^2HHCO_2H (2)

Eight volunteers (five female, three male; weight 55–105 kg; ages 25–50) ingested $[\beta,\beta\text{-}^2H_2]$-*p*-tyramine hydrochloride (1) (125 mg). Urine was collected from 0–3 hours and from 3–24 hours and analysed by mass spectrometry for $[\beta,\beta\text{-}^2H_2]$-*p*-tyramine (1) (conjugated and unconjugated), $[\alpha,\alpha\text{-}^2H_2]$- and $\alpha\text{-}[^2H_1]$-*p*-hydroxyphenylacetic acid (2) (conjugated and unconjugated), $[\beta\text{-}^2H_1]$-*p*-octopamine (3) (unconjugated), and $[\alpha\text{-}^2H_1]$-*p*-hydroxymandelic acid (4) (unconjugated). Analysis was as follows: *p*-tyramine, by direct probe selected ion monitoring (SIM) of its dansyl derivative using a $[^2H_4]$-labelled internal standard; (2) and (4) by capillary GC–high resolution SIM of the methyl esters of the pentafluoropropionyl derivatives, using $[^2H_4]$-labelled internal standards; (3) by direct probe SIM of the acetyl bis-dansyl derivative using a $[^2H_3]$-labelled internal standard. 71–72% of the ingested dose was accounted for by these metabolites with more than 90% of this being unconjugated $[^2H_2]$-(2) (59 mg) and conjugated (1) (9 mg). 59% of the metabolite label was excreted in 0–3 hours [50% of (1) and 70% of (2)]. Small amounts of (4) (*ca.* 100 μg) and (3) (*ca.* 5 μg) were found in the unconjugated form in the urine. Interestingly $[^2H_1]$-(2) was also found (3.3 mg, unconjugated), showing that some exchange of deuterium had occurred in its formation.

Reference

A. A. Boulton and B. A. Davis, The metabolism of ingested deuterium-labelled *p*-tyramine in normal subjects, *Biomed. Environ. Mass Spectrom.*, 1987, **14**, 207.

Methamphetamine

Use/occurrence:	Analeptic
Key functional groups:	Secondary alkylamine
Test system:	Rat, guinea pig (intraperitoneal, 45 mg kg^{-1})

Structure and biotransformation pathway

In vitro studies showed that the *N*-hydroxy metabolites (1) and (2) of amphetamine (3) and methamphetamine (4) reacted with formaldehyde, acetaldehyde, and propionaldehyde in a Schiff's base condensation to give aldehyde adducts.

One adduct, *N*-[(1-methyl-2-phenyl)ethyl]ethanimine *N*-oxide (5) was shown by GC–MS to be formed *in vitro* when (4) was incubated with liver S9 (9000 g) supernatant from rats and guinea pigs. Compound (5) was also detected by GC in the urine of rats and guinea pigs dosed intraperitoneally with (4) but represented only $1.0 \pm 0.1\%$ (rats) or $0.6 \pm 0.3\%$ of the dose in 0–24 hour urine.

It is postulated that the *N*-hydroxy metabolite (1) is formed via two independent pathways which are initiated by either hydroxylation of nitrogen (pathway *a*) or hydroxylation of a carbon atom of the *N*-methyl group (pathway *b*).

Reference

T. Baba, H. Yamada, K. Oguri, and H. Yoshimura, 'A new metabolite of methamphetamine; evidence for formation of *N*-[(1-methyl-2-phenyl)ethyl]-ethanimine *N*-oxide, *Xenobiotica*, 1987, **17**, 1029.

Dimethylamphetamine

Use/occurrence:	Stimulant
Key functional groups:	Alkylphenyl, dimethylaminoalkyl
Test system:	Rat (oral, 20 mg kg^{-1}), man (oral, 10 mg)

Structure and biotransformation pathway

After oral administration of dimethylamphetamine (1) to rat and man, urinary metabolites were extracted into organic solvent, separated by TLC, and identified by GC–MS. Metabolites excreted in three days after administration of (1) amounted to about 57% in rat and 53–56% in man. About 12% of the dose was recovered in rat urine as unchanged (1). The major metabolite excreted in rat urine was dimethylamphetamine N-oxide (2), accounting for about 30% of the dose. N-Demethylation to produce methylamphetamine (3; 6% of the dose) and amphetamine (4; 0.3%) was a minor pathway. The *para*-hydroxylated metabolites of dimethyl-amphetamine, (5), methamphetamine, (6), and amphetamine, (7),

235

each represented less than 5% of the dose. Urinary excretion of the metabolites of (1) by man was very similar to that in rat except that the proportion of aromatic hydroxylated metabolites was lower in man. Excretion of (5) and (6) represented 2–3% of the dose in total and (7) was not detected.

Reference

T. Inoue and S. Suzuki, The metabolism of dimethylamphetamine in rat and man, *Xenobiotica*, 1987, **17**, 965.

Deprenyl

Use/occurrence:	Monoamine oxidase inhibitor
Key functional groups:	Alkylphenyl, alkyl tertiary amine, ethynyl
Test system:	Rat liver, lung, and kidney microsomes

Structure and biotransformation pathway

The biotransformation of deprenyl (1) was investigated in rat liver, lung, and kidney microsomes. Metabolites were determined by GC. Metabolism of (1) was lower in lung and kidney microsomes than in liver microsomes. Liver microsomes produced methamphetamine (2), amphetamine (3), and nor-deprenyl (4). Only (2) and (4) were produced by lung microsomes and (2) only by kidney microsomes. Production of (2) and (3) by liver microsomes was increased by pretreatment of animals with phenobarbitone but not by 3-methylcholanthrene. However, phenobarbitone did not affect metabolism of (1) by the lung and reduced the activity in the kidney. The metabolism of deprenyl was decreased in both lung and kidney microsomes by pretreatment with 3-methylcholanthrene. Sex- and strain-linked differences in the metabolism of (1) were noted.

Reference

T. Yoshida, T. Oguro, and Y. Kuroiwa, Hepatic and extrahepatic metabolism of deprenyl, a selective monoamine oxidase (MAO)B inhibitor of amphetamine in rats: sex and strain differences, *Xenobiotica*, 1987, **17**, 957.

Chloramphenicol

Use/occurrence: Antibacterial

Key functional groups: Chloroacetamide, nitrophenyl

Test system: Male human volunteer (oral, 500 mg)

Structure and biotransformation pathway

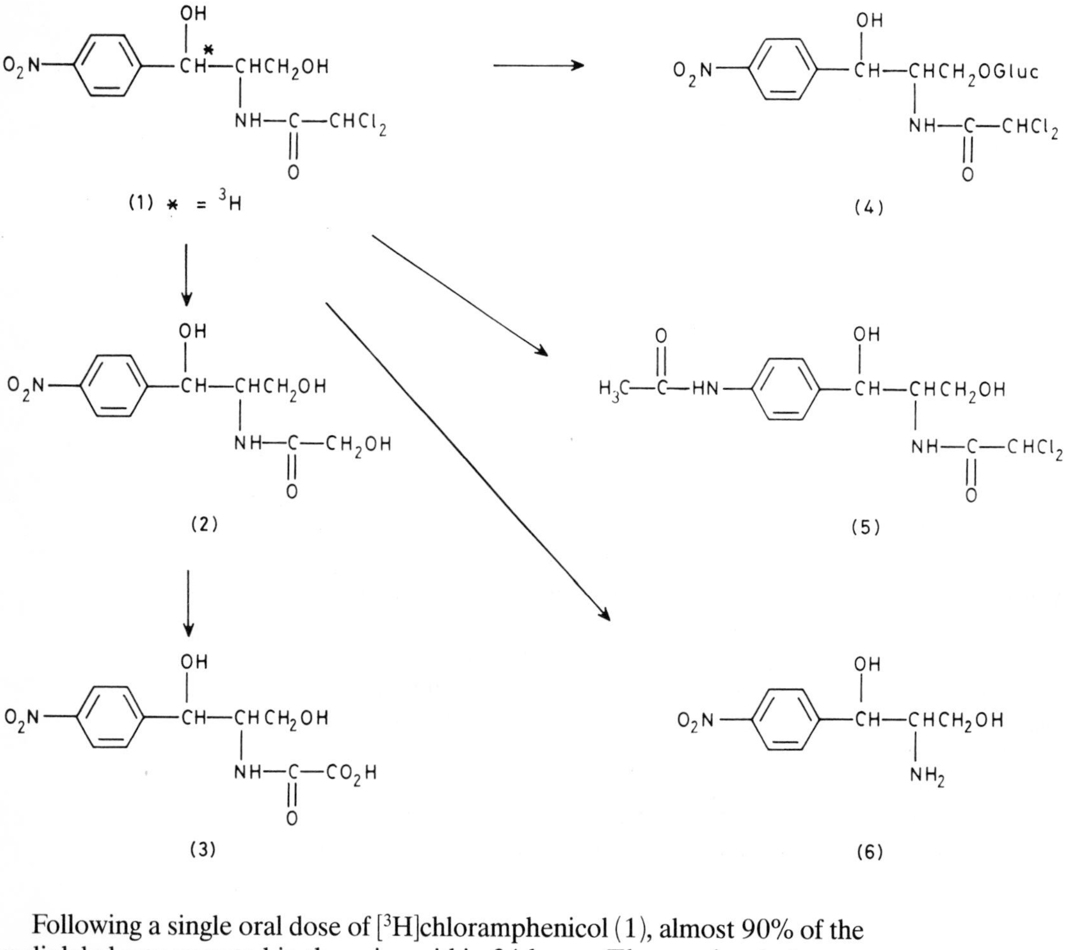

Following a single oral dose of [^{3}H]chloramphenicol (1), almost 90% of the radiolabel was excreted in the urine within 24 hours. The previously known metabolites (2), (4), (5), and (6), together with unchanged (1) and the novel oxamic acid metabolite (3), were identified by HPLC and MS comparison with synthetic standards. These compounds accounted for almost all the urinary radioactivity. The glucuronide conjugate (4) was the major metabolite in urine (51% dose) followed by the oxamic acid (13% dose). The other metabolites and unchanged (1) each accounted for 4–7% dose. The

238

arylamine which would have been intermediate in the formation of the arylamide (5) could not be identified in urine (<0.1% dose).

Reference

D. E. Corpet and G. F. Bories, [³H]Chloramphenicol metabolism in human volunteer: oxamic acid as a new major metabolite, *Drug Metab. Dispos.*, 1987, **15**, 925.

Diethylpropion

Use:	Anorectic drug
Key functional groups:	Aryl ketone, dialkylaminoalkyl
Test system:	Human (oral, 25 and 75 mg)

Structure and biotransformation pathway

Re-investigation of a method previously used to quantify the basic urinary metabolites of diethylpropion showed that, contrary to expectation, the parent drug and the metabolites (1) and (2) had not decomposed during the analysis procedure and consequently had interfered with the determination of the amino-alcohols (3), (4), and (5). A new extraction and GC method for the analysis of diethylpropion and metabolites (1)–(5) was therefore developed and used to investigate the urinary metabolites of diethylpropion hydrochloride in man. Acidic urine was maintained in the subjects using ammonium chloride sustained-release pellets. The diethylpropion and its basic metabolites found in urine accounted for about 70% of the administered dose. The major metabolites were the mono dealkylated product (1) (33–37% of the dose) and the alcohol (3) (13–20%). Metabolites (2) and (5), for which only a combined analysis was possible, and metabolite (4) accounted for 8–14% and 3–4% of the dose respectively. N-Dealkylation was a more important pathway of metabolism than ketone reduction. Administration of a sustained-release formulation of diethylpropion (75 mg) caused delayed uptake and elimination but the proportions of the metabolites were essentially identical.

Greater amounts of diethylpropion and metabolite (1), and lower amounts of metabolites (4) and (2) plus (5) were obtained in this study compared with those in a previous investigation, and the total recoveries of drug and metabolites were 20% lower.

Reference

A. H. Beckett and M. Stanojčić, Re-evaluation of the metabolism and excretion of diethylpropion in non-sustained and sustained release formulations, *J. Pharm. Pharmacol.*, 1987, **39**, 409.

Penbutolol

Use/occurrence:	β-Adrenoceptor blocking agent
Key functional groups:	Secondary alkyl alcohol, alkyl aryl ether, cycloalkyl, tertiary alkyl amine
Test system:	Human patients (dose not stated)

Structure and biotransformation pathway

The aim of the work reported in this paper was to isolate and characterize intact conjugates of penbutolol (1).

The conjugates (3)–(6) were isolated from urine and identified by enzymic deconjugation studies and by FAB-MS of the intact conjugates. ^{1}H NMR spectra and EI-MS of TMS derivatives were obtained for some of the conjugates. From the limited quantitative data presented it appeared that the glucuronide (3) of the parent compound was the major conjugate while the

242

novel dehydro compound (4) was a minor metabolite. The formation of the aliphatic glucuronide (5) of the phase I metabolite (2) in preference to the potential 4′-aromatic glucuronide was a suprising feature of the results. On *in vitro* incubation of (2) with glucuronyl-transferases from rat, rabbit, and bovine liver microsomes, the aromatic glucuronide was formed. It was stated that further investigation would be needed to clarify whether human glucuronyl-transferases were responsible for this regioselectivity, or whether the parent compound glucuronide (3) was formed first and then oxidized to (5). The dehydro structure of (4) had not been previously reported and it was suggested that it was formed *via* an as yet unidentified intermediate metabolite.

Reference

K. H. Lehr, P. Damm, H. W. Fehlhaber, and P. Hajdù, Isolation, identification, and *in vitro* synthesis of conjugates of penbutolol and its metabolites, *Arzneim-Forsch.*, 1987, **37** (II), 1222.

Metoprolol

Use/occurrence: β-Adrenergic blocker

Key functional groups: Alkyl aryl ether, secondary alkyl alcohol, isoalkylamino, methoxyethyl

Test system: Human (oral, 100 mg)

Structure and biotransformation pathway

244

Three novel metabolites of the β-adrenergic blocking drug metoprolol (1) have been identified in human urine following oral administration of metoprolol tartrate. These metabolites were derivatized in the urine by addition of phosgene, extracted (from basic and acidic solutions), trimethylsilylated, and subjected to capillary GC–MS. Electron impact spectra showed the new metabolites to be degradation products of α-hydroxymetoprolol (2) with the structures (3), (4), and (5). Compound (2) is a previously identified metabolite accounting for about 10% of the dose of (1). All of these metabolites (2)–(5) showed an ion at M/Z 336 in their EI spectra resulting from cleavage at the benzylic position, and this ion was used for selective ion monitoring. Metabolites (2), (3), and (5) were found in the basic urine extract and metabolite (4) in the acidic extract. Also in the acidic extract were the previously identified metabolites (6) and (7). Metabolite (3) was quantitated by GC with nitrogen detection and was found to account for only 0.25% of the dose of (1). This technique of using phosgene to derivatize aminopropanols to oxazolidinones appears generally applicable to other β-adrenoceptor antagonists.

Reference

K. J. Hoffman, O. Gyllenhaal, and J. Vessman, Analysis of α-hydroxy metabolites of metoprolol in human urine after phosgene/trimethylsilyl derivatization, *Biomed. Environ. Mass Spectrom.*, 1987, **14**, 543.

Propranolol

Use/occurrence:	*β*-Blocker
Key functional groups:	Secondary alkyl alcohol, isoalkylamino, alkyl aryl ether, naphthalene
Test system:	Rat and guinea pig liver microsomes

Structure and biotransformation pathway

$$R = CH_2CH(OH)CH_2NHCH(CH_3)_2$$
$$R^1 = CH_2CH(OH)CH_2NH_2$$
$$GS = glutathionyl$$

The 4- and 5-hydroxylated metabolites (3) and (4) and the desisopropyl metabolite (2) were produced from *S*-propranolol (1) in liver microsomes from both rats and guinea pigs which had been pretreated with 3-methyl-cholanthrene. Addition of rat liver cytosol and glutathione to the systems decreased the formation of the 5-hydroxylated metabolite (4), had no effect on the formation of metabolites (2) and (3), and resulted in the formation of a new water-soluble metabolite. Evidence that this metabolite was a glutathione adduct was obtained from three approaches: (*a*) treatment with *γ*-glutamyltranspeptidase resulted in formation of a compound with a different HPLC retention time, which was not formed in control experiments

246

in the absence of enzyme; (b) when [^{14}C]glutathione and [^{3}H]-S-propranolol were used the ratio of the two isotopes in the metabolite indicated a 1:1 adduct; (c) thermospray LC-MS showed fragment ions characteristic of glutathione conjugates although no molecular ions were observed by this technique or FAB-MS or californium plasma desorption. The finding that the glutathione conjugate was formed at the expense of (4) suggested that it was formed *via* the 5,6-epoxide (5). Evidence for this was obtained from incubation of a pseudoracemic mixture of (1) and [2,4-^{2}H$_2$]-R-propranolol. Thermospray LC-MS of the resulting diastereomers of the glutathione adduct showed no evidence of loss of deuterium in the spectrum of the relevant adduct. Further HPLC analysis partially separated the glutathione adduct from (1) into two peaks. Acid hydrolysis of the combined adducts resulted in the formation of two products which co-chromatographed (HPLC) with 5-hydroxy- and 6-hydroxy-propranolol, indicating that both possible regioisomers of the conjugate (6) and (7) had been formed. It was not possible to separate the 6-hydroxy compound from the 4-hydroxy compound (3) but it was considered unlikely on the basis of the other results that the glutathione would have been attached to the A ring of (1).

Reference

H. A. Sasame, D. J. Liberato, and J. R. Gillette, The formation of a glutathione conjugate derived from propranolol, *Drug Metab. Dispos.*, 1987, **15**, 349.

Propranolol

Use/occurrence: β-Blocker

Key functional groups: Alkyl aryl ether, secondary alkyl alcohol, isoalkylamino, naphthalene

Test system: Human liver microsomes

Structure and biotransformation pathway

(1)

$* = {}^{14}C$

(2)

The possibility that propranolol exerts its inhibitory effect on P-450 mediated metabolism by irreversibly binding to P-450 was investigated. The binding of propranolol to microsomal proteins was found to be relatively non-specific. The relationship between protein binding and oxidative metabolism was also studied. The hydroxylation of propranolol to (2) and the *N*-desisopropylation to (1) was inhibited by 1 mM phenacetin. Phenacetin was also found to reduce the extent of protein binding, whereas 1 mM glutathione, debrisoquine, or antipyrine did not inhibit the two routes of metabolism. Metabolites were determined by HPLC assay.

Reference

L. Shaw, M. S. Lennard, G. T. Tucker, N. D. S. Bax, and H. F. Woods, Irreversible binding and metabolism of propranolol by human liver microsomes: relationship to polymorphic oxidation, *Biochem. Pharmacol.*, 1987, **36**, 2283.

Miscellaneous Aromatics

Vitamin K$_1$

Use/occurrence:	Vitamin
Key functional groups:	Naphthaquinone
Test system:	Serum (*in vitro*) (bovine, calf, sheep, pig, horse, rat, chick, rabbit, guinea pig); plasma (*in vitro*) (bovine, calf)

Structure and biotransformation pathway

(1) R = $(CH_2CH_2CH_2\overset{CH_3}{\underset{|}{C}}H)_3CH_3$

(2)

Vitamin K$_1$ (1) is converted by plasma or serum from a variety of species into vitamin K$_1$ chromenol (2). Bovine and calf plasma and serum, and rabbit serum, are notably effective for this metabolic transformation. Yields up to 200 pmol min^{-1} ml^{-1} were obtained for incubations of vitamin K$_1$, at concentrations of 26 nmol ml^{-1} at 37 °C for 2 hours under nitrogen. The structure of the compound was established by comparisons of mass and ultraviolet spectra, and HPLC retention times, with synthetic vitamin K$_1$ chromenol. Vitamin K$_1$ chromenol is unstable in aqueous solution and reacts with nucleophiles (*e.g.* amines), and higher rates of its formation were observed for incubations carried out under nitrogen and in the absence of strongly nucleophilic buffers. The possibility of vitamin K$_1$ chromenol formation *in vivo* was not ascertained in these investigations.

Reference

M. J. Fasco, A. C. Wilson, R. G. Briggs, and J. F. Gierthy, Identification of vitamin K$_1$ chromenol — a novel metabolite of vitamin K$_1$ formed *in vitro* by a component in blood, *Arch. Biochem. Biophys.*, 1987, **252**, 501.

4'-O-Tetrahydropyranyladriamicin (THP-adriamycin)

Use/occurrence: Antibiotic, anticancer agent

Key functional groups: Acetal, alkyl ketone, anthraquinone, glycoside

Test system: Human (intravenous, 30 mg m^{-2}, increased by increments of 10 mg m^{-2})

Structure and biotransformation pathway

Animal studies suggested that THP-adriamycin (1) had an improved therapeutic index over adriamycin (2), and preliminary clinical studies showed a short half-life and higher concentrations in whole blood than in plasma, implying rapid cellular uptake. THP-adriamycin (1) was metabolized by cancer patients to adriamycin (2), THP-adriamycinol (3), and adriamycinol (4). The parent compound had a terminal half-life of 13 ± 1.6 hours in plasma and 12 ± 3 hours in cells where peak concentrations were reached after 5 minutes. Adriamycin (2), the major metabolite, had a terminal half-life of 33 ± 10 hours in plasma whereas in blood cells it was > 72 hours, the observation period. Its formation *in vivo* appeared to be influenced by pH and temperature. Thus, if (1) is active, with preclinical studies suggesting reduced cardiotoxicity of adriamycin, it might be used therapeutically to avoid peak concentrations of the latter.

Reference

A. A. Miller and C. G. Schmidt, Clinical pharmacology and toxicity of 4'-O-tetrahydropyranyladriamycin, *Cancer Res.*, 1987, **47**, 1461.

Marcellomycin, Daunorubicin

Use/occurrence:	Antibiotics, anti-cancer agents
Key functional groups:	Anthraquinone, glycoside
Test system:	Xanthine oxidase, rat liver cytosol

Structure and biotransformation pathway

Pharmacology of anthracycline antibiotics in humans has generally been studied by following fluorescent metabolites, but recovery is not total, indicating the physiological production of non-fluorescent metabolites. Under anaerobic conditions rat liver homogenate, microsomes, or cytosol will metabolize marcellomycin (1) to its 7-deoxyaglycone, 7-deoxypyrromycinone (2), which is subsequently converted into a non-fluorescent metabolite. This latter reaction also requires anaerobic conditions and an electron donor, but does not occur with purified NADPH cytochrome P-450 reductase, the main enzyme responsible for formation of 7-deoxyaglycones. The reaction did occur in the presence of xanthine oxidase. NADH or NADPH were equally effective electron donors for rat liver cytosol. Whereas xanthine was the best co-factor for xanthine oxidase, it was inactive with rat liver cytosol and no xanthine oxidase activity was detected. Xanthine oxidase also induced formation of a non-fluorescent metabolite when incubated with 7-deoxydaunorubicin aglycone (3) but other metabolic pathways such as reduction of the carbonyl side-chain complicate the study of this compound.

Tentative identification of the non-fluorescent metabolite from 7-deoxypyrromycinone (2) suggested that it was the dihydroquinonic derivative of the parent deoxyaglycone.

It was concluded that several enzymes including xanthine oxidase are capable of reducing 7-deoxyaglycones of anthracycline antibiotics to produce non-fluorescent dihydroquinonic derivatives.

Reference

P. Dodion, A. L. Bernstein, B. M. Fox, and N. R. Bachur, Loss of fluorescence by anthracycline antibiotics: effects of xanthine oxidase and identification of the non-fluorescent metabolites, *Cancer Res.*, 1987, **47**, 1036.

Adriamycin

Use/occurrence:	Antibiotic
Key functional groups:	Anthraquinone, glycoside
Test system:	Rat, rabbit hepatocytes

Structure and biotransformation pathway

(1)   (2)

(3)   (4)

The biotransformation of adriamycin (1) in hepatocytes from rats and rabbits, under aerobic conditions, was investigated and the metabolites were quantified by specific HPLC assays. Adriamycin was the major species within the cells of both rat and rabbit hepatocytes. In the rat little adriamycinol (2) was detected, but there were significant levels of the aglycones deoxyadriamycin (3) and deoxyadriamycinol (4). In contrast in the rabbit hepatocytes (2) and (4) predominated whereas the levels of (3) were low. The relative formation of (3) and (4) indicates that removal of the glycoside from (2) may be more effective than from (1). There was no evidence for the formation of conjugates and it was suggested that conjugate synthesis is extremely slow compared with the other metabolic transformations of the anthracyclines.

Reference

D. A. Gewirtz and S. Yanovich, Metabolism of adriamycin in hepatocytes isolated from the rat and rabbits, *Biochem. Pharmacol.*, 1987, **36**, 1793.

256

Deoxynivalenol

Use/occurrence:	Mycotoxin
Key functional groups:	Epoxide, trichothecene
Test system:	Rat (oral, 10 mg kg^{-1})

Structure and biotransformation pathway

B$_1$ $\longrightarrow$ Q$_1$

CH$_3$ ... OH ... (1) $\longrightarrow$ CH$_3$ ... OH ... (2)

[^{14}C]-3α,7α,15-trihydroxy-12,13-epoxytrichothec-9-en-8-one (deoxy-nivalenol) (1) was prepared (via the biosynthesis of its 3-acetyl derivative) and administered orally to rats (10 mg kg^{-1}, 3.2 μCi kg^{-1}). Urine and extracts of faeces were examined by HPLC on a reverse phase column using neutral or acidic solvent systems. Urine contained 25.4 ± 1.7% of the dose in the first 96 hours. Under neutral HPLC conditions five major peaks were observed including *ca.* 50% as polar materials and *ca.* 25% as unchanged (1). Under acidic HPLC conditions two major peaks were observed, these being (1) (impure) and 3α,7α,15-trihydroxytrichothec-9,12-dien-8-one (2) (*ca.* 10% of urinary radioactivity). In 24–48 hour faeces *ca.* 5% of the radioactivity was (1) and *ca.* 13% was (2), with *ca.* 75% present as polar materials. The identity of (2) was confirmed by GC–MS of its trimethylsilyl derivative. Tissue retention of radioactivity was limited although some was found in the adrenals and lower gastrointestinal tract 96 hours after dosing.

Reference

B. G. Lake, J. C. Phillips, D. G. Walters, D. L. Bayley, M. W. Cook, L. V. Thomas, J. Gilbert, J. R., Startin, N. C. P. Baldwin, B. W. Bycroft, and P. M. Dewick, Studies on the metabolism of deoxynivalenol in the rat, *Food Chem. Toxicol.*, 1987, **25**, 589.

T-2 toxin

Use/occurrence:	Fungal toxin (*Fusarium*)
Key functional groups:	Alkyl ester, isobutyl, epoxide, methyl carboxylate, trichothecene
Test system:	Liver microsomes and 9000g supernatant (rat, mouse, guinea pig, rabbit, pig, cow, chicken); intestinal microsomes (rabbit)

Structure and biotransformation pathway

(2) R^1 = OH ; R^2 = $(CH_3)_2CHCH_2OCO$

(3) R^1 = $OCOCH_3$; R^2 = $(CH_3)_2C(OH)CH_2OCO$

(4) R^1 = OH ; R^2 = $(CH_3)_2C(OH)CH_2OCO$

The metabolism of T-2 toxin (1) to the deacetylated product (2) by liver microsomal esterase has previously been reported. Evidence also exists that hydroxylation of the isovaleryl chain occurs, as cows excrete (3) and (4) in their urine. This metabolism has now been demonstrated to be dependent on cytochrome P-450 monooxygenase.

Incubation of (1) with hepatic 9000g supernatant or microsomes from untreated mice yielded the deacetylated metabolite (2) and 3′-hydroxy-T-2-toxin (3). Pretreatment of the animals with phenobarbital, polychlorinated biphenyl, or 3-methylcholanthrene induced cytochrome P-450; microsomal metabolism of (1) now yielded 2–3 times more (3) and also 3′-hydroxy deacetylated T-2 toxin (4). SKF-525A or 7,8-benzoflavone pretreatment inhibited the production of (3) and (4) by mouse microsomes. Addition of an esterase inhibitor, eserine, to incubation mixtures caused a great reduction in the amount of (2) formed and no (4) could be detected. Similar results were obtained with rat microsomes, although for unknown reasons 3-methylcholanthrene pretreatment decreased the extent of hydroxylation of (1). Species comparison of the metabolism was carried out using microsomes from noninduced mouse, rat, pig, chicken, rabbit, cow, and guinea pig. All species deacetylated (1) to (2), and 3'-hydroxylation to (3) occurred with all microsomes (except those from rabbit), guinea pig being the most effective species. Production of (4) interestingly did occur in the rabbit. Rabbit intestinal microsomes gave no production of (3) or (4) from (1) although some deacetylation was observed.

All metabolites were identified and analysed by GC of dichloromethane extracts of the microsomal incubations.

Reference

J. Kobayashi, T. Horikoshi, J. C. Ryu, F. Tashiro, K. Ishii, and Y. Ueno, The cytochrome P-450 dependent hydroxylation of T-2 toxin in various animal species, *Food Chem. Toxicol.*, 1987, **25**, 539.

T-2 toxin

Use/occurrence:	Fungal toxin (*Fusarium*)
Key functional groups:	Alkyl ester, isobutyl, epoxide, methyl carboxylate, trichothecene
Test system:	Rat liver S-9 fraction (phenobarbital pretreated)

Structure and biotransformation pathway

(1) R = CH₃
(2) R = CH₂OH

This paper described the HPLC isolation and identification by GC–MS and ¹H and ¹³C NMR of a new relatively minor metabolite of T-2 toxin (1), which had previously been shown to be produced by hepatic microsomes from rats, mice, and chickens. This new metabolite (2) was formed by hydroxylation of a terminal methyl group (C-4′ position) on the butyryl ester side-chain. Results of a rat skin toxicity bioassay test showed that (2) was nearly equal in toxicity to the parent compound (1) and that hydroxylation at the C-4′ position was not a detoxification reaction.

Reference

C. A. Knupp, D. G. Corley, M. S. Tempesta, and S. P. Swanson, Isolation and characterization of 4′-hydroxy T-2-toxin, a new metabolite of the trichothecene mycotoxin T-2, *Drug Metab. Dispos.*, 1987, **15**, 816.

Zearalenone

Use/occurrence:	Fungal constituent
Key functional groups:	Alkyl ketone, aryl carboxylate, lactone
Test system:	Intestinal mucosa, intestinal microsomes (sow)

Structure and biotransformation pathway

(2)　　　　　　　　　(1)　　　　　　　　　(3)

Glucuronide　　　　　　Glucuronide

Zearalenone, 6-(10-hydroxy-6-oxo-*trans*-undec-1-enyl)-β-resorcylic acid lactone (1),· is a *Fusarium* mycotoxin, which is capable of producing reproductive disorders in swine. It is metabolized in pigs to the glucuronide conjugates of (1) and of its reduced metabolite α-zearalenol (2). The current study involves an assessment of the ability of intestinal mucosa to carry out this metabolism.

Zearalenone was incubated at a concentration of 0.012 mM with sow intestinal mucosa homogenate. In the presence of NADPH (1) was reduced to α-zearalenol (2) and β-zearalenol (3), which were extracted and analysed by HPLC. The β-isomer (3) was the major product [*i.e.* 0.25 nmol mg^{-1} protein h^{-1} for duodenum mucosa, compared with 0.11 nmol mg^{-1} protein h^{-1} for (2)]. Conjugation of (1) with glucuronic acid was faster (11.3 nmol mg^{-1} protein h^{-1} for duodenum mucosa). The isoenzyme responsible for this conjugation was not the same as that which glucuronidated α-naphthol.

It thus appears that (1) is conjugated in the intestinal mucosa before it may be reduced, which may explain the apparent lack of the glucuronide of β-zearalenol as an *in vivo* metabolite.

Reference

M. Olsen, H. Pettersson, K. Sandholm, A. Visconti, and K..H. Kiessling, Metabolism of zearalenone by sow intestinal mucosa *in vitro*, *Food Chem. Toxicol.*, 1987, **25**, 681.

Aflatoxin B₁

Use/occurrence:	Mycotoxin
Key functional groups:	Cyclopentenone
Test system:	Post-mitochondrial fraction of human autopsy or biopsy liver and monkey necropsy liver

Structure and biotransformation pathway

Aflatoxin Q_1, a hydroxy metabolite of aflatoxin B_1 was produced by 18 out of 22 human liver samples and by 5 of 7 species of non-human primate investigated. Q_1 was one of a number of chloroform-soluble metabolites produced and was of variable quantitative importance. Human specimens from both sexes, ages 20–80 years, and with a variety of hepatic histological and historical influences, produced Q_1. Although Q_1 has been shown to be less toxic and mutagenic than B_1 in several *in vitro* tests, it was considered that its persistent formation in the *in vitro* systems investigated in this study indicated that it will be a metabolite of toxicological significance *in vivo*.

Reference

D. M. Yourtee, T. A. Bean, and C. L. Kirk-Yourtee, Human aflatoxin B₁ metabolism: an investigation of the importance of aflatoxin Q₁ as a metabolite of hepatic post-mitochondrial fraction, *Toxicol. Lett.*, 1987, **38**, 213.

(+)-*trans*-Δ^8-Tetrahydrocannabinol

Use/occurrence:	Model compound
Key functional groups:	Methylcyclohexene, n-pentyl
Test system:	Mouse (intraperitoneal, 50 mg kg^{-1}), mouse liver microsomes

Structure and biotransformation pathway

Compound	Relative concentration In vitro	In vivo	R^1	R^2	R^4	R^4	R^5	R^6	R^7
1'-OH	+	+	CH$_3$	H	OH	H	H	H	H
2'-OH	+	−	CH$_3$	H	H	OH	H	H	H
3'-OH	+ +	−	CH$_3$	H	H	H	OH	H	H
4'-OH	+ +	+	CH$_3$	H	H	H	H	OH	H
5'-OH	+	−	CH$_3$	H	H	H	H	H	OH
7α-OH	+ + +	−	CH$_3$	α-OH	H	H	H	H	H
7β-OH	+ +	−	CH$_3$	β-OH	H	H	H	H	H
11-OH	+ + +	+ + +	CH$_2$OH	H	H	H	H	H	H
3',7-Di-OH (?)	+	−	CH$_3$	OH	H	H	OH	H	H
4',7-Di-OH (?)	+	−	CH$_3$	OH	H	H	H	OH	H
1',11-Di-OH	−	+ +	CH$_2$OH	H	OH	H	H	H	H
2',11-Di-OH	−	+	CH$_2$OH	H	H	OH	H	H	H
3',11-Di-OH	+	+	CH$_2$OH	H	H	H	OH	H	H
4',11-Di-OH	+	+	CH$_2$OH	H	H	H	H	OH	H
7β,11-Di-OH	+ +	−	CH$_2$OH	β-OH	H	H	H	H	H
7α,11-Di-OH	+	−	CH$_2$OH	α-OH	H	H	H	H	H
11-OAc	+ + +	−	CH$_2$OAc	H	H	H	H	H	H
11-CO$_2$H	−	+ + +	CO$_2$H	H	H	H	H	H	H
1'-OH,11-CO$_2$H	−	+	CO$_2$H	H	OH	H	H	H	H
3'-OH,11-CO$_2$H	−	+ +	CO$_2$H	H	H	H	OH	H	H
4'-OH,11-CO$_2$H	−	+ +	CO$_2$H	H	H	H	H	OH	H

+, compound detected, number of +s indicates relative abundance; −, not detected

The $(+)$-*trans*-Δ^8-tetrahydrocannabinol isomer has very little psycho-activity. The objective of the study under review was to examine the metabolism of this inactive $(+)$-isomer to see whether any major differences occurred between its metabolism and the reported metabolism under the same conditions of the active $(-)$-isomer, which differs in the stereochemistry of the ring fusion. For the *in vivo* study, animals were killed one hour after dosing and the livers were removed and extracted. Following clean-up, these liver extracts and extracts from *in vitro* incubations were examined by GC–MS. Identification rested on the GC–MS properties of TMS, $[^2H_9]$-TMS, and methyl derivatives and the products of reduction with lithium aluminium deuteride. No reference compounds were available, but GC–MS properties of metabolites were compared with those of the corresponding known metabolites of $(-)$-*trans*-Δ^8-tetrahydrocannabinol. Twenty-three metabolites were assigned structures. In addition to those shown in the table, an 8,9-epoxide and its derived glycol were detected in the *in vitro* incubation. The major metabolic route was allylic hydroxylation at C-11, followed *in vivo* by oxidation to the corresponding acid. [This was also a major pathway for the active $(-)$-isomer.] Further hydroxylation at C-7 and in the aliphatic side-chain led to a series of disubstituted metabolites. Major differences between the metabolism of the $(+)$-isomer under investigation and its $(-)$-isomer concerned the positions of hydroxylation in the side-chain and relative amounts of side-chain and alicyclic substitution. Acetylation of the 11-hydroxy group *in vitro* was unique to the $(+)$-isomer. However, differences in metabolism observed between the $(+)$- and $(-)$-isomers related mainly to those metabolites in the $(-)$-series not having significant biological activity and it was not considered that differences in metabolism would contribute significantly to the absence of psychoactivity shown by the $(+)$-isomer.

Reference

D. J. Harvey and H. J. Marriage, Metabolism of $(+)$-*trans*-Δ^8-tetrahydrocannabinol in the mouse *in vitro* and *in vivo*, *Drug Metab. Dispos.*, 1987, **15**, 914.

Galanthamine

Use/occurrence:	Alkaloid, antiparalysis agent
Key functional groups:	Allylic alcohol, methoxyphenyl, N-methylalkylamine
Test system:	Rat and rabbit liver homogenate (10 000g supernatant)

Structure and biotransformation pathway

Metabolism of (1) in rat liver homogenate was slow and no more than 5% biotransformation occurred. In co-factor enriched rabbit liver homogenate, (1) was oxidized to the ketone (2) which was then reduced to the epimer (3). All of these reactions were shown to be reversible but in the reduction of (2) the *epi*-isomer was favoured in the ratio 8:1. The oxidation of (3) was shown to proceed at a greater rate than the oxidation of (1). Other possible metabolic pathways, N- and O-dealkylation, were reported to be negligible.

Reference

D. Mihailova, M. Velkov, and Z. Zhivkova, *In vitro* metabolism of galanthamine hydrobromide (nivalin) by rat and rabbit liver homogenate, *Eur. J. Drug Metab. Pharmacokinet.*, 1987, **12**, 25.

Naltrexonium methiodide

Use:	Narcotic antagonist
Key functional groups:	Alkyl quaternary ammonium, cyclohexanone, *N*-cyclo-propylmethyl
Test system:	Rat (intravenous, 4 mg kg^{-1})

Structure and biotransformation pathway

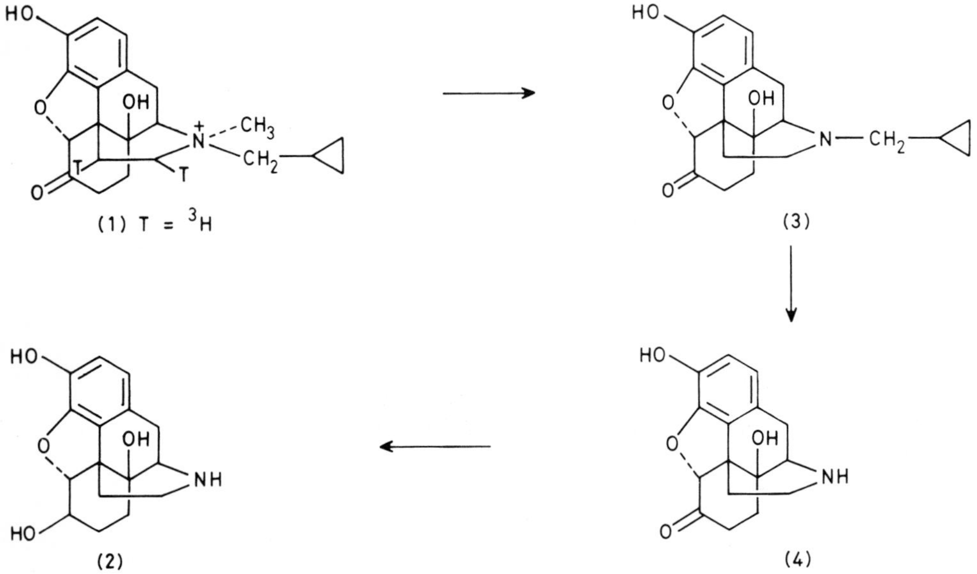

(1) T = ^{3}H

(3)

(2)

(4)

The plasma of male rats given [15,16-^{3}H]naltrexonium methiodide (1) was analysed for metabolites. Thin layer chromatographic analysis of ethyl acetate extracts showed the presence of 7,8-dihydro-14-hydroxynor-morphine (2), naltrexone (3), and 7,8-dihydro-14-hydroxynormorphinone (4). These metabolites were found in the ratio 68:32:5 and together accounted for only about 2% of the radioactivity in plasma, the remainder being naltrexonium methiodide. Naltrexone was also detected in brain but only at early times after administration of the drug.

Reference

A. L. Misra, R. B. Pontani, and N. L. Vadlamani, Intravenous kinetics and metabolism of [15,16-^{3}H]naltrexonium methiodide in the rat, *J. Pharm. Pharmacol.*, 1987, **39**, 225.

Dextromethorphan

Use/occurrence:	Antitussive agent
Key functional groups:	Methoxyphenyl, *N*-methyl-alkylamine
Test system:	Human volunteers (oral, 50 mg)

Structure and biotransformation pathway

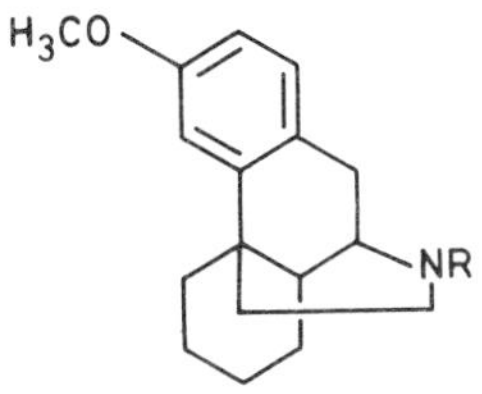

(1) R = CH₃
(2) R = H
(5) R = COCH₃

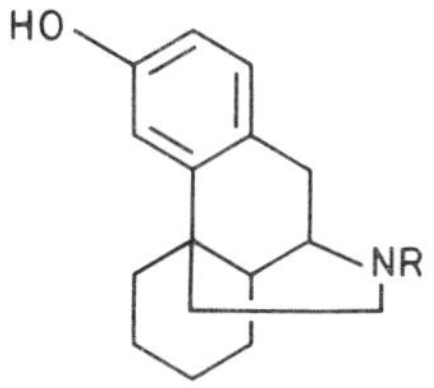

(3) R = CH₃
(4) R = H

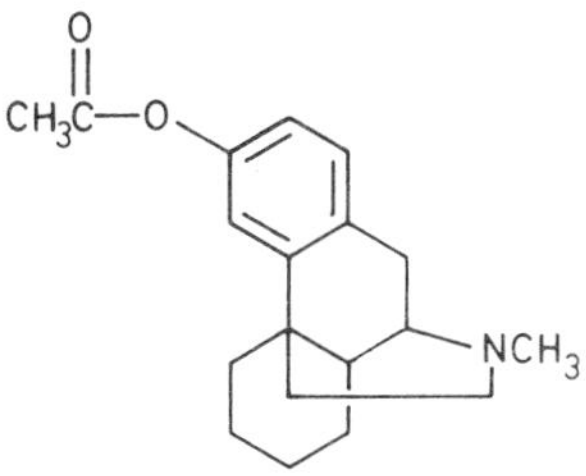

(6)

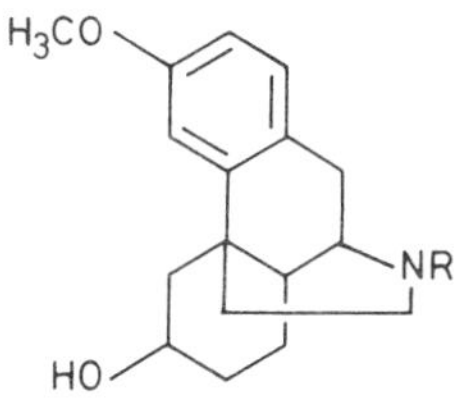

(7) R = CH₃
(18) R = H

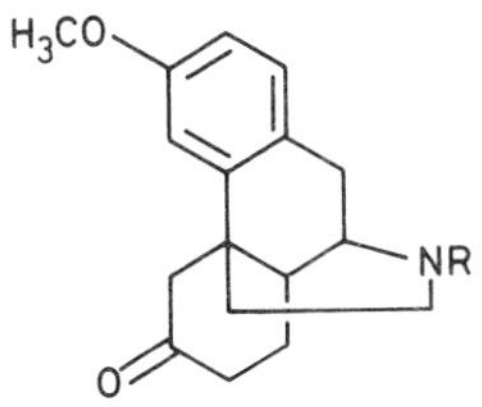

(8) R = CH₃
(19) R = H

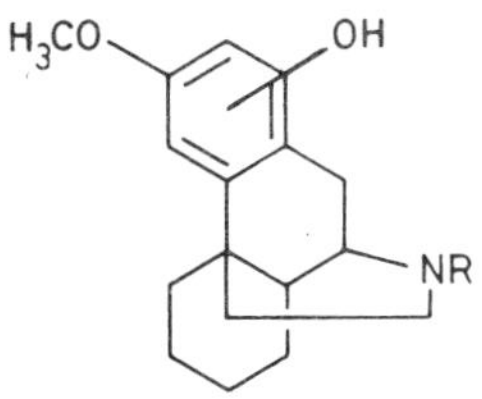

(9) R = CH₃
(16) R = H

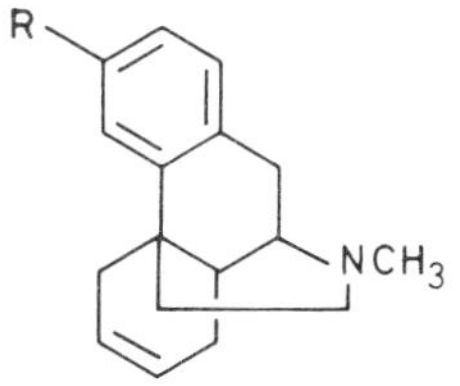

(10) R = CH₃O
(14) R = OH

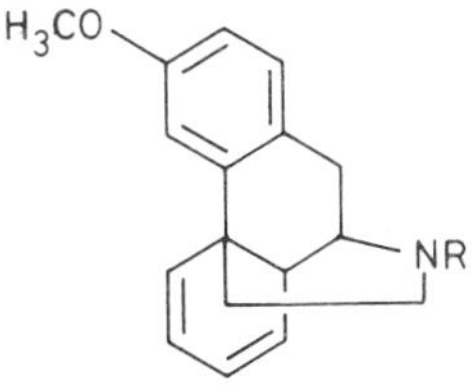

(11) R = CH₃
(17) R = H

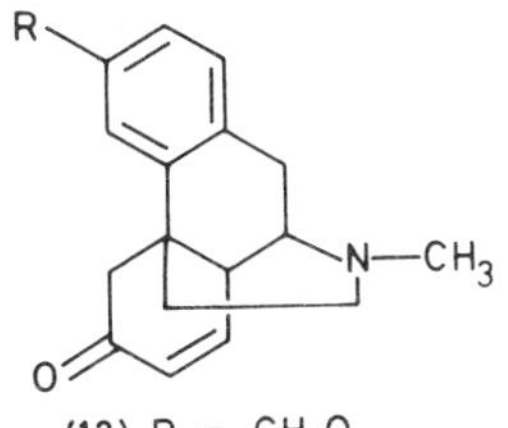

(12) R = CH₃O
(15) R = OH

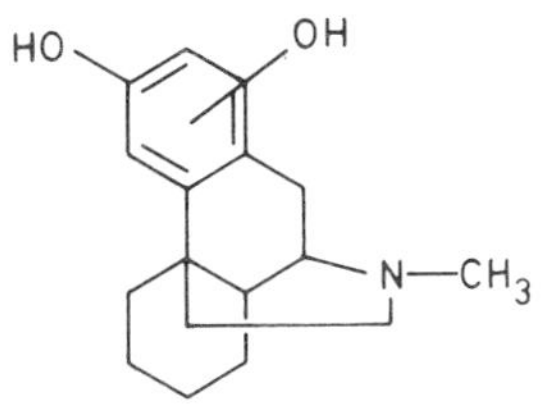

(13)

Urine collected during 24 hours after a single dose was incubated with glucuronidase/sulphatase and extracted with dichloromethane/isopropanol at pH 3 and pH 10. Extracts were analysed by GC–MS before and after acetylation. The study used non-radioactive methodology and no quantitative data were presented. Structural elucidation was based on interpretation of mass spectra, comparison with spectra of reference compounds or reference spectra, and formation of acetyl derivatives. Nineteen components including the parent dextromethorphan (1) were detected in extracts. Metabolites (1)–(9) and (12) could be identified without derivatization. Metabolites (13)–(16), (18), and (19) could only be identified as acetyl derivatives. Acetyl derivatives of metabolites (3), (4), (7), (8), and (9) were also formed. Compounds (10), (11), and (17) were apparently only detected in acetylated extracts, although not as acetyl derivatives.

Metabolites (2)–(4) were previously known, but the additional fifteen compounds were reported to be previously unknown. The metabolic pathways leading to the identified metabolites were *O*- and *N*-demethylation, *O*- and *N*-acetylation, and hydroxylation of the phenyl ring and the saturated ring system with subsequent further oxidation. Compounds (10), (11), (14), and (17) were considered to be artefacts formed by acetic anhydride induced dehydration of the corresponding alcohols, *e.g.* (7)→(10).

Reference

C. Koppel, J. Tenczer, and K. Ibe, Urinary metabolism of dextromethorphan in man, *Arzneim.-Forsch.*, 1987, **37** (II), 1304.

4-Desacetylvinblastine monoclonal antibody conjugate (LY256787)

Use/occurrence:	Antitumour agent
Key functional groups:	Vinca alkaloid, monoclonal antibody
Test system:	Rats (intravenous, 10 and 100 mg kg^{-1}), rhesus monkey (intravenous, 40 mg kg^{-1})

Structure and biotransformation pathway

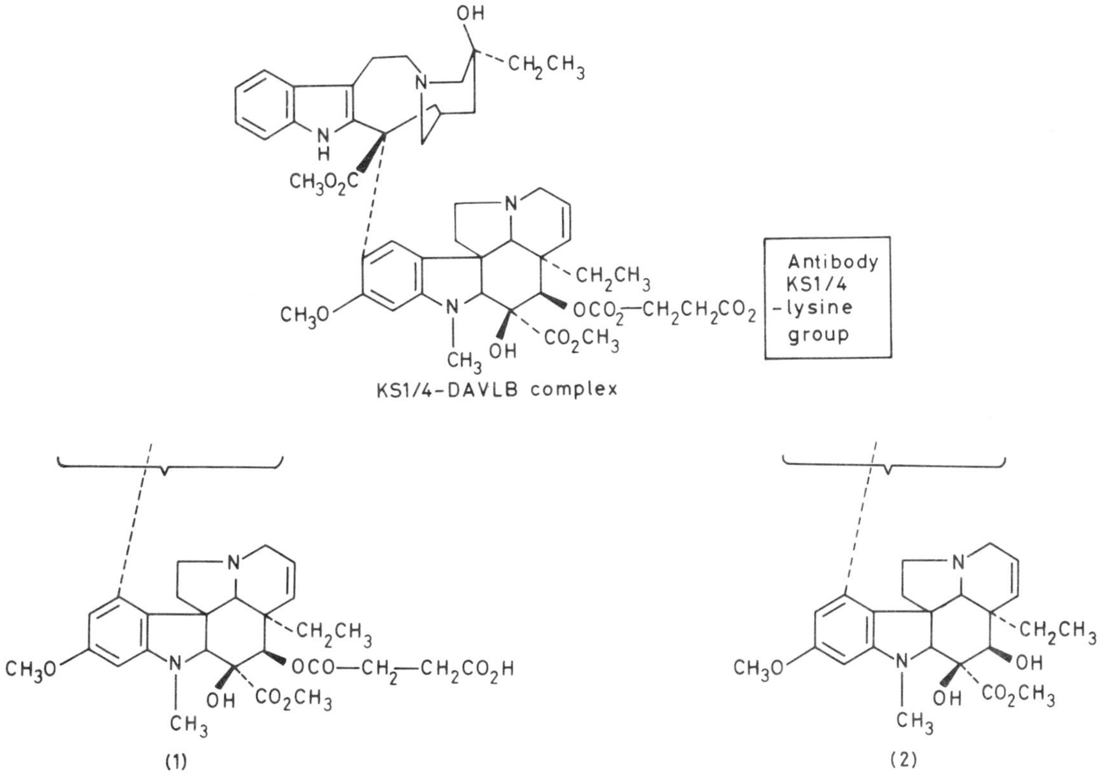

KS1/4-DAVLB complex

(1)

(2)

The conjugate KS1/4-DAVLB of the murine monoclonal antibody KS1/4 with the vinca alkaloid 4-desacetylvinblastine (DAVLB) was prepared both uniformly labelled with tritium in the alkaloid moiety and ^{35}S-labelled in the KS1/4 moiety. On average 4–6 molecules of the alkaloid were combined with one molecule of KS1/4. The [^{3}H]conjugate was administered intravenously to rats and monkeys and terminal plasma half-life values were measured as radioequivalents and as functionally immunoreactive antibody conjugate. The half-life values determined radiometrically were 145 and 62 hours in rats after 10 and 100 mg kg^{-1} doses respectively and 90 hours in monkeys after a 40 mg kg^{-1} dose. Comparable results were obtained when functionally immunoreactive conjugate concentrations were determined by an enzyme-linked immunosorbent assay technique. The ratio of [^{35}S]:[^{3}H] in the rats dosed with 100 mg kg^{-1} [^{35}S]KS1/4-[^{3}H]DAVLB remained reasonably constant during 336 hours after dosing. Less than 1% of the total vinca alkaloid equivalents present in the plasma could be extracted as free vinca species; the major vinca alkaloid species at early time points were hemisuccinate derivatives of DAVLB (1), whereas at later times DAVLB (2) and an *N*-oxide were also present in similar amounts. The position of *N*-oxidation was not specified. The major pathway of elimination was via the faeces. The ratio of [^{35}S]:[^{3}H]DAVLB was substantially less than that in the plasma. Analysis of the radioactivity present in the bile by size exclusion HPLC showed that most of the tritium present in the bile was associated with material of lower molecular weight than KS1/4-DAVLB. Thus the KS1/4-DAVLB circulates mainly in the intact form in the plasma and is catabolized in the liver with subsequent excretion via the bile.

Reference

M. E. Spearman, R. M. Goodwin, and D. Kau, Disposition of the monoclonal antibody–vinca alkaloid conjugate KS1/4-DAVLB (LY256787), in Fisher-344 rats and rhesus monkeys, *Drug Metab. Dispos.*, 1987, **15**, 640.

Tannic acid

Use/occurrence: Plant constituent

Key functional groups: Aryl carboxylic acid

Test system: Rat liver microsomes

Structure and biotransformation pathway

The metabolism of tannic acid (1), a polymeric glycoside derivative of gallic acid, by rat liver microsomes has been studied in an attempt to find a product which might be responsible for the hepatocarcinogenicity of (1). Following the incubation of (1) (0.5 mg ml^{-1}) with a microsomal suspension and NADPH, three metabolites (2), (3), and (4) were separated from the supernatant by silica TLC. A further metabolite (5) was bound to the microsomal fraction and was isolated by TLC of an SDS solution of this fraction. Metabolite structures were determined by IR, UV, NMR, MS, and elemental analysis, and the carcinogenic potential tested by the microsomal degranula-

271

tion technique. Metabolite (3) [1,1,2-trimethyl-1-ethanyl-2-ylidenetris(3,4,5-trihydroxybenzoate)] showed high activity in this test and was postulated to be responsible in part for the carcinogenicity of (1). The authors suggest that both C-1 and C-2 in (3) could behave electrophilically after further metabolism and thus give rise to the compound's genotoxic activity.

Reference

M. M. Gupta and H. M. Dani, Characterization of tannic acid metabolites formed *in vitro* by rat liver microsomes and assay of their carcinogenicity by the microsomal degranulation technique, *Chem. Biol. Interact.*, 1987, **63,** 39.

Heterocycles

5-Hydroxymethyl-2-furaldehyde

Use/occurrence:	Impurity in sugar containing foodstuff
Key functional groups:	Furan, aldehyde
Test system:	Rat (oral and intravenous, 0.08–330 mg kg^{-1})

Structure and biotransformation pathway

HOCH$_2$ —[furan]— CHO   * = ^{14}C

HOCH$_2$ —[furan]— CO$_2$H   (1)

HOCH$_2$ —[furan]— CONHCH$_2$CO$_2$H   (2)

In rats 85% of the radioactivity associated with an oral dose of [^{14}C]-5-hydroxymethyl-2-furaldehyde was excreted in urine within 8 hours. A similar excretion followed intravenous administration. The major metabolites were 5-hydroxymethyl-2-furoic acid (1), which was identified by UV, MS, NMR, and HPLC comparison with a reference standard, and its glycine conjugate (2) which was identified by NMR, and by MS. The latter was consistent with a published spectrum for this substance. The relative proportion of the metabolites formed was dose-dependent. At low doses similar proportions of each metabolite were excreted whereas at higher doses the free acid (1) was the major metabolite. Tissue distribution studies suggest that glycine conjugation occurs in the kidney.

Reference

J.-E. Germond, G. Philippossian, U. Richli, I. Bracco, and M. J. Arnaud, Rapid and complete urinary elimination of [^{14}C]-5-hydroxymethyl-2-furaldehyde administered orally or intravenously to rats, *J. Toxicol. Environ. Health*, 1987, **22**, 79.

Furazolidone

Use/occurrence: Antimicrobial

Key functional groups: Nitrofuran, oxazolidinone

Test system: Swine liver microsomes

Structure and biotransformation pathway

(2)   (1)   (3)

$R = CH=N-N-C$

(4)

Furazolidone, N-(5-nitro-2-furfurylidene)-3-amino-2-oxazolidone (1), is an antimicrobial, used in veterinary practice, but with mutagenic and carcinogenic properties. Residues must therefore be investigated in edible animal tissues. Residues derived from $[^{14}C]$-(1) were detected in tissues 14 days after withdrawal of medication which were apparently bound covalently to cellular macromolecules. Metabolic conversion of (1) by swine liver microsomes has been studied to elucidate the biotransformation pathway and identify intermediates which may bind to cellular protein.

Furazolidone was metabolized aerobically (2.55 nmol mg^{-1} protein min^{-1}) and anaerobically (3.25 nmol mg^{-1} protein min^{-1}), with aerobic covalent binding to microsomal protein of 0.29 nmol mg^{-1} protein min^{-1}. Addition of cysteine to the incubation decreased protein binding, suggesting involvement of thiol groups. At least 50% of the total metabolites resulted from reduction of (1), indicated by formation of a conjugate, 3-(4-cyano-3-β-hydroxyethyl-mercapto-2-oxobutylideneamino)-2-oxazolidone (2), after addition of mercaptoethanol to the incubation mix. This result suggested that the open-chain acrylonitrile derivative is the reactive intermediate of (1) which may be involved in covalent binding to protein. Two known minor metabolites were also identified: 3-(4-cyano-2-oxobutylideneamino)-2-oxazolidone (3) and 2,3-dihydro-3-cyanomethyl-2-hydroxy-5-nitro-1α,2-di(2-oxo-oxazolidin-3-yl)iminomethylfuro[2,3-b]furan (4).

Reference

L. H. M. Vroomen, J. P. Groten, K. van Muiswinkel, A. van Velduizen, and P. J. van Bladeren, Identification of a reactive intermediate of furazolidone formed by swine liver microsomes, *Chem. Biol. Interact.*, 1987, **64**, 167.

4-Amino-5-ethylthiophene-3-carboxylic acid methyl ester (RO 22-0654)

Use/occurrence:	Anti-obesity agent
Key functional groups:	Alkyl ester, aminothiophene, aryl carboxylate
Test system:	Rats (intragastric, 100 mg kg^{-1}; intraduodenal, 50 mg kg^{-1}), dogs (intragastric, 30 mg kg^{-1})

Structure and biotransformation pathway

(1) $*$ = ^{14}C   (2)   (3)

After intragastric administration of (1) to male rats, 77% of the dose was excreted in urine in 24 hours with a further 6% in faeces. Urine was extracted with hexane and the extracts concentrated prior to analysis by TLC and HPLC. Metabolites were isolated by HPLC and their structures identified by NMR and MS. Unchanged (1) represented 1% of the dose in urine but 4-amino-5-ethylthiophene-3-carboxylic acid (2; free and conjugated forms) represented 57.2% of the dose. In rats dosed intraduodenally high levels of (2) were found in both systemic and portal blood within 5 minutes.

Dogs dosed intragastrically with (1) excreted 0.3% of dose as (1), 30.8% as (2), and 6.8% as 5-ethyl-4-(methylamino)thiophene-3-carboxylic acid (3). From intravenous and intragastric administration of (1) and (2) to dogs it was concluded that the two compounds are pharmacokinetically equivalent and that (1) may be a prodrug for (2). Additional minor metabolites and conjugates were detected in the urine of rats and dogs but not identified.

Reference

F.-J. Leinweber, A. J. Szuna, A. C. Loh, T. H. Williams, G. J. Sasso, I. Bekersky, E. Bagglioni, and J. Tristcari, Metabolism of the anti-obesity agent 4-amino-5-ethyl-thiophene-3-carboxylic acid methyl ester hydrochloride in rats and dogs, *Xeno-biotica*, 1987, **17**, 1405.

2-(2-Thienyl)allylamine

Use/occurrence:	Dopamine hydroxylase inactivator
Key functional groups:	Allyl amine, thiophene
Test system:	Rat (oral and intravenous, 100 mg kg^{-1})

Structure and biotransformation pathway

HPLC analysis of 0–18 hour urine from rats administered oral or intravenous doses of [^{14}C]thienylallylamine showed only one major metabolite. The metabolite was isolated by solvent extraction and analysed by GC–MS after formation of a trimethylsilyl derivative. Confirmation of the metabolite as 2-(2-thienyl)propionic acid (3) was obtained by comparison with synthetic reference compound. The postulated mechanism of formation involves deamination and oxidation to yield the carboxylic acid (2) followed by the action of enoyl reductase to give the terminal metabolite. Formation and reduction of an α,β-unsaturated acid derivative has been postulated as the mechanism for the stereospecific interconversion of enantiomeric 2-aryl-propionic acids. Thus it would be expected that the metabolite would be optically active due to stereospecific reduction but this was not determined.

Reference

W. P. Gordon, J. R. McCarthy, and S. Y. Chang, Metabolism of 2-(2-thienyl)allylamine hydrochloride in the rat: identification of a novel metabolite, *Biochem. Biophys. Res. Commun.*, 1987, **145**, 575.

Asparagusic acid

Use/occurrence:	Natural product
Key functional groups:	Disulphide, dithiolane
Test system:	Human (oral)

Structure and biotransformation pathway

Male and female volunteers ate 500 g asparagus in the evening and collected urine the following morning. Those samples of urine having odorous (volatile) components were analysed by head space capillary GC–MS. Six discrete compounds were identified: dimethyl sulphide, dimethyl disulphide, bis(methylthio)methane, dimethyl sulphoxide, dimethyl sulphone, and methanethiol. When asparagusic acid (1) was given to volunteers who produced the odorous urine the same volatiles were detected in similar proportions in the urine. It was suggested that (1) is reduced to the dithiol (2) which may be methylated followed by oxidation of the ring carbons, liberating methanethiol. Subsequent oxidation, methylation, and dimerization would yield the detected compounds.

Reference

R. H. Waring, S. C. Mitchell, and G. R. Fenwick, The chemical nature of the urinary odour produced by man after asparagus ingestion, *Xenobiotica*, 1987, **17**, 1363.

N-Methylpyrrolidine

Use/occurrence:	Drug metabolite
Key functional groups:	Pyrrolidine
Test system:	Dog (intravenous, 2 mg kg^{-1}), rat (intravenous, 2–10 mg kg^{-1})

Structure and biotransformation pathway

(1) * = ^{14}C   (2)   (3)

   N-Methylpyrrolidine (1) is a potential intermediate metabolite of the cephalosporin antibiotic cefepime (3). Following intravenous administration of [^{14}C]-(1) to rats or dogs, most of the radioactivity (rat 81% dose, dog 91% dose) was excreted in the urine within 24 hours. HPLC analysis of urine showed a single major metabolite (86–93% urinary radioactivity), which was identified by HPLC–MS and ^{1}H NMR comparison with a reference standard as the *N*-oxide (2). A further minor metabolite (1–3% urinary radioactivity) was not identified. Unchanged (1) accounted for 4–13% urinary radioactivity. After intravenous administration of [*Me*-^{14}C]cefepime (3) to rats, the unchanged compound accounted for 86% of the radiolabel recovered in urine while the *N*-oxide (2) and the pyrrolidine (1) accounted for 12% and 2% respectively. Oxidation of (1) to (2) by rat and human liver microsomes was also demonstrated.

Reference

S. T. Forgue, P. Kari, and R. Barbhaiya, *N*-Oxidation of *N*-methylpyrrolidine released *in vivo* from cefepime, *Drug Metab. Dispos.*, 1987, **15**, 808.

Nicotine (from smoking)

Use/occurrence:	Cigarette constituent
Key functional groups:	Pyridine, pyrrolidine
Test system:	Human cigarette smokers

Structure and biotransformation pathway

(2)   (1)   (6)

(3)   (4)   (5)

The urine of cigarette smokers was analysed for nicotine (1) and its metabolites cotinine (2), nicotine *N'*-oxide (3), nornicotine (4), *N*-methyl-nicotinium ions (5) and *trans*-3'-hydroxycotinine (6). The latter compound, which was a known metabolite, but not previously quantitated, was the major constituent. The extent of excretion of nicotine and metabolites was correlated in nine smokers with the nicotine content of the cigarettes. For a smoker of 19 cigarettes a day with 1.35 mg mainstream nicotine per cigarette the urinary excretion of nicotine was 14.0 mg day^{-1}. The excretions of cotinine, nicotine *N'*-oxide, nornicotine, *N*-methylnicotinium ions and *trans*-3'-hydroxycotinine were 12.2, 3.4, 1.1, 1.0, and 49.5 mg nicotine equivalents respectively. The average serum level of *trans*-3'-hydroxycotinine was 69 ± 34 ng ml^{-1} in those individuals, compared with a cotinine level of 155 ± 91 ng ml^{-1}.

The analytical methods employed were HPLC for *N*-methylnicotinium ions and GC for the remaining metabolites. *trans*-3'-Hydroxycotinine was measured by nitrogen detection of its acetate derivative.

Reference

G. B. Neurath, M. Dünger, D. Orth, and F. G. Pein, *trans*-3'-Hydroxycotinine as a main metabolite in urine of smokers, *Int. Arch. Occup. Environ. Health*, 1987, **59**, 199.

Cotinine

Use/occurrence:	Metabolite of nicotine
Key functional groups:	*N*-Methylpyrrolidone, pyridine
Test system:	Guinea pig (intraperitoneal)

Structure and biotransformation pathway

(1)   (2)

Following intraperitoneal injection of $[G\text{-}^3H]\text{-}S\text{-}(-)$-cotinine to guinea pigs the radioactivity was almost quantitatively recovered in urine within 24 hours. Analysis of the urine by reverse phase HPLC showed the presence of cotinine and four metabolites. Two of these were identified as 3-hydroxy-cotinine (1; 56% of the urinary radioactivity) and cotinine *N*-oxide (2; 18%) by chromatographic comparison with reference standards. Cotinine itself accounted for <1% of the radioactivity. The other two metabolites, which together contributed 20% of the urinary radioactivity, could not be identified. Analysis by cation exchange HPLC showed that $S\text{-}(-)$-cotinine is neither converted into *N*-methylated metabolites in the guinea pig nor does it undergo reduction to $S\text{-}(-)$-nicotine or any of its methylated metabolites.

Reference

K. C. Cundy and P. A. Crooks, Biotransformation of primary nicotine metabolites II. Metabolism of [^{3}H]-S-$(-)$-cotinine in the guinea pig: determination of *in vivo* urinary metabolites by high performance liquid radiochromatography, *Xenobiotica*, 1987, **17**, 785.

Phenytoin

Use/occurrence:	Anti-epileptic drug
Key functional groups:	Alkylphenyl, hydantoin, prochiral carbon
Test system:	Phenotyped human volunteers (oral, 100 mg)

Structure and biotransformation pathway

(1) † prochiral carbon   (2)

The stereochemical course of the known aromatic hydroxylation pathway of metabolism was investigated in subjects with and without genetic drug hydroxylation deficiencies for mephenytoin or debrisoquine. The enantiomers of the hydroxylated metabolite (2) in urine were separated by chiral ligand exchange HPLC. Irrespective of the various drug-hydroxylation phenotypes, (2) was the major urinary metabolite of (1) and its *S*-enantiomer accounted for $95 \pm 4\%$ of the total present. However, a statistically significant difference in the *S/R* ratio of (2) was found between extensive (14 ± 5) and poor (73 ± 17) metabolizer phenotypes of mephenytoin. No such correlation was observed with debrisoquine hydroxylation phenotypes. Evidence was obtained to exclude substrate specific glucuronidation of the enantiomers of (2).

Reference

S. Fritz, W. Lindner, I. Roots, B. M. Frey, and A. Küpfer, Stereochemistry of aromatic phenytoin hydroxylation in various drug hydroxylation phenotypes in humans, *J. Pharm. Exp. Ther.*, 1987, **241**, 615.

Antipyrine

Use/occurrence:	Hepatic probe drug, analgesic, antipyretic
Key functional groups:	*N*-Phenyl, pyrazolin-5-one
Test system:	Venda Africans

Structure and biotransformation pathway

The metabolic profile of antipyrine (1) is well established, the major products being 4-hydroxyantipyrine (2), norantipyrine (3), and 3-hydroxymethylantipyrine (4). The drug is frequently used as a probe of liver mixed function oxidase activity. In order to determine whether there are genetic differences in the capacity to metabolize antipyrine, the present study involved the determination of the metabolite profile in a group (20) of Venda Africans who had adopted a Western way of life and diet. Results were then compared with previously reported data on Caucasians and Orientals living in Canada, and on American subjects. Venda Africans metabolized antipyrine to 4-hydroxyantipyrine (26.2% of dose), norantipyrine (7.4%), and 3-hydroxymethylantipyrine (13.3%). This metabolic profile differs quantitatively from that of the Caucasian and Oriental subjects, suggesting a genetic difference between these populations. Such a conclusion was not supported by comparison of metabolite-formation rate constants. The metabolites were identified and quantified by reverse phase HPLC.

Reference

De K. Sommers, J. Moncrieff, and J. C. Avenant, Antipyrine metabolism in the Venda, *Human Toxicol.*, 1987, **6**, 127.

Benznidazole

Use/occurrence: Antiprotozoal

Key functional groups: Acetamide, nitroimidazole

Test system: Mouse, mouse tumour, mouse liver fractions (oral, 650 mg kg^{-1}, 0.083–1 mM *in vitro*)

Structure and biotransformation pathway

Benznidazole (1) is reduced to its amine metabolite (2) *in vivo* and *in vitro*. *In vivo* benznidazole was extensively metabolized with only 5% of the dose excreted unchanged in 0–24 hour urine. The amine was a minor metabolite accounting for 3% of the dose in 0–24 hour urine. The amine was characterized by co-chromatography on HPLC. *In vitro*, 85% of the total nitro-reductase activity in liver was associated with microsomes. The major microsomal enzymes responsible for the reduction of benznidazole were characterized as NADPH cytochrome P-450 reductase and cytochrome P-450 on the basis of inhibitor studies. The cytosolic enzyme aldehyde oxidase plays a minor role in the nitroreduction. The relative rates of the nitroreduction in different tissues *in vivo* may be a function of oxygen concentration.

Reference

M. I. Walton and P. Workman, Nitroimidazole bioreductive metabolism, *Biochem. Pharmacol.*, 1987, **36**, 887.

Metronidazole

Use/occurrence: Antiprotozoal agent

Key functional groups: Nitroimidazole

Test system: Rat liver 10 000g supernatant

Structure and biotransformation pathway

Metronidazole (1) was incubated with a rat liver supernatant for two hours and a metabolite isolated which was shown to be identical to the product formed from oxidation of metronidazole with hydrogen peroxide. The structure was assigned as the N-3 oxide (2) based on NMR and UV spectra and the observation that the metabolite could be reduced back to metronidazole with sulphur dioxide. However, there is a possibility that the metabolite could be the N-1 oxide.

Reference

E. E. Essien, J. I. Ogonor, H. A. B. Coker, and M. M. Bamisile, Metabolic N-oxidation of metronidazole, *J. Pharm. Pharmacol.*, 1987, **39**, 843.

Midaglizole

Use/occurrence:	Hypoglycaemic
Key functional groups:	Imidazoline, phenylethyl, pyridine
Test system:	Male beagle dog (oral, 14 mg kg^{-1})

Structure and biotransformation pathway

[^{14}C]Midaglizole (1) was administered to male beagle dogs and urine collected for 24 hours. Urine was extracted by various column chromatographic procedures and the metabolites were characterized by IR, UV, NMR, and MS. Unchanged (1) was the major radioactive component (39% urine radioactivity) isolated from dog urine. TLC indicated the presence of about

287

30 metabolites of which six were characterized: (2), 2-[2-phenyl-2-(2-pyridyl)ethyl]-2-imidazole; (3), 2-[2-phenyl-2-(2-pyridyl)ethyl]-4,5-imidazolinedione; (4), 3-phenyl-3-(2-pyridyl)propionimidamide; (5), 3-phenyl-3-(2-pyridyl)propionamide; (6), 3-phenyl-3-(2-pyridyl)propion-imidoylaminoacetic acid; (7), 3-phenyl-3-(2-pyridyl)propionimidoylamino-acetamide.

The metabolites (2), (4), and (6) represented 5, 16, and 13% respectively of the urinary radioactivity. Oxidation on the imidazoline ring and subsequent ring opening were the major routes of biotransformation. Various inter-mediates were postulated as indicated in the scheme. Two of the terminal metabolites of the process were the amides (5) and (7).

Reference

M. Nakaoka and H. Haksui, Identification of the metabolites of a new hypoglycaemic agent, midaglizole, in dogs, *Xenobiotica*, 1987, **17**, 1329.

Pyrazole

Use/occurrence: Model compound

Key functional groups: Pyrazole

Test system: Rat liver microsomes

Structure and biotransformation pathway

(1)　　　　　　　　　　(2)

4-Hydroxypyrazole (2) is a major metabolite in the urine of mice and rats administered pyrazole (1), and is also known to be a potent scavenger of the hydroxyl radical. The production of (2) from the interaction of (1) with hydroxyl radical has now been studied in three systems (microsomes, xanthine oxidase, and ascorbate dependent oxidations). The production of hydroxyl radicals was catalysed by the addition of ferric-EDTA, and the formation of (2) was determined by HPLC with electrochemical or UV detection. All three systems yielded (2) and furthermore the reaction was inhibited by other hydroxyl radical scavengers such as ethanol. Microsomes also produced (2) in the absence of the stimulators of hydroxyl radical generation (ferric-EDTA plus azide) in a time dependent manner, although the extent of reaction was only *ca.* 20% of that when the radical stimulators were present.

Reference

S. Puntarulo and A. I. Cederbaum, Production of 4-hydroxypyrazole from the interaction of the alcohol dehydroxygenase inhibitor pyrazole with hydroxyl radical, *Arch. Biochem. Biophys.*, 1987, **255**, 217.

TA-1801

Use/occurrence:	Hypolipidemic agent
Key functional groups:	Alkyl ester, aryl acetic acid, furan, chlorophenyl, oxazole
Test system:	Rat, rabbit, dog (oral, 50 mg kg^{-1})

Structure and biotransformation pathway

Rat, rabbit, and dog respectively excreted 38%, 65%, and 5% dose in urine and 48%, 25%, and 82% dose in faeces during 48 hours after administration. The four metabolites (1)–(4) were formed in all three species in varying proportions. The acid (1), resulting from ester hydrolysis, was important in all three species and was excreted mainly in the faeces of rats and dogs and in the urine of rabbits. It was identified by GC–MS and TLC comparison with a synthetic standard. Its glucuronide conjugate (2) was a significant metabolite in rat (10% dose) and rabbit (8% dose) urine, but was of very minor importance in dog urine. The structure of this metabolite was assigned by enzymic deconjugation and by GC–MS after methylation and trimethylsilylation. A number of isomers of the derivatized metabolite were separated by GC which were assumed to arise from rearrangement of the ester glucuronide. Cleavage of the furan ring yielded metabolites (3) and (4) which together in urine accounted for 22% dose (rat), 13% dose (rabbit), and 3% dose (dog). Metabolites (3) and (4) were not detected in faeces. They were identified by GC–MS and ¹H NMR comparison with synthetic standards.

Reference

T. Kobayashi, H. Ando, J. Sugihara, and S. Harigaya, Metabolism of ethyl 2-(4-chlorophenyl)-5-(2-furyl)oxazole-4-acetate, a new hypolipidemic agent, in the rat, rabbit, and dog, *Drug Metab. Dispos.*, 1987, **15**, 262.

Roxatidine

Use/occurrence:	H-2 Receptor antagonist
Key functional groups:	Methyl carboxylate, piperidine
Test system:	Rat and dog liver homogenates

Structure and biotransformation pathway

A mixture of roxatidine acetate (1) and its [^{2}H$_{10}$]piperidine analogue were incubated with rat and dog liver 9000g supernatant and rat liver microsomes. Metabolites were isolated by extraction with ethyl acetate, purified by preparative TLC and analysed by GC–MS after formation of the trimethyl-silyl derivatives. Both roxatidine (2) and the hydroxylated compound (3) were formed by rat and dog liver supernatant but the amide (4) was only formed by the rat incubate. The amount of (3) formed was about 2.5-fold higher in the dog. There was an apparent isotope effect in the formation of the metabolites (3) and (4) since in all cases amounts of the protonated metabolite were higher (H/D ratios 1.34–1.69). During formation of (3) and (4) three and four deuterium atoms respectively were lost from the [^{2}H$_{10}$]-compound.

Reference

S. Iwamura, K. Shibata, Y. Kawabe, K. Tsukamoto, and S. Honma, The metabolism of roxatidine acetate hydrochloride in rat and dog liver homogenates, *J. Pharma-cobio-Dyn.*, 1987, **10**, 229.

Roxatidine acetate hydrochloride (RA)

Use/occurrence:	H-2 Receptor antagonist
Key functional groups:	Alkyl amide, alkyl aryl ether, methyl carboxylate, piperidine
Test system:	Rat (oral, 30 mg kg^{-1}), dog (oral, 10 mg kg^{-1})

Structure and biotransformation pathway

Co-administration of [^{14}C]-RA and [^{2}H$_{10}$]-RA labelled with deuterium in the piperidine ring expedited the detection and identification by GC–MS of metabolites in enzyme treated urine. Synthetic standards were available for all metabolites except (2), (12), and (13), and concentrations of major metabolites in urine were determined by selective ion monitoring using the deuterium-labelled compounds as internal standards. No unchanged drug was detected in rat or dog urine and evidence of structure was presented for 15 metabolites. Deacetylation to yield (1) was the initial process in the formation of all metabolites. Metabolite (1) exhibited the same pharmacological activity as the parent compound. Other biotransformation processes fell into three groups: (*a*) hydrolysis of the acetamide group in (1) to yield the amine (8), followed by side-chain oxidation to hydroxyl and acid metabolites, (6), (7), (10), or acetylation (9); (*b*) *O*-dealkylation of the entire side-chain to yield (11) from (1) and (12)–(15) in combination with other processes; (*c*) hydroxylation of the aromatic and piperidine rings to yield (2)–(5), and (13)–(15), followed by methylation (12).

Major metabolites in both species (values as % dose, rat followed by dog) were (1) (17%, 31%), (10) (15%, 13%), (11) (44%, 9%), and (8) (7%, 10%). Metabolite (4) was not detected in the dog but accounted for 2% dose in rat. These major metabolites were formed *via* processes (*a*) and (*b*) above. Ring hydroxylation was a minor pathway and total hydroxylation of both aromatic and piperidine rings amounted to about 3% dose in both species. Hydroxylation of the piperidine ring to yield metabolites (3) and (4) involved the unexpected loss of three or four deuterium atoms respectively from the ring for which no explanation was proposed.

Reference

S. Honma, S. Iwamura, R. Kobayashi, Y. Kawabe, and K. Shibata, The metabolism of roxatidine acetate hydrochloride, *Drug Metab. Dispos.*, 1987, **15**, 551.

Alfentanil

Use/occurrence: Synthetic opioid

Key functional groups: Alkyl aryl amide, alkyl carboxamide, methyl ether, piperidine

Test system: Rat (intravenous, 0.16 mg kg^{-1}), dog (intravenous, 0.05 mg kg^{-1})

Structure and biotransformation pathway

Approximately 75% of single intravenous doses of [^{3}H]alfentanil (1) were excreted in urine during 96 hours by rats and dogs. Faecal excretion accounted for about 24% dose (rat) or 8% dose (dog). In cannulated rats, about 24% dose was excreted in bile during 24 hours. Biotransformation pathways of (1) involved *N*-dealkylation at the piperidine and amide nitrogens, *O*-demethylation, aromatic hydroxylation, and conjugation (mainly glucuronidation). Metabolites were identified by GC–MS and chromatographic comparison with reference standards in most cases. Conjugates were assigned on the basis of enzymatic deconjugation. The position of the hydroxy group in (3) was not established, but it was assumed to be as shown on the basis of the structure of (6). Quantitative differences in metabolite profiles in rat and dog were large. The major metabolite in the rat was the *N*- and *O*-dealkylated compound (5), which was present in both urine and faeces and accounted for 32% dose. (5) and its conjugate were also the major metabolites in bile. Other important metabolites in the rat were conjugates of (6) [14% dose; (6) itself accounted for <1% dose]; (3) and its conjugates (12% dose) and (4) (7% dose). In the dog, the glucuronide of (3) was the major urinary metabolite (21% dose). The free aglycone accounted for an additional 5% dose. Metabolite (5) accounted for only 3% dose. A number of minor metabolites were not characterized. There was some indication that formation of (7) by migration of the propionyl group was non-enzymatic and occurred during the isolation procedure.

Reference

W. Meuldermans, J. Hendrickx, W. Lauwers, R. Hurkmans, E. Swysen, J. Thijssen, Ph. Timmerman, R. Woestenborghs, and J. Heykants, Excretion and biotransformation of alfentanil and sufentanil in rats and dogs, *Drug Metab. Dispos.*, 1987, **15**, 905.

Sufentanil

Use/occurrence:	Synthetic opioid
Key functional groups:	Alkyl aryl amide, alkyl carboxamide, methyl ether, piperidine
Test system:	Rat (intravenous, 0.16 mg kg^{-1}), dog (intravenous, 0.004 mg kg^{-1})

Structure and biotransformation pathway

Approximately 38% and 60% respectively of single intravenous doses of [^{3}H]sufentanil (1) were excreted in urine by rats and dogs during 96 hours. Faecal excretion accounted for about 62% dose (rat) or 42% dose (dog). In cannulated rats about 68% dose was excreted in bile during 24 hours. Routes

297

of metabolism involved *N*-dealkylation of the piperidine and amide nitrogen, *O*-demethylation, aromatic hydroxylation, and conjugation (mainly glucuronidation). Metabolites were identified by GC–MS and chromatographic comparison with reference standards in most cases. Conjugates were assigned on the basis of enzymic deconjugation. The position of the hydroxy group in (3) was not established, but it was assumed to be as shown, on the basis of the structure of (6). There were quantitative differences between metabolite profiles from the two species. In the rat, the *N*- and *O*-dealkylated metabolite (5) was the most important metabolite (22% dose) and was present in both urine and faeces. Other important metabolites in the rat were (4) (11% dose), (3) and its conjugate (8% dose together), and the conjugate of (6) (6% dose). In the dog the major metabolites were (4) (19% dose), (3) and its conjugate (15% dose together), and (5) (12% dose). A number of minor metabolites from both species were unidentified and rat bile contained a complex pattern of metabolites most of which were not characterized. There was some indication that formation of (7) by migration of the propionyl group was non-enzymic and occurred during the isolation procedure.

Reference

W. Meuldermans, J. Hendrickx, W. Lauwers, R. Hurkmans, E. Swysen, J. Thijssen, Ph. Timmerman, R. Woestenborghs, and J. Heykants, Excretion and biotransformation of alfentanil and sufentanil in rats and dogs, *Drug Metab. Dispos.*, 1987, **15**, 905.

3-n-Octyl- and *N*-octyl-pyridoglutethimide

Use/occurrence:	Aromatase inhibitors
Key functional groups:	Octyl, piperidine dione, pyridine
Test system:	Rat liver microsomes

Structure and biotransformation pathway

R(CH₂)₆CH₂CH₃ → R(CH₂)₆—C—CH₃ (with =O) + R(CH₂)₆—CHCH₃ (with OH)

(1)   (2)   (3)

R =

(3-octyl)   or   (*N*-octyl)   (4)

After incubation of the *N*- and 3-n-octyl analogues (1) of pyridoglutethimide (4) with liver microsomes from rats pretreated with sodium phenobarbital, organic extracts were examined by TLC. The metabolites were isolated and subjected to EI and CI mass spectrometry. The 7-oxo-octyl (2) and 7-hydroxyoctyl (3) metabolites were identified. This study showed that microsomal metabolism occurred primarily in the alkyl side-chain. Unlike pyridoglutethimide, the octyl analogues did not appear to form *N*-oxidized products.

Reference

A. Seago, M. H. Baker, J. Houghton, C.-S. Leung, and M. Jarman, Metabolism and pharmacokinetics of the *N*- and *C*-n-octyl analogues of pyridoglutethimide [3-ethyl-3(4-pyridyl)piperidine-2,6-dione]: novel inhibitors of aromatase, *Biochem. Pharmacol.*, 1987, **36**, 573.

1-Methyl-4-phenyl-1,2,3,6-tetrahydro-pyridine (MPTP)

Use/occurrence:	Model compound
Key functional groups:	*N*-Methyltetrahydro-pyridine
Test system:	Isolated rat hepatocytes

Structure and biotransformation pathway

MPTP (1) is a xenobiotic which destroys dopaminergic neurons of the substantia nigra, thus mimicking the clinical symptoms of Parkinson's disease. Biotransformation of (1) is catalysed via monoamine oxidase B to 1-methyl-4-phenyl-2,3-dihydropyridinium ion, MPDP$^+$ (2) and the fully oxidized derivative 1-methyl-4-phenylpyridinium ion, MPP$^+$ (3). It has been shown that both intact liver and brain mitochondria concentrate (3) in an energy dependent process.

In this study freshly isolated hepatocytes have been used as a model system to study the relationship between metabolism and toxicity of (1). MPTP is converted into (2) and (3) by hepatocytes, but only (3) accumulates in the cells. Exposure of cells to (2) also results in accumulation of (3) with more rapid conversion than after exposure to (1). Exposure of cells to (3) requires a long lag phase before toxicity is apparent, with slow accumulation presumably reflecting its limited access to the cells. Cytotoxicity of (2) is dose dependent and preceded by complete depletion of intracellular ATP. Pargyline, a monoamine oxidase inhibitor blocks conversion of (1) into (2) and (3) but does not affect accumulation in cells of (3) after exposure to (2). The authors therefore conclude that (1) is toxic to hepatocytes after monoamine oxidase metabolism via (3), which accumulates in the cells, causing ATP depletion and cell death.

Reference

D. DiMonte, G. Ekstrom, T. Shinka, M. T. Smith, A. J. Trevor, and N. Castagnoli, Role of 1-methyl-4-phenylpyridinium ion formation and accumulation in 1-methyl-4-phenyl-1,2,3,6-tetrahydropyridine toxicity to isolated hepatocytes, *Chem. Biol. Interact.*, 1987, **62**, 105.

Areca-specific alkaloids, mainly arecoline

Use/occurrence:	In betel quid, cholinergic, anthelmintic
Key functional groups:	Methyl carboxylate, *N*-methyltetrahydropyridine
Test system:	Human betel quid chewers

Structure and biotransformation pathway

The chewing of betel quid appears to be associated with oral cavity cancer. One possible causative factor is 3-(methylnitrosamino)propionitrile (MNPN) (2), which is one of the nitrosation products of the major alkaloid arecoline (1) of the most abundant constituent of betel quid (Areca nut). MNPN has now been found in the saliva of betel quid chewers at concentrations of 0.5–11.4 μg l^{-1}. No MNPN could be detected in control saliva. The compound was identified by GC–TEA and by comparison of its mass spectrum with that of the synthetic compound. The mechanism of its formation from arecoline is unknown. MNPN is metabolically activated to a compound which can methylate guanine at the O^6 and *N*-7 positions, and is a carcinogen in rats.

Reference

B. Prokopczyk, A. Rivenson, P. Bertinato, K. D. Brunnemann, and D. Hoffmann, 3-(Methylnitrosamino)propionitrile: occurrence in saliva of betel quid chewers, carcinogenicity, and DNA methylation in F-344 rats, *Cancer Res.*, 1987, **47**, 467.

Use/occurrence:	Anxiolytic
Key functional groups:	Pyridine
Test system:	Rat (oral, 925 mg kg^{-1})

Structure and biotransformation pathway

The N-oxide (2) was isolated as a metabolite of (1) from rat urine after hydrolysis with β-glucuronidase. The identification was confirmed by NMR and MS by comparison with the authentic reference compound. There was no precise quantitation of the amount of metabolite excreted.

Reference

J. M. Morand, M. Cussac, M. Kaouadji, and J. Alary, Etude metabolique d'un aza-chalcol: 10184 CERM isolement d'un metabolite urinaire N-oxide chez le rat et synthese, *J. Pharm. Belg.*, 1987, **42**, 229.

Nilvadipine

Use/occurrence:	Calcium antagonist
Key functional groups:	Alkyl ester, dihydropyridine, nitrophenyl
Test system:	Rat (intravenous, 1 mg kg^{-1}; oral, 10 mg kg^{-1}), dog (intravenous, 0.1 mg kg^{-1}; oral, 1 mg kg^{-1})

Structure and biotransformation pathway

Six metabolites of [^{14}C]nilvadipine (1) were isolated by HPLC from dog urine and urine and bile from rat. The metabolites were identified by NMR and MS analysis by comparison with authentic reference compounds. Unchanged (1) was not detected. The metabolic profiles in rat and dog excreta were qualitatively similar. The main metabolites arose from oxidation of the 1,4-dihydropyridine ring (2)–(6), hydrolysis of the 5-isopropyl ester (3), (5), (6), and (7) or 3-methyl ester (2) and (4), and/or hydroxylation of the 6-methyl (5) or methyl group of the isopropyl ester (4). The expected metabolite (8), arising from oxidation of the dihydropyridine ring without subsequent metabolism, was not detected. Reduction of the 3-nitro group to produce the primary amine (6) was a minor pathway. Metabolites (3) and (5) accounted for 34.3% and 17.7% of the radioactivity in rat urine. A total of 17 metabolites were detected in rat bile of which (2) was the major (14.8% of the radioactivity). The major metabolites in dog urine were (3; 32.1%), (2; 12.4%), and (4; 9.2%).

Reference

S. Terashita, Y. Tokuma, T. Fujiwara, Y. Shiokawa, K. Okumura, and H. Noguchi, Metabolism of nilvadipine, a new dihydropyridine calcium antagonist, in rats and dog, *Xenobiotica*, 1987, **17**, 1415.

2,6-Dimethyl-4-phenylpyridine-3,5-dicarboxylic acid diethyl ester

Use/occurrence: Model compound

Key functional groups: Alkyl ester, aryl carboxylate

Test system: Rat liver microsomes

Structure and biotransformation pathway

Rat liver microsomes were shown to catalyse the removal of an ethyl group from the pyridinedicarboxylic acid ester (1). It was shown that de-ethylation involved cytochrome P-450 and resulted in the formation of the monoester (3) and acetaldehyde. Rate studies using the diestester with deuterated ethyl groups showed an apparent kinetic isotope effect of about 8. This high intrinsic isotope effect indicated C—H cleavage as a rate-limiting step and together with the formation of acetaldehyde provided evidence for formation of the hydroxylated intermediate (2). The same reaction was also shown to occur with the isopropyl or methyl esters although at a lower rate.

Reference

F. P. Guengerich, Oxidative cleavage of carboxylic esters by cytochrome P-450, *J. Biol. Chem.*, 1987, **262**, 8459.

HI-6

Use:	Cholinesterase reactivator
Key functional groups:	Alkyl ether, aryl carboxamide, oxime, pyridinium
Test system:	Rat (intravenous, 50 mg kg^{-1})

Structure and biotransformation pathway

(1)   (2)   (3)

Rats were given an intravenous dose of HI-6 dihydrochloride (1). The 0–24 hour urine contained HI-6 together with two metabolites (2) and (3). The metabolites were identified by MS, IR, UV, and NMR. There were no differences between the spectra of the metabolites and those of authentic reference compounds. The metabolites together accounted for 20% of the dose, 60% being excreted unchanged. It was suggested that metabolite (2) could be therapeutically active because of the presence of the intact pyridinium aldoxime group.

Reference

D. A. Ligtenstein, E. R. J. Wils, S. P. Kossen, and A. G. Hulst, Identification of two metabolites of the cholinesterase reactivator HI-6 isolated from rat urine, *J. Pharm. Pharmacol.*, 1987, **39**, 17.

N-Benzyl-N′-picolinylpiperazine

Use/occurrence:	Antidepressant
Key functional groups:	Aryl carboxamide, benzyl amine
Test system:	Rat (oral, 200 mg kg^{-1})

Structure and biotransformation pathway

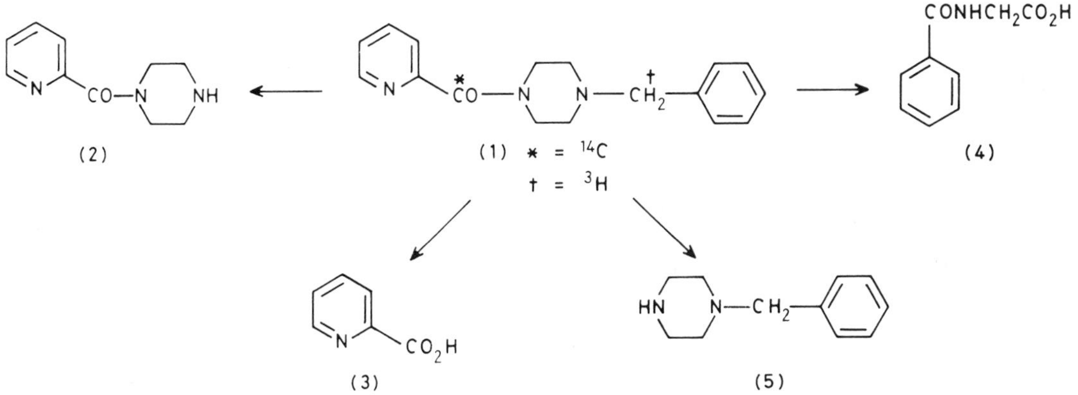

(2) ← (1) $* = {}^{14}C$ $\dagger = {}^{3}H$ → (4)

(3)

(5)

Metabolites of the antidepressant (1) were investigated in urine after administration of single oral doses. A 0–48 hour urine sample contained 40% of the dose and metabolites were isolated by Sephadex G-15 column chromatography and identified by GC–MS after formation of trimethylsilyl derivatives. The two major meabolites were picolinylpiperazine (2) and picolinic acid (3), each representing about 13% of the dose. A further major metabolite was hippuric acid formed via debenzylation of (1) or (5), the latter being detected as a minor metabolite (<2% dose).

Reference

A. Guttmann, J. Nagy, and K. Magyar, The metabolism of EGYT-475, a new antidepressant agent in rats, *Pol. J. Pharmacol. Pharm.*, 1987, **39**, 123.

Use: Urapidil, antihypertensive; IP-66, hypotensive and adrenolytic; Enciprazine, anxiolytic; MJ7378, anti-anxiety

Key functional groups: Methoxyphenyl, N-alkyl-piperazine

Test system: Rat (oral, 50 μmol kg^{-1})

Structure and biotransformation pathway

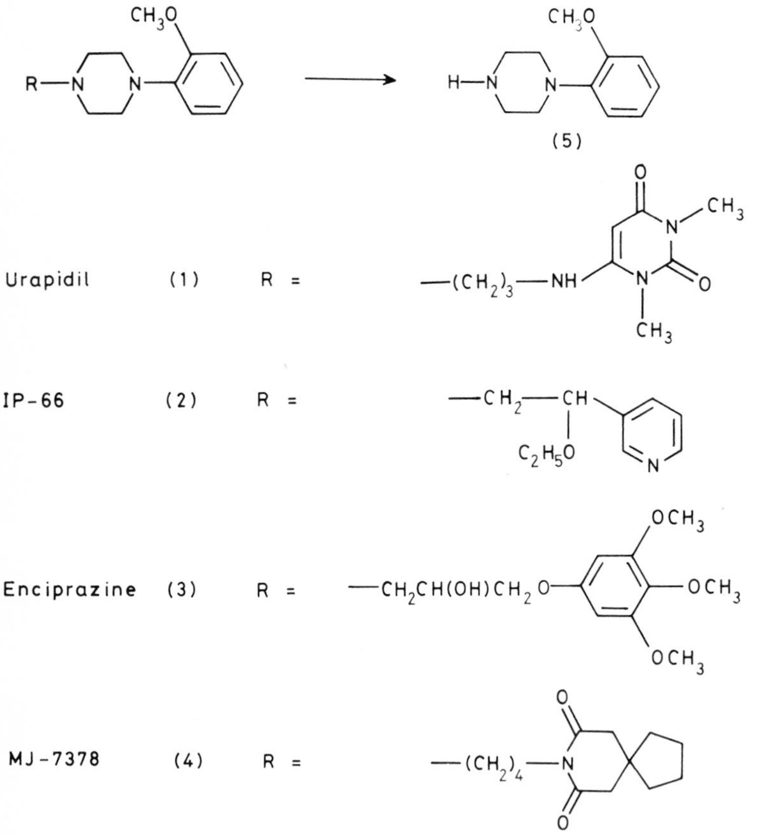

Urapidil (1)
IP-66 (2)
Enciprazine (3)
MJ-7378 (4)

Male rats were given equimolar doses of compounds (1)–(4) and urine was collected for 0–24 hours. Irrespective of which of the drugs was administered, 1-(o-methoxyphenyl)piperazine (5) was present in urine. This metabolite was detected by GC and confirmed by GC–MS, and is known to have potent hypotensive action in man.

Reference

E. Benfenati, S. Caccia, and F. Della Vedova, 1-(*o*-Methoxyphenyl)piperazine is a metabolite of drugs bearing a methoxyphenylpiperazine side-chain, *J. Pharm. Pharmacol.*, 1987, **39**, 312.

Phenobarbital

Use/occurrence:	Anticonvulsant
Key functional groups:	*N*-Alkylimide, alkylphenyl, barbiturate
Test system:	Human volunteers (oral, 300 mg)

Structure and biotransformation pathway

	R^1	R^2
(1a) (Phenobarbital) , (2a)	H	H
(1b) (Mephobarbital) , (2b)	CH_3	H
(1c),(2c)	C_2H_5	H
(1d),(2d)	CH_3	CH_3
(1e),(2e)	C_2H_5	C_2H_5

GC–MS comparison with synthetic standards was used to identify *O*-methylcatechol metabolites of phenobarbital (1a) and its *N*-alkylated derivatives (1b–e) in *β*-glucuronidase treated urine. Synthesis of the standard compounds was described. The *N*-alkyl barbiturates additionally formed *O*-methylcatechol derivatives in which *N*-dealkylation had occurred. For example, following administration of the dimethyl compound (1d), metabolites (2a), (2b), and (2d) were detected. In all cases the *O*-methylcatechol derivatives were minor metabolites, accounting for <0.1–4% dose. Metabolite (2a) was detected in larger quantities after administration of the *N*-methyl derivatives (1b) and (1d) than after administration of (1a). The *N*-de-ethylated metabolite (2a) was of very minor importance in the metabolism of the monoethyl compound (1c). No quantitative data for the diethyl compound (1e) were given.

Reference

A. M. Treston, A. Philippides, N. W. Jacobsen, M. J. Eadie, and W. D. Hooper, Identification and synthesis of *O*-methylcatechol metabolites of phenobarbital and some *N*-alkyl derivatives, *J. Pharm. Sci.*, 1987, **76**, 496.

Phenobarbitone

Use:	Sedative
Key functional groups:	Alkylphenyl, barbiturate
Test system:	Hamster liver microsomes

Structure and biotransformation pathway

(1)   (2)   (3)   (4)

The metabolism of [^{14}C]phenobarbitone (1) was investigated in liver microsomes prepared from male hamsters either with or without pretreatment with Aroclor 1254. The extent of metabolism was low (*ca.* 3%) and was not dependent on phenobarbitone concentration in the range 0.097–5.0 μmol substrate/3.5 ml of microsomal suspension. Only two metabolites were characterized; these were *p*-hydroxyphenobarbitone (2) and either *m*-hydroxyphenobarbitone (3) or phenylethylmalondiamide (4). These metabolites were identified by TLC.

Reference

A. Seago and J. W. Gorrod, The *in vitro* metabolism of [^{14}C]pentobarbitone and [^{14}C]phenobarbitone by hamster liver microsomes, *J. Pharm. Pharmacol.*, 1987, **39**. 84.

Pentobarbitone

Use:	Sedative
Key functional groups:	Barbiturate, pentyl
Test system:	Hamster liver microsomes

Structure and biotransformation pathway

(1)   (2)   (3)   (4)

The metabolism of $[^{14}C]$pentobarbitone (1) was investigated in liver microsomes from male hamsters either with or without pretreatment with Aroclor 1254. Metabolism was dose dependent in the range 0.054–5.0 μmol pentobarbitone/3.5 ml of microsomal suspension. At the lowest concentration less than 50% of the substrate remained after 1 hour whereas greater than 90% remained at the highest concentration. The major metabolite formed was 3′-hydroxypentobarbitone (2). 3-Oxopentobarbitone (3) and 5-ethyl-5(1′-methyl-3′-carboxypropyl)barbituric acid (4) were present as minor metabolites. All metabolites were identified by TLC comparison with authentic samples.

Reference

A. Seago and J. W. Gorrod, The *in vitro* metabolism of $[^{14}C]$pentobarbitone and $[^{14}C]$phenobarbitone by hamster liver microsomes, *J. Pharm. Pharmacol.*, 1987, **39**, 84.

Flucytosine

Use/occurrence:	Antifungal agent
Key functional groups:	Cytosine, fluorocytosine
Test system:	Human male patient (intravenous, 2.5 g at 8 hour intervals)

Structure and biotransformation pathway

Glucuronide (2)

(1)

(4)

$H_2NCH_2CHCO_2H$ with F substituent (3)

F^-

^{19}F NMR was used for analysis of flucytosine (1) and its metabolites in biological fluids of a patient with cryptococcal meningitis. In urine, in addition to unchanged (1) and the previously reported metabolite α-fluoro-β-alanine (3), three new metabolites were detected: a glucuronide (2) of the parent compound, the ring-hydroxylated metabolite (4), and fluoride ion. The same metabolites, except fluoride ion, were found in plasma. In cerebrospinal fluid, only unchanged compound and glucuronide were observed. Total urinary excretion during an 8 hour period between two injections was 100.4% of an injected dose. Unchanged (1) was the major excretory product accounting for 96.1% of the total. The conjugate (2) and the hydroxy metabolite (4) accounted for 2.7% and 1.2% respectively of excreted material. Fluoride ion (0.3%) and the fluoroalanine (3) (0.1%) were minor metabolites. Identification of these minor metabolites and the parent (1) rested on correspondence of the ^{19}F NMR signals with standards. The glucuronide was characterized by the disappearance of its ^{19}F NMR signal when urine was incubated with β-glucuronidase. The ^{19}F NMR signal was unaffected when urine was incubated without enzyme or when the enzyme was inhibited with saccharo-1,4-lactone. No new signals were observed in urine following enzymic incubation indicating that the conjugate had been hydrolysed to the parent compound. The hydroxy metabolite (4) was characterized by correspondence of its ^{19}F NMR signal with the signal of a synthetic standard, and by comparison of the variation of the chemical shift of this signal with pH.

Reference

J. P. Vialaneix, M. C. Malet-Martino, J. S. Hoffmann, J. Pris, and R. Martino, Direct detection of new flucytosine metabolites in human biofluids by ^{19}F nuclear magnetic resonance, *Drug Metab. Dispos.*, 1987, **15**, 718.

5-Fluorouracil

Use/occurrence: Anti-cancer drug

Key functional groups: Fluorouracil, uracil

Test system: Human patients (intravenous, 500 mg m^{-2})

Structure and biotransformation pathway

(1) * = ^{3}H   (2)   (3)

Aliquots of bile containing highest concentrations of radioactivity (0.75–6 hours) after intravenous administration of [^{3}H]-5-fluorouracil (1) were used to isolate an unkown metabolite. The metabolite was isolated and purified by HPLC and identified by FAB–MS as a conjugate of 2-fluoro-β-alanine (2) with endogenous cholic acid. The structural assignment as (3) was confirmed by comparison with an authentic reference compound and showing that treatment with cholyglycine hydrolase resulted in formation of cholic acid and 2-fluoroalanine. This represents a new biotransformation pathway of an exogenous amino acid competing with endogenous glycine and taurine to produce an abnormal bile acid. This metabolite is probably formed owing to the high liver concentrations of 2-fluoro-β-alanine the major metabolite of the fluorouracil or a high affinity with cholyl-CoA-amino acid N-acyltransferase. A further unknown metabolite was not identified but it was postulated that it could be a similar conjugate with chenodeoxycholic acid.

Reference

D. J. Sweeney, S. Barnes, G. D. Heggie, and R. B. Diasio, Metabolism of 5-fluorouracil to an N-cholyl-2-fluoro-β-alanine conjugate: previously unrecognised role for bile acids in drug conjugation, *Proc. Natl. Acad. Sci. USA* , 1987, **84**, 5439.

1-Phthalidyl-5-fluorouracil

Use/occurrence:	Antineoplastic agent (pro-drug)
Key functional groups:	Fluorouracil, phthalidyl
Test system:	Rat hepatocytes and liver homogenate 105 000g supernatant

Structure and biotransformation pathway

When the 5-fluorouracil derivative (1) was incubated with isolated hepatocytes more than 50% was hydrolysed in 30 minutes. 5-Fluorouracil (2) and 2-carboxybenzaldehyde (3) were the metabolites formed by a hydrolase. The aldehyde was converted by a cytosolic reductase into α-hydroxymethylbenzoic acid (4) which formed phthalide (5) in acidic solution. The hydrolase and reductase were induced by pretreatment of animals with phenobarbital.

Reference

K. Tonda, S. Kamata, and M. Hirata, Metabolism of 1-phthalidyl-5-fluorouracil in rat liver and enzyme induction by phenobarbital, *Xenobiotica*, 1987, **17**, 759.

2′,3′-Dideoxycytidine

Use/occurrence:	Antiviral agent
Key functional groups:	Cytidine, cytosine, nucleoside
Test system:	Mouse (oral and intravenous bolus, 100 mg kg^{-1} or intraperitoneal infusion), rhesus monkey (intravenous infusion, 27 mg kg^{-1})

Structure and biotransformation pathway

(1)   * = ^{3}H   (2) X = Mono-, di-, or tri-phosphate

2′,3′-Dideoxycytidine was predominantly excreted via the urine as unchanged drug, although a minor metabolite (1) was detected in the urine of monkeys but not of mice. The oral absorption of 2′,3′-dideoxycytidine in the mouse was rapid with plasma levels approaching those seen after intravenous administration within 45 minutes. Relatively high concentrations of drug were found in mouse kidney, pancreas, and liver. The drug in the tissues was present mainly as the parent nucleoside, with its pharmacologically active anabolite (2) accounting for only a small fraction of the retained drug.

Reference

J. A. Kelley, C. L. Litterst, J. S. Roth, D. T. Vistica, D. G. Poplack, D. A. Cooney, M. Nadkarni, F. M. Balis, S. Broder, and D. G. Johns, The disposition and metabolism of 2′,3′-dideoxycytidine, an *in vitro* inhibitor of human T-lymphotrophic virus type III infectivity, in mice and monkeys, *Drug Metab. Dispos.*, 1987, **15**, 595.

Pyrazinamide

Use/occurrence:	Antitubercular agent
Key functional groups:	Aryl carboxamide, pyrazine
Test system:	Rat (oral, 150 mg kg^{-1}), man (oral, 12.5 mg kg^{-1})

Structure and biotransformation pathway

(2)   (1) * = ^{14}C   (3)   (4)

A new metabolite of pyrazinamide (1) was isolated from both human and rat urine and identified by NMR and MS as the hydroxylated derivative (2). Confirmation of identity was obtained by comparison of the methylated derivative with authentic 5-methoxypyrazinamide. Only small amounts of pyrazinamide were detected in rat urine (3% dose) but other major metabolites were pyrazinoic acid (3) (35% dose) and the hydroxylated acid (4) (25% dose). Hydroxylated pyrazinamide accounted for about 14% dose.

Reference

L. W. Whitehouse, B. A. Lodge, A. W. By, and B. H. Thomas, Metabolic disposition of pyrazinamide in the rat: identification of a novel *in vivo* metabolite common to both rat and human, *Biopharm. Drug Dispos.*, 1987, **8**, 307.

Pyrazinamide

Use/occurrence:	Antitubercular agent
Key functional groups:	Aryl carboxamide, pyrazine
Test system:	Human (oral, 3 g) and *in vitro* (human liver cytosol)

Structure and biotransformation pathway

In the *in vivo* study 5-hydroxypyrazinamide (2) was identified as a novel metabolite of pyrazinamide in the urine of volunteers. The metabolites (1) and (3) have been described previously. The identity of the novel metabolite (2) was confirmed by mass spectrometry. For the *in vitro* study a xanthine oxidase rich fraction of human liver cytosol was prepared. This fraction was found to catalyse both the oxidation pyrazinamide and pyrazinoic acid (1) to the corresponding 5-hydroxy metabolites (2) and (3). Kinetic studies suggested that *in vitro* xanthine oxidase was responsible for this oxidation. Metabolites *in vivo* and *in vitro* were measured by an HPLC method.

References

T. Yamamoto, Y. Moriwaki, S. Takahashi, T. Hada, and K. Higishano, 5-Hydroxy-pyrazinamide a human metabolite of pyrazinamide, *Biochem. Pharmacol.*, 1987, **36**, 2415.

T. Yamamoto, Y. Moriwaki, S. Takahashi, T. Hada, and K. Higishano, *In vitro* conversion of pyrazinamide into 5-hydroxypyrazinamide and that of pyrazinoic acid into 5-hydroxypyrazinoic acid by xanthine oxidase from human liver, *Biochem. Pharmacol.*, 1987, **36**, 3317.

N-(4,6-Dimethyl-2-pyrimidinyl)benzene-sulphonamide (desaminosulfamethazine, DAS)

Use/occurrence:	Animal metabolite of sulfamethazine
Key functional groups:	Methyl pyrimidine, sulphonamide
Test system:	Rat (oral, in the diet, approximately equivalent to 100 mg kg^{-1})

Structure and biotransformation pathway

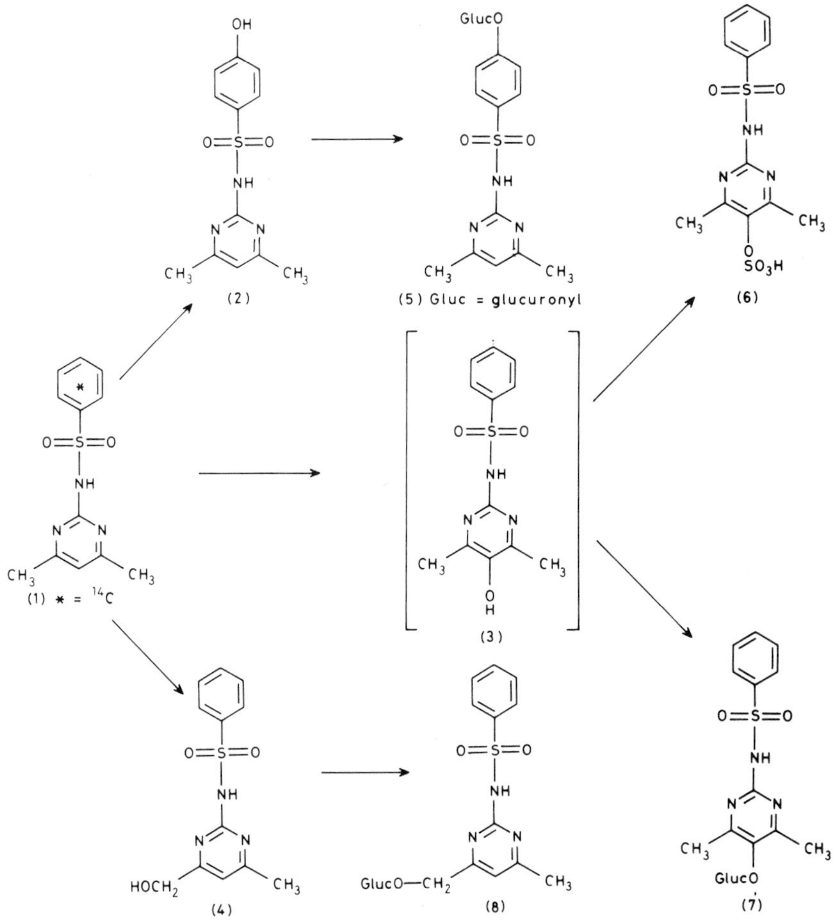

(2)

(5) Gluc = glucuronyl

(6)

(1) ∗ = ^{14}C

(3)

(4)

(8)

(7)

DAS (1) is an unusual deaminated metabolite of the antibiotic growth promoter sulfamethazine, whose formation is reported to be accelerated by high dietary nitrite.

Powdered diet (5 g), containing 20 mg [^{14}C]-(1) was consumed within 30 minutes by rats. During 96 hours after a single administration the rats excreted 64% dose in urine and 22% in faeces. Compounds accounting for most of the radiolabel in the 0–24 hour urine were isolated and identified by IR, NMR, and MS. The results showed that (1) was metabolized by hydroxylation at three different positions in the molecule, each being followed by varying degrees of conjugation. The major pathway was hydroxylation at the 5-position on the pyrimidine ring. Very little (< 1%, results as % urinary ^{14}C) of the free aglycone (3) was excreted in urine; most of this intermediate was conjugated and excreted as the sulphate (6; 45%) or glucuronide (7; 16%). Hydroxylation at the 4-position of the benzene ring was quantitatively the second most important pathway and urine contained both hydroxylated metabolite (2; 9%) and its glucuronide conjugate (5; 10%). Hydroxylation also occurred on one of the methyl groups attached to the pyrimidine ring to yield (4; 9%) and its glucuronide conjugate (8; 1%). At least four minor urinary metabolites were not identified and it was postulated that these were the result of metabolism at more than one site on the molecule.

Reference

G. D. Paulson and V. J. Feil, The disposition of *N*-(4,6-dimethyl-2-pyrimidinyl)-benzene [*U*-^{14}C]sulphonamide in the rat, *Drug Metab. Dispos.*, 1987, **15**, 671.

Sulphamethazine

Use: Antibiotic drug

Key functional groups: Aryl amine, methyl pyrimidine, sulphonamide

Test system: Rat (oral, 60 mg kg^{-1})

Structure and biotransformation pathway

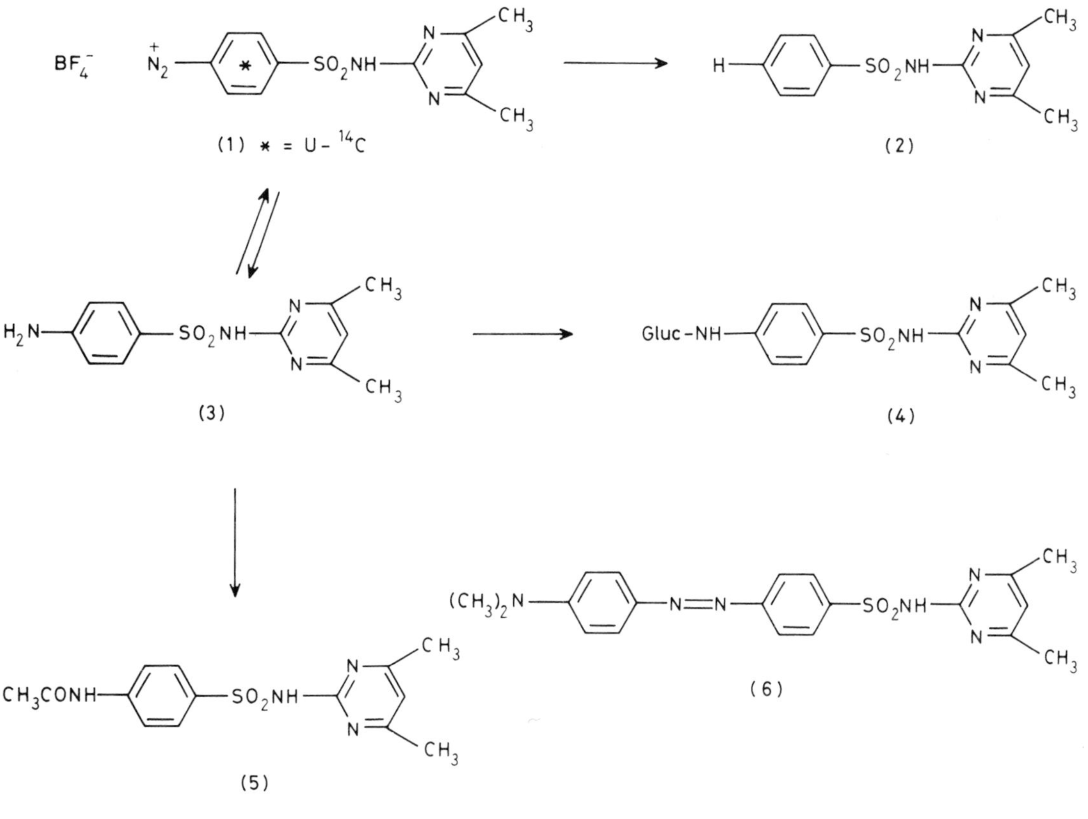

Six hours after the administration of [^{14}C]sulphamethazinediazonium tetra-fluoroborate (1) to rats 72% of the radioactivity remained in the gut and less than 4% had been eliminated in urine. The remainder was distributed in a range of tissues. Although the radioactivity in blood, liver, and skeletal muscle could be almost quantitatively extracted with methanol, 28% and 32% of the radioactivity in the stomach and intestine respectively could not be recovered under the same conditions. The major metabolite present in tissues and the stomach was desaminosulphamethazine (2) (38–63% of the tissue radioactivity) but in the small intestine and urine this metabolite accounted for a smaller proportion of the radioactivity (<6%). Sulphametha-zine (3) (4–14%) and its N-glucuronide (4) and N-acetyl metabolites (5) (each less than 5%) were present as minor metabolites in tissues, stomach, small

321

intestine, and urine. When the stomach contents of rats which were simultaneously given [14C]sulphamethazine, sodium nitrite, and dimethylaniline were analysed three compounds were identified. These were the parent substance (3), the desamino metabolite (2), and the dimethylaminophenyl adduct (6). The similarity in the disposition and metabolism of [14C]sulphamethazinediazonium tetrafluoroborate (1) and [14C]sulphamethazine (2) (reported previously) suggests that a large proportion of sulphamethazine is diazotized in the rat stomach, and this is confirmed by the results of the experiment using dimethylaniline to trap the diazonium intermediate. The diazonium intermediate (1) was also shown to possess weak mutagenic activity in bacterial assays using *Salmonella typhimurium*.

Reference

G. D. Paulson and V. J. Feil, Formation of a diazonium cation intermediate in the metabolism of sulphamethazine to desaminosulphamethazine in the rat, *Xenobiotica*, 1987, **17**, 697.

Indeloxazine

Use/occurrence:	Cerebral activator
Key functional groups:	Alkyl aryl ether, indene, morpholine
Test system:	Rat (oral, 10–50 mg kg^{-1} and intravenous, 3 mg kg^{-1})

Structure and biotransformation pathway

Metabolites of indeloxazine (1) were investigated in samples of plasma, urine, and bile from rats administered oral or intravenous doses.

Metabolites were characterized by IR, NMR, and MS. The morpholinone (2) was the major plasma metabolite but was not detected in urine. The indanediol (3) was the major urine metabolite and was also present in plasma.

Both *cis*- and *trans*-indane-1,2-diols were synthesized and the metabolite was shown to co-chromatograph with the *trans*-isomer. Two phenolic metabolites present as glucuronides in urine were identified after enzyme hydrolysis. One metabolite (6) resulted from aromatic hydroxylation in the indene ring and the other (7) was formed from cleavage of the arylether link. Two other minor urinary metabolites (4) and (5) were identified as indene diols. It was demonstrated that the metabolite (5) was excreted in urine following administration of (2) and that both (4) and (5) were excreted in urine after administration of the diol (3). A further metabolite (8) was formed by degradation of the morpholine ring but the amide (2) was shown not to be an intermediate. Bile collected during 72 hours after dosing contained about 50% of the dose as unknown polar metabolites which were not hydrolysed by β-glucuronidase/ sulphatase.

Reference

H. Kamimura, Y. Enjoji, H. Sasaki, R. Kawai, H. Kaniwa, K. Niigate, and S. Kageyama, Disposition and metabolism of indeloxazine hydrochloride, a cerebral activator, in rats, *Xenobiotica*, 1987, **17**, 645.

Molsidomine

Use/occurrence:	Anti-anginal agent
Key functional groups:	Imine carboxylate, morpholine, sydnone
Test system:	Man (healthy volunteers, oral, 2 mg)

Structure and biotransformation pathway

After oral administration of [^{14}C]molsidomine to man, the majority of radioactivity (93% of dose) was excreted in urine. Pharmacologically active 3-morpholinosydnonimine (1) formed by de-ethoxycarbonylation of molsidomine accounted for about 20% of the dose. N-Cyanomethylamine-N-(2'-hydroxyethyl)glycine (4) and (N-cyanomethylamino-2-aminoethoxy)acetic acid (5) formed by oxidative metabolic cleavage of the morpholine ring accounted for 33% and 16% of the dose respectively. The proportion of the metabolites increased with time relative to (1). Unstable degradation products of (1), N-nitroso-N-morpholineaminoacetonitrile (2), and N-cyanomethylaminomorpholine (3) were also excreted in urine accounting for 5% and 8% of the dose respectively. [^{14}C]Thiocyanate, probably resulting from free cyanide formed from nitrile-containing metabolites of molsidomine, was also detected in urine (<5% dose). Metabolites were characterized by co-chromatography (HPLC) with reference compounds and had all been identified previously in animal studies.

Reference

I. D. Wilson, J. M. Fromson, H. P. A. Illing, and E. Schraven, The metabolism of [^{14}C]-N-ethoxycarbonyl-3-morpholinosydnonimine (molsidomine) in man, *Xenobiotica*, 1987, **17**, 93.

Caprolactam

Use/occurrence:	Nylon monomer
Key functional groups:	Lactam
Test system:	Rat (oral, 3% in diet)

Structure and biotransformation pathway

Male Sprague–Dawley rats were fed caprolactam (3% in diet estimated to be 385 mg day^{-1}) for 2–3 weeks. The urine contained four ninhydrin-positive components (A–D) (19% of dose), which were resolved on an amino acid analyser. None of these components were present in control urine. The major metabolite (D) coeluted with 6-amino-γ-caprolactone (2) and was excreted at 63 mg day^{-1}. Metabolite C (8.8% of the four components) co-eluted with 6-aminohexanoic acid (3) (6.3 mg day^{-1}). Metabolite A was concluded to be 6-amino-4-hydroxyhexanoic acid (4), and metabolite B (3.7%) was unidentified. Metabolites D and A were shown to be in equilibrium with each other under acidic conditions, and together totalled 87.5% of the four metabolites. The metabolites were purified by preparative ion exchange chromatography and subjected to further spectroscopic analysis (IR, NMR, MS). Metabolite D was confirmed to be (2), and was shown to have a negative optical rotation, suggesting an *S*-configuration. Rats consuming 6-amino-hexanoic acid did not excrete (2) or (4), suggesting that it was caprolactam itself that was hydroxylated to yield a 4-hydroxy derivative, which is hydrolysed and rearranges to (4) and (2) respectively, possibly under the influence of the acidic conditions of the urine.

Reference

L. K. Kirk, B. A. Lewis, D. A. Ross, and M. A. Morrison, Identification of ninhydrin-positive caprolactam metabolites in the rat, *Food Chem. Toxicol.*, 1987, **25**, 233.

Oxazepam

Use/occurrence:	Antidepressant
Key functional groups:	Benzodiazepine
Test system:	Mouse (oral, 22 mg kg^{-1})

Structure and biotransformation pathway

(2)　　　　　(1) * = ^{14}C　　　　　(3)

Glucuronide　　　　　Glucuronide

Following oral administration of [2-^{14}C]oxazepam (1) to male mice, 27% dose was recovered in urine and 58% in faeces during five days. The radioactivity in faeces was not investigated. The major radiolabelled components in the 0–48 hour urine were unchanged (1) (*ca.* 19%, all results as % urinary radioactivity), its glucuronide conjugate (37%), the ring-contracted quinazoline carboxylic acid (2) (14%), and the glucuronide conjugate of 4'-hydroxyoxazepam (3) (*ca.* 10%). Unconjugated (3) was a minor metabolite. Sulphate conjugates of (3) and 4'-hydroxy-3'-methoxyoxazepam (structure not shown) were very minor metabolites (*ca.* 1.5% urinary metabolites together). Identifications of metabolites were based on two-dimensional thin layer co-chromatography with reference compounds. The acid (2) was additionally characterized by GC–MS following methylation. The results were compared with the previously reported metabolism of (1) in the rat.

Reference

S. F. Sisenwine, C. O. Tio, A. L. Liu, and J. F. Politowski, The metabolic fate of oxazepam in mice, *Drug Metab. Dispos.*, 1987, **15**, 579.

Diazepam

Use/occurrence:	Tranquilizer
Key functional groups:	Benzodiazepine
Test system:	Cultured hepatocytes from rat, rabbit, dog, guinea pig, and man

Structure and biotransformation pathway

Metabolite profiles obtained from HPLC analysis of culture media after incubation of diazepam indicated that substantial species differences existed corresponding to known differences in the *in vivo* metabolite profiles of diazepam. The disappearance of diazepam was rapid in rat, dog, and guinea pig hepatocytes, but slow in human hepatocytes. In the rat the major assignable metabolites were nordiazepam (3) and the 4'-hydroxy metabolite (5), the latter being mainly present as a glucuronide conjugate, but only about 40% of the metabolites corresponded chromatographically with authentic standards after enzymic hydrolysis. In the dog about 85% of the metabolites corresponded to standards of which (3) was the dominant metabolite. No 4'-

hydroxy metabolites were detected. In the guinea pig about 80% of the diazepam metabolites could be assigned. Compound (3) was again the major metabolite and oxazepam (4) and the 4'-hydroxy metabolites (5) and (6) were also formed and extensively glucuronidated. In the rabbit essentially all of the metabolized diazepam was recovered as either (3) or (5), the latter mainly as a glucuronide. In human hepatocytes the rate of metabolism was low and nordiazepam (3) and temazepam (2) accumulated in the culture medium. Intrinsic clearances were also determined for (2), (3), and (4) in hepatocytes from each species. The intrinsic clearance of nordiazepam (3) exhibited the most marked species variation and it appeared that species exhibiting a high intrinsic clearance for (3) were also those that exhibited significant hydroxylation at the 4' site of the molecule.

Reference

R. J. Chenery, A. Ayrton, H. G. Oldham, P. Standring, S. J. Norman, T. Seddon, and R. Kirby, Diazepam metabolism in cultured hepatocytes from rat, rabbit, dog, guinea pig, and man, *Drug Metab. Dispos.*, 1987, **15**, 312.

Guanethidine

Use:	Antihypertensive drug
Key functional groups:	*N*-Alkylcycloalkylamine, guanidine
Test system:	Human liver microsomes

Structure and biotransformation pathway

(1)

Incubation of guanethidine with human liver microsomes resulted in oxidation at the cyclic aliphatic tertiary nitrogen to give the *N*-oxide (1). This product was identified by HPLC comparison with a reference standard. The optimum pH for this biotransformation was 8.5. The addition of the cytochrome P-450 inhibitors proadifen and 2,4-dichloro-6-phenylphenoxyethylamine to the incubation medium caused a less than 20% reduction in the rate of *N*-oxide formation. The results demonstrated that the flavin-containing monooxygenase and not cytochrome P-450 is the enzyme mainly responsible for *N*-oxidation of guanethidine in human liver microsomes.

Reference

M. E. McManus, D. S. Davies, A. R. Boobis, P. H. Grantham, and P. J. Wirth, Guanethidine *N*-oxidation in human liver microsomes, *J. Pharm. Pharmacol.*, 1987, **39**, 1052.

Proglumetacin

Use/occurrence:	Anti-inflammatory drug
Key functional groups:	Alkyl amide, alkyl ester, aryl amide, chlorobenzoyl, indole, methoxyphenyl, piperazine
Test system:	Rat (oral, 27 mg kg^{-1})

Structure and biotransformation pathway

Identification of metabolites in this study relied on TLC comparison with synthetic standards. Proglumetacin (1) was shown to act as a prodrug for indomethacin (4), which was formed in the stomach and was the major radio-labelled component in peripheral blood and in major organs. Radioactivity

332

was excreted *via* both urine (34% dose) and faeces (61% dose). The extensive faecal elimination was probably mainly a result of biliary excretion of radioactivity. Indomethacin (4) was the major metabolite in faeces (15% dose), followed by (6) (10% dose), and unchanged (1) (8% dose). Metabolites (2), (3), and (5) were also detected. Compounds (6) and (5) were the major metabolites in urine but (2), (3), and (4) were also present. Significant proportions of urinary and faecal radioactivity were associated with unidentified metabolites. Enzymic incubations indicated that glucuronide conjugates were not present.

Reference

F. Makovec, R. Christé, W. Peris, and I. Setnikar, Pharmacokinetics and metabolism of [^{14}C]proglumetacin after oral administration in the rat, *Arzneim. Forsch.*, 1987, **37 (II)**, 806.

Dimethyltryptamine, 5-Methoxydimethyl-tryptamine

Use/occurrence:	Psychomimetic agents
Key functional groups:	Dimethylaminoalkyl, indole, methoxyphenyl
Test system:	Rat (intraperitoneal, 10 mg kg^{-1})

Structure and biotransformation pathway

X = H , Dimethyltryptamine
X = CH$_3$, 5-Methoxydimethyltryptamine

(1)

(2)

Previous studies have shown that for psychomimetic indole alkylamines oxidative deamination is a major route of metabolism. The products of this metabolism by monoamine oxidase are indoleacetic acid metabolites. In this study the effect of monoamine oxidase inhibitors (pargyline and iproniazid) on the excretion of the indole alkylamines and their characteristic metabolites (1) and (2) was studied. In the absence of treatment with monoamine oxidase inhibitors urinary excretion of the *N*-oxide metabolites (1) accounted for 6–7% of the dose (both compounds). Iproniazid treatment increased the excretion of the *N*-oxide of 5-methoxydimethyltryptamine to 49% of the dose. A lesser effect was seen for dimethyltryptamine where excretion of the *N*-oxide increased to *ca.* 20% of the dose followng treatment of the rats with either pargyline or iproniazid. Amounts of parent compound or the monomethyl metabolites (2) excreted were quite low (representing *ca.* 1% of the dose or less), treatment with monoamine oxidase inhibitors increased this proportion. Metabolites were analysed by HPLC using a cation exchange column.

Reference

B. R. Sitaram, L. Lockett, G. L. Blackman, and W. R. McLeod, Urinary excretion of 5-methoxy-*N,N*-dimethyltryptamine, *N,N*-dimethyltryptamine, and their *N*-oxides in the rat, *Biochem. Pharmacol.*, 1987, **36**, 2235.

3-Methylindole

Use/occurrence:	Product of tryptophan degradation in ruminants
Key functional groups:	Indole, methylindole
Test system:	Mouse (intraperitoneal, $350-900$ mg kg^{-1})

Structure and biotransformation pathway

X = H , 3-Methylindole
X = D , Deutero-3-methylindole

(1)

Glutathione depletion and toxicity

The pneumotoxicity of 3-methylindole was compared with that of deutero-3-methylindole. The depletion of pulmonary glutathione was also measured following administration of both chemicals. The deutero analogue was found to be less toxic and furthermore caused less depletion of pulmonary glutathione than 3-methylindole. The imine methide (1) is the proposed primary reactive intermediate. This intermediate is thought to be responsible for the pulmonary glutathione depletion. Formation of this intermediate would involve C–H bond breakage and thus deuteration of the 3-methyl group was expected to reduce the toxicity.

Reference

J. C. Huijzer, J. D. Adams, and G. S. Yost, Decreased pneumotoxicity of deuterated 3-methylindole: bioactivation requires methyl C–H bond breakage, *Toxicol. Appl. Pharmacol.*, 1987, **90**, 60.

(−)-Indolactam V

Use/occurrence:	Tumour promoter
Key functional groups:	Indole, lactam, N-methyl-arylamine
Test system:	Rat liver microsomes

Structure and biotransformation pathway

(1)

(2)   (3)   (4)

The indole alkaloid (−)-indolactam V (1) is a skin tumour promoter related in structure to the teleocidin class of compounds isolated from *Streptomyces*. Metabolism of (1) by microsomes from phenobarbital pretreated rats yielded three products which were separated by HPLC. Compound (2) was demonstrated by co-chromatography and MS to be the desmethyl derivative. Compounds (3) and (4) were formed in similar amounts and were also found in the control (boiled microsomes) to the extent of *ca.* 25% of the amount in the active microsomal incubation, presumably because of an autoxidative process. The structures were determined by co-chromatography and MS of their t-butyldimethylsilyl derivatives to be the diastereoisomers of the 2-oxo compounds. The metabolites had lower tumour-promoting activity than (1).

Reference

N. Hagiwara, K. Irie, H. Tokuda, and K. Koshimizu, The metabolism of indole alkaloid tumour promoter, (−)-indolactam V, which has the fundamental structure of teleocidins, by rat liver microsomes, *Carcinogenesis*, 1987, **8**, 963.

α-[(Dimethylamino)methyl]-2-(3-ethyl-5-methyl-4-isoxazolyl)-1H-[3-¹⁴C]indole-3-methanol (DIIM)

Use/occurrence:	Hypoglycaemic
Key functional groups:	Dialkylisoxazole, dimethylaminoalkyl, indole, alkyl secondary alcohol
Test system:	Humans (oral, 50 mg, 200 mg)

Structure and biotransformation pathway

(2)   (1) * = ¹⁴C   (3)

[¹⁴C]-DIIM (1) was administered to 15 male volunteers at one of two dose levels (50 mg, 200 mg). Urinary excretion of radioactivity was dose dependent $76.9 \pm 9.9\%$ for 50 mg but $66.0 \pm 6.2\%$ at 200 mg. Faecal excretion was comparable at both dose levels. Analysis of urine indicated that $54.1 \pm 4.8\%$ of total urinary radioactivity after the 50 mg dose was DIIM compared with $49.7 \pm 6.3\%$ after the 200 mg dose. The major metabolites were detected in urine. Hydroxy-DIIM (2) was only a minor metabolite in urine, but two major metabolites were identified as diastereoisomeric glucuronide conjugates of DIIM (3). These glucuronides represented 16–17% of the dose in urine. The metabolites were extracted by XAD-2, separated by HPLC, and their structures determined by NMR and MS.

Reference

F. L. S. Tse, B. A. Orwig, J. M. Jaffe, and J. G. Dain, Pharmacokinetics and metabolism of α-[(dimethylamino)methyl]-2-(3-ethyl-5-methyl-4-isoxazolyl)-1H-indole-3-methanol, a hypoglycaemic agent, in man, *Xenobiotica*, 1987, **17**, 1259.

Jacobine, Jacoline, Senecionine, Seneciphylline

Use/occurrence:	Hepatotoxic alkaloids
Key functional groups:	Alkyl ester, pyrrolizidine
Test system:	Rat liver microsomes (phenobarbital pretreated)

Structure and biotransformation pathway

The test compounds (not radiolabelled) were extracted from the plant species *Senecio jacobaea* (tansy ragwort) and purified by recrystallization and preparative HPLC. An analytical standard of the *N*-oxide was prepared from each compound. Following incubation of each compound, metabolites were quantified and characterized by HPLC and comparison with these synthetic standards and authentic 6,7-dihydro-7-hydroxy-1-hydroxymethyl-5*H*-pyrrolozine (DHP; 5). Confirmatory evidence of structure was obtained by MS in most cases. Microsomal metabolism of all four alkaloids resulted in the formation of *R*-DHP. The *N*-oxides of jacobine (1), senecionine (3), and seneciphylline (4) were also formed but the *N*-oxide of jacoline (2) was not detected (<0.3% added substrate). The rate of formation (results as nmol min^{-1} mg^{-1} protein) of DHP was also lowest for (2) (0.36), compared with values of 1.1–1.5 for the other three alkaloids. Reported values for rates of formation of the *N*-oxides of (1), (3), and (4) were in the range 1.4–3.1. Rates of formation of both *N*-oxide and DHP increased in the order (1)<(4)<(3). Some minor unidentified metabolites were formed from all four alkaloids. No detectable metabolism of any of the *N*-oxides occurred when the individual compounds were incubated with microsomes. The authors concluded that the *N*-oxides and DHP were formed by the action of different enzymes, possibly different cytochrome P-450 iso-enzymes. The results were discussed in relation to published work on other pyrrolizidine alkaloids and with regard to the hypothesis that pyrrole formation is involved in the mechanism of toxicity, whereas *N*-oxidation is a detoxification pathway.

Reference

H. S. Ramsdell, B. Kedzierski, and D. R. Buhler, Microsomal metabolism of pyrrolizidine alkaloids from *Senecio jacobaea*, *Drug Metab. Dispos.*, 1987, **15**, 32.

1,2-Dihydro-2,2,4-trimethylquinoline (TMQ)

Use/occurrence:	Antioxidant in rubber industry
Key functional groups:	Alkyl aryl amine, dihydroquinoline
Test system:	Rat (oral, 2, 20, and 200 mg kg^{-1}; intravenous, 20 mg kg^{-1}), *in vitro* rat liver cell fractions

Structure and biotransformation pathway

Absorption, distribution, metabolism, and excretion were essentially independent of dose in the range studied. Approximately 60–70% of an intravenous dose was excreted via urine and 20–30% via the faeces. More than 99% of the dose was in the form of metabolites. Urine contained two major

340

metabolites (4) and (5 or 6) and ten minor metabolites. The two major metabolites were identified by NMR and MS. *In vitro* metabolite (1) and a glutathione conjugate of unspecified structure were the major metabolites found in the 10 000g supernatant. The proportion of the glutathione conjugate was increased by addition of 2.5 mM glutathione to the incubation mixture. Cytosol alone was almost completely inactive. Metabolite (1) was only tentatively identified and metabolites (2) and (3) were postulated intermediates.

Reference

Y. M. Ioanno, L. T. Burka, J. M. Sanders, M. P. Moorman, and H. B. Matthews, Absorption, distribution, metabolism, and excretion of 1,2-dihydro-2,2,4-trimethylquinoline in the male rat, *Drug Metab. Dispos.*, 1987, **15**, 367.

Chloroquine

Use/occurrence:	Antimalarial drug
Key functional groups:	Secondary alkyl aryl amine, dialkylaminoalkyl, quinoline
Test system:	Human volunteers (oral, 150 mg), human patients (oral, 2×150 mg week^{-1}), human overdose cases (oral, 4 g or 10 g)

Structure and biotransformation pathway

Three previously unknown metabolites (2), (3), and (4) were detected by GC–MS in acid-hydrolysed urine from most subjects, although they amounted to less than 2% of the unchanged chloroquine (1) concentration. The results gave no evidence for dose dependent formation of these metabolites. It was proposed that (2) was formed by acetylation of the previously known metabolite didesethylchloroquine. It is also conceivable that an acetyl group could arise by C-oxidation of an ethyl group. Possible mechanisms for formation of (3) and (4) were discussed which involved oxidative cleavage of the diethylamino group and cyclization of the resulting aldehyde or derived carboxylic acid. However, formation of these components as artefacts during analysis cannot be precluded.

Reference

C. Koppel, J. Tenczer, and K. Ibe, Urinary metabolism of chloroquine, *Arzneim.-Forsch.*, 1987, **37 (I)**, 208.

Primaquine

Use/occurrence:	Antimalarial drug
Key functional groups:	Alkyl amine, chiral carbon
Test system:	Isolated perfused rat liver

Structure and biotransformation pathway

(1) † chiral carbon   (2)

The disposition of the two enantiomers of primaquine (1) was studied in the isolated perfused rat liver using an HPLC assay which permitted the measurement of (1) and the carboxylic acid metabolite (2). Stereoselectivity in the hepatic elimination efficiency of the enantiomers of (1) and stereoselective deamination to give (2) were observed.

Reference

D. D. Nicholl, G. Edwards, S. A. Ward, M. L. E. Orme, and A. M. Breckenridge, The disposition of primaquine in the isolated perfused rat liver. Stereoselective formation of the carboxylic acid metabolite, *Biochem. Pharmacol.*, 1987, **36**, 3365.

6,7-Dimethoxy-4-(4'-chlorobenzyl)iso-quinoline

Use/occurrence:	Muscle relaxant
Key functional groups:	Benzyl, chlorophenyl, methoxyphenyl, isoquinoline
Test system:	Male albino rat (intravenous, 2.34 mg kg^{-1}, 30 mg kg^{-1}; oral, 30 mg kg^{-1})

Structure and biotransformation pathway

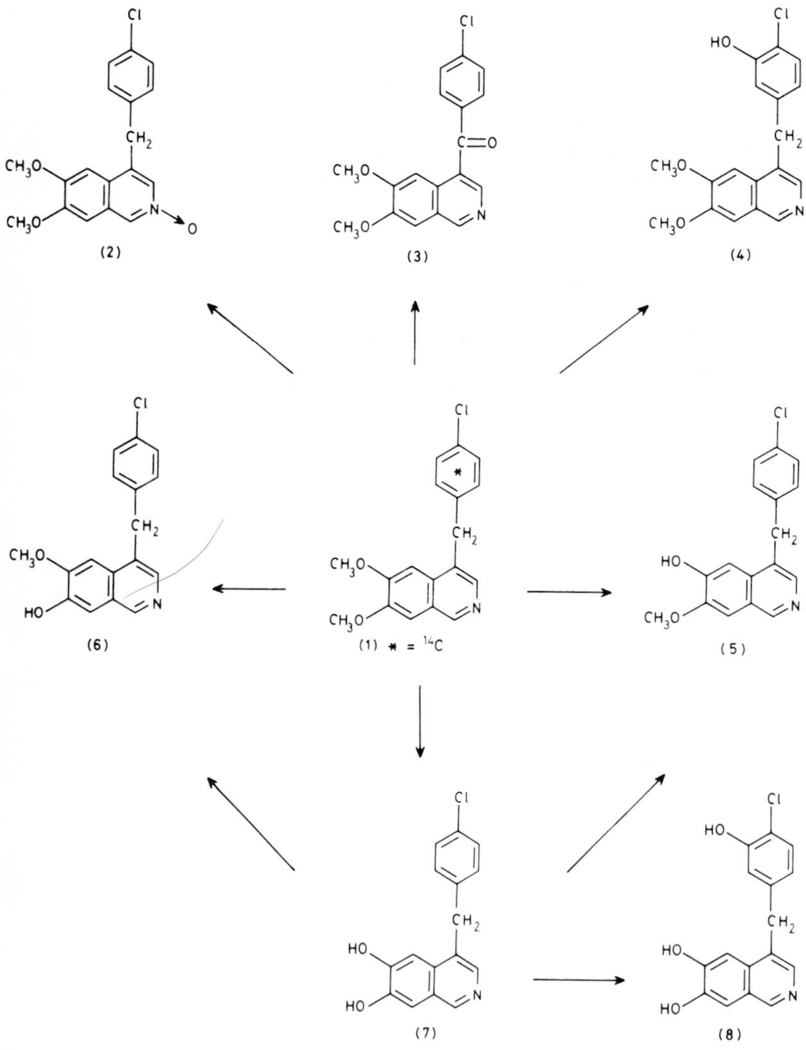

(2)

(3)

(4)

(6)

(1) ✳ = ^{14}C

(5)

(7)

(8)

344

Following intravenous administration of (1), the parent compound and metabolites were excreted in urine (23% of dose) and faeces (72%). Bile was also collected from bile duct-cannulated rats for 2 hours after an intravenous injection of (1). Solvent extraction of excreta (before and after enzymic/acidic hydrolysis) was used followed by TLC to identify seven metabolites by comparison with reference compounds. The compounds identified were the *N*-oxide (2), ketone (3), 3'-hydroxy (4), 6-hydroxy (5), 7-hydroxy (6), 6,7-dihydroxy (7), and 6,7,3'-trihydroxy (8) metabolites. *In vitro* and *in vivo* procedures were used to confirm that liver catechol-*O*-methyl transferase methylated (7) to give (5). Glutathione depletion (*in vivo*) by diethyl maleate produced an increase in liver covalent binding of metabolites and a decrease in the biliary excretion of benzyl ring-hydroxylated metabolites. An epoxide intermediate of these latter metabolites followed by glutathione conjugation was therefore implicated.

Reference

A. L. Servin, D. Wicek, M. P. Oryszczyn, C. Jacquot, J. P. Lussiana, H. Christinaki, and C. Viel, Metabolism of 6,7-dimethoxy-4-(4'-chlorobenzyl)isoquinoline, II. Role of liver catechol *O*-methyltransferase and glutathione, *Xenobiotica*, 1987, **17,** 1381.

Acenocoumarol

Use:	Anticoagulant
Key functional groups:	Alkyl ketone, coumarin, nitrophenyl
Test system:	Rat (subcutaneous, 3 mg kg^{-1})

Structure and biotransformation pathway

The R-and S-enantiomers of acenocoumarol were separately administered to bile duct-cannulated male rats. Irrespective of the enantiomer given about 50% of the dose was eliminated via bile and 20% via urine within 24 hours. Three metabolites were found both free and conjugated in bile and urine together with some unchanged acenocoumarol. Two of the metabolites were 6- and 7-hydroxyacenocoumarol, (1) and (2) respectively, and these were identified by HPLC. The third metabolite was not identified. Conjugates were identified after hydrolysis with β-glucuronidase/aryl sulphatase.

A greater proportion of the R-isomer (6%) was excreted in bile than the S-isomer ($<1\%$) though similar amounts were excreted in urine (ca. 2%). Both the 6- and 7-hydroxy metabolites and their conjugates were eliminated mainly in bile and accounted for 35–40% and 10–12% of the dose respectively. A greater proportion of the S-isomer of either metabolite was eliminated in a free form compared with the R-enantiomer. The unidentified metabolite was formed in a greater amount from the S-isomer than from the R-form and was also eliminated mainly in bile.

Reference

H. H. W. Thijssen and L. G. M. Baars, The biliary excretion of acenocoumarol in the rat: stereochemical aspects, *J. Pharm. Pharmacol.*, 1987, **39**, 655.

Fenbendazole

Use/occurrence:	Anthelmintic
Key functional groups:	Aryl thioether, benzimidazole, methyl carbamate
Test system:	Goat (oral, 5 mg kg^{-1})

Structure and biotransformation pathway

Analytical methodology using direct insertion probe mass spectrometry (MS) and tandem mass spectrometry (MS–MS) has now been developed to identify and quantify fenbendazole (1) and four of its metabolites (2)–(5) in faecal samples. Although these metabolites had previously been identified in other biological matrices and methods are available for their analysis by HPLC, the novel approach allows a specific, selective, and more rapid determination of these compounds. The mass spectrometric analyses were carried out using EI selective ion monitoring using mebendazole (M/Z 295) as internal standard. The ions monitored were (1) M/Z 299, oxfendazole (2) M/Z 315, fenbendazole sulphone (3) M/Z 331, p-hydroxyfenbendazole (4) M/Z 283, aminofenbendazole (5) M/Z 241. Corrections were made for ions common to more than one metabolite; $e.g.$ M/Z 315 present in the spectra of both (2) and (4). The analytical approach was evaluated on spiked goat faeces samples and then applied to the faeces of six goats treated orally with 5 mg kg^{-1} (1). 44.7% of the dose was excreted as unchanged (1), 1.1% as (2), 0.3% as (3), and 19.4% as (4). No (5) could be detected. The results were com-

347

parable to those obtained by the established HPLC method. MS–MS was also used to reduce background in the selective ion monitoring analyses.

Reference

S. A. Barker, L. C. Hsieh, T. R. McDowell, and C. R. Short, Qualitative and quantitative analysis of the anthelmintic febendazole and its metabolites in biological matrices by direct exposure probe mass spectrometry, *Biomed. Environ. Mass Spectrom.*, 1987, **14**, 161.

Thiabendazole

Use: Anthelmintic drug

Key functional groups: Benzimidazole, thiazole

Test system: Mice (oral, 1300 mg kg^{-1})

Structure and biotransformation pathway

(1) * = $U - {}^{14}C$

[^{14}C]Thiabendazole (1) was administerd orally to mice on day 11 of pregnancy using olive oil or gum arabic as vehicle. Absorption and excretion occurred more rapidly with olive oil, but irrespective of the vehicle 60–62% of the dose was eliminated in urine and 34–37% in faeces over 7 days. The metabolites were isolated and identified using a combination of TLC, HPLC, GC–MS, and NMR. Four major components were present in urine corresponding to the unchanged drug (12–15% of the urinary radioactivity), 5-hydroxythiabendazole (2) (22–24%), and the glucuronide (4) and sulphate (3) conjugates of 5-hydroxythiabendazole (28–29% and 30–31% respectively). Thiabendazole and 5-hydroxythiabendazole were also present in faeces together with more polar metabolites which were not identified. *N*-Methylthiabendazole (5) was detected as a minor metabolite (<0.1%) in both the urine and faeces.

The teratogenic potential of thiabendazole and its metabolites were tested using a mouse limb bud cell culture system. Thiabendazole and its 5-hydroxy metabolite were of equal potency but *N*-methylthiabendazole was less potent. The glucuronide and sulphate conjugates had little activity in this test.

Reference

T. Tsuchiya, A. Tanaka, M. Fukuoka, M. Sato, and T. Yamaha, Metabolism of thiabendazole and teratogenic potential of its metabolites in pregnant mice, *Chem. Pharm. Bull.*, 1987, **35**, 2985.

Astemizole

Use/occurrence:	H$_1$-antagonist
Key functional groups:	Benzimidazole, methoxyphenyl
Test system:	Cultured rat hepatocytes

Structure and biotransformation pathway

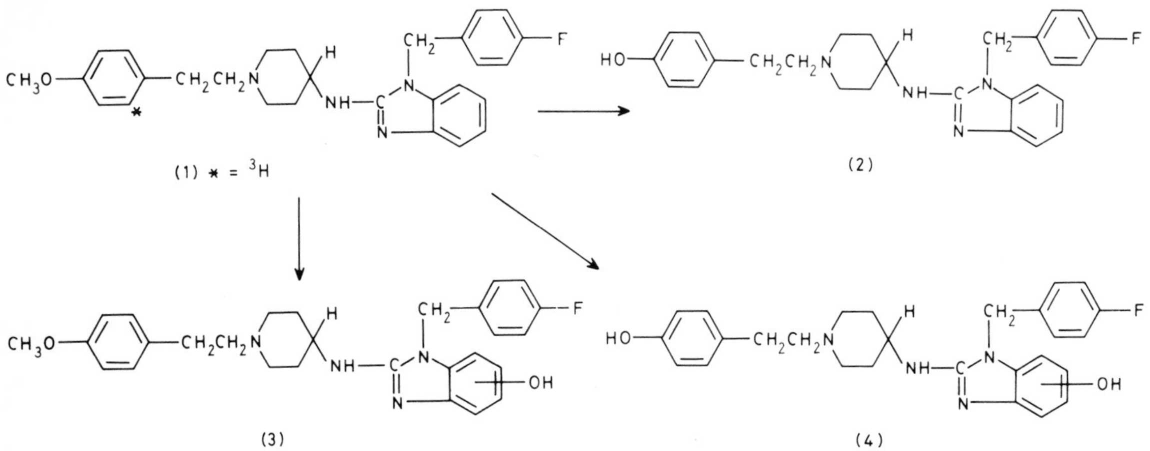

(1) ✻ = ^{3}H

(2)

(3)

(4)

Metabolites produced by incubation of (1) in cultured rat hepatocytes were determined by HPLC. Desmethyl (2), hydroxy (3), and hydroxydesmethyl (4) metabolites were formed. Unaltered (1) was not detectable by 24 hours. At 1 hour the metabolites were present mainly in the unconjugated form whereas after 24 hours the glucuronides predominated. The metabolites produced by hepatocytes in this study correspond to the principal metabolites formed *in vivo* by rat, dog, and man.

Reference

C. Waterkeyn, P. Laduron, W. Meuldermans, A. Trouet, and Y.-J. Schneider, Uptake, subcellular distribution and biotransformation of ^{3}H-labelled astemizole in cultured rat hepatocytes, *Biochem. Pharmacol.*, 1987, **36**, 4129.

Isomazole

Use/occurrence:	Vasodilator
Key functional groups:	Methylsulphinyl
Test system:	Rat (oral, 20 mg kg^{-1})

Structure and biotransformation pathway

Rats were dosed orally with [^{18}O]isomazole (1) (86.7% ^{18}O) and the plasma concentrations of [^{18}O]- and [^{16}O]-isomazole were measured in blood samples taken up to 24 hours after dosing. Samples were analysed by HPLC–MS selected ion monitoring of the isotopic molecular ions. There was a progressive decrease in the proportion of ^{18}O retained in the circulating isomazole from 95% at 30 minutes to 47% at 6 hours. Previous work had shown that the drug is metabolized to the sulphone and sulphide, both contributing to the pharmacological activity. Although the parent drug was the major entity in plasma this study showed that a large proportion of this was produced by recycling of the sulphide metabolite (1).

Reference

J. R. Bernstein and R. B. Franklin, *In vivo* reduction and reoxidation of isomazole: studies with ^{18}O-labelled isomazole, *Med. Sci. Res.*, 1987, **15**, 569.

Benzarone

Use/occurrence:	Thrombolytic agent
Key functional groups:	Arylalkyl, aryl ketone, benzofuran, phenol
Test system:	Rat (oral, 2 mg kg^{-1}), dog (oral, 0.5 mg kg^{-1}), man (oral, 100 mg)

Structure and biotransformation pathway

In rat and dog only 14% and 7% of an oral dose was excreted in the urine but most of the remaining material was absorbed and excreted in bile. In man more than 70% of the dose was eliminated in urine. Components in human urine were present as conjugates (mainly glucuronides). The two major aglycones were the hydroxylated metabolites (2) and (5) which represented 26% and 8% dose respectively. Small amounts (2–3% dose) of benzarone and metabolites (3) and (4) were also detected. The same conjugated components were also present in human plasma and benzarone and metabolites (2) and (3) were the major components in faeces. Metabolites were identified by MS and NMR where necessary. Metabolism was apparently much less extensive in the rat and dog with benzarone (free and conjugated) representing more than 70% of the total material in bile, urine, and faecal extracts from dogs and in bile from rats.

Reference

S. G. Wood, B. A. John, L. F. Chasseaud, R. Bonn, H. Grote, K. Sandrock, A. Darragh, and R. F. Lambe, Metabolic fate of the thrombolytic agent benzarone in man: comparison with rat and dog, *Xenobiotica*, 1987, **17,** 881.

Hydralazine

Use/occurrence:	Antihypertensive drug
Key functional groups:	Hydrazine, phthalazine, pyridazine
Test system:	Rat liver microsomes

Structure and biotransformation pathway

(1)

(2)

(3)

(4)

(5)

In the presence of NADPH, hydralazine was metabolized by rat liver microsomes to three major oxidation products, phthalazine (3), phthalazinone (4), and a dimer (5). Under similar conditions radioactivity derived from [^{14}C]hydralazine was found to be covalently bound to microsomes. Metabolite formation and covalent binding increased following pretreatment with phenobarbital. In contrast pretreatment with 3-methylcholanthrene or piperonyl butoxide slightly decreased both metabolite formation and covalent binding. Metabolites were determined by HPLC assay. In addition ESR studies showed the formation of nitrogen radicals under similar conditions to those required for covalent binding and metabolite production. These observations support the hypothesis that the oxidation of hydralazine generates reactive metabolites, which may contribute to the side effects seen in individuals with a low capacity for *N*-acetylation on hydralazine therapy.

Reference

L. B. LaCagnin, H. D. Colby, N. S. Dalal, and J. P. O'Donnell, Metabolic activation of hydralazine by rat liver microsomes, *Biochem. Pharmacol.*, 1987, **36**, 2667.

1-Aminophthalazine, 1-Chlorophthalazine

Use/occurrence:	Model compounds
Key functional groups:	Phthalazine, pyridazine
Test system:	Partially purified rabbit liver aldehyde oxidase, bovine milk xanthine oxidase

Structure and biotransformation pathway

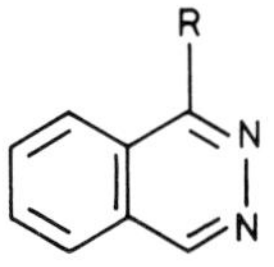

(1) R = NH$_2$ or Cl

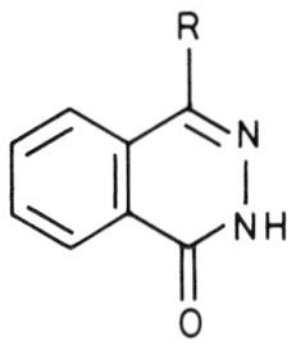

(2)

1-Aminophthalazine, (1; R = NH$_2$) was found to be a substrate for aldehyde oxidase but not bovine milk xanthine oxidase for which it was a competitive inhibitor. The single product of metabolism by aldehyde oxidase, 4-amino-1-(2H)-phthalazinone (2; R = NH$_2$) was identified by IR and MS but not quantified.

1-Chlorophthalazine (1; R = Cl) was a substrate for both aldehyde oxidase and bovine milk xanthine oxidase and also gave the phthalazinone (2; R = Cl) as the only metabolite.

Reference

C. Johnson, C. Beedham, and J. G. P. Stell, Reaction of 1-amino- and 1-chlorophthalazine with mammalian molybdenum hydroxylases *in vitro*, *Xenobiotica*, 1987, **17**, 17.

1-Methylxanthine, 3-Methylxanthine, 7-Methylxanthin

Use/occurrence:	Model compounds and metabolites of caffeine, theobromine, and theophylline
Key functional groups:	*N*-Methylpurine, xanthine
Test system:	Rat (intraperitoneal, serial doses of 10, 22, 68, 78 mg kg^{-1} at 20 minute intervals), *in vitro* in perfused rat liver

Structure and biotransformation pathway

1-Methylxanthine (R^1 = CH_3, R^2, R^3 = H)
3-Methylxanthine (R^2 = CH_3, R^1, R^3 = H)
7-Methylxanthine (R^3 = CH_3, R^1, R^2 = H)

(1)

In vivo 1-methylxanthine was rapidly metabolized to the corresponding uric acid (1). 7-Methylxanthine was also metabolized to its corresponding uric acid (1) at a lower rate, but the corresponding uric acid of 3-methylxanthine was not found. All three monomethylxanthines are active adenosine antagonists *in vitro*. Both 3-methylxanthine and 7-methylxanthine were shown to be active *in vivo*, whereas 1-methylxanthine was inactive, presumably due to its rapid metabolism. The major metabolite of 1-methylxanthine in perfused rat liver was shown to be 1-methyluric acid (1) and xanthine oxidase was shown to be the enzyme responsible for this metabolism. Methylurates and other potential metabolites were assayed by HPLC.

Reference

L. A. Reinke, M. Nakamura, L. Logan, H. D. Christensen, and J. M. Carney, *In vivo* and *in vitro* 1-methylxanthine metabolism in the rat, *Drug Metab. Dispos.*, 1987, **15**, 295.

Theobromine

Use/occurrence:	Stimulant, naturally occurring in cocoa
Key functional groups:	*N*-Methylpurine, xanthine
Test system:	Rat (oral, 5 mg kg^{-1}), *in vitro* rat liver cell fractions

Structure and biotransformation pathway

3,7-Dimethyluric acid (4) was the major metabolite of theobromine (1) produced by rat liver microsomes with higher rates of metabolism in both 3-methylcholanthrene and phenobarbital induced microsomes. Monoclonal antibodies to rat cytochrome P-450 inhibited this metabolism. Addition of cytosol or physiological concentrations of glutathione resulted in a dose dependent conversion of (4) into the ring-opened derivative, 6-amino-5-(*N*-methylformylamino)-1-methyluracil (6). Cytosol alone was catalytically inert and it was suggested that these results provided a mechanistic basis for the occurrence of (6) as the major urinary metabolite.

The major urinary metabolite after oral dosing was (6) with lesser amounts of (4), 3-methylxanthine (2), 7-methylxanthine (5), 7-methyluric acid (8), and traces of dimethylallantoin (7). Metabolites were measured by HPLC with use of authentic standards.

Reference

C. A. Shively and E. S. Vesell, *In vivo* and *in vitro* biotransformation of theobromine by phenobarbital- and 3-methylcholanthrene-inducible cytochrome P-450 monooxygenases in rat liver, *Drug Metab. Dispos.*, 1987, **15**, 217.

Caffeine

Use/occurrence:	Stimulant, naturally occurring in tea and coffee
Key functional groups:	*N*-Methylpurine, xanthine
Test system:	Rat perfused liver

Structure and biotransformation pathway

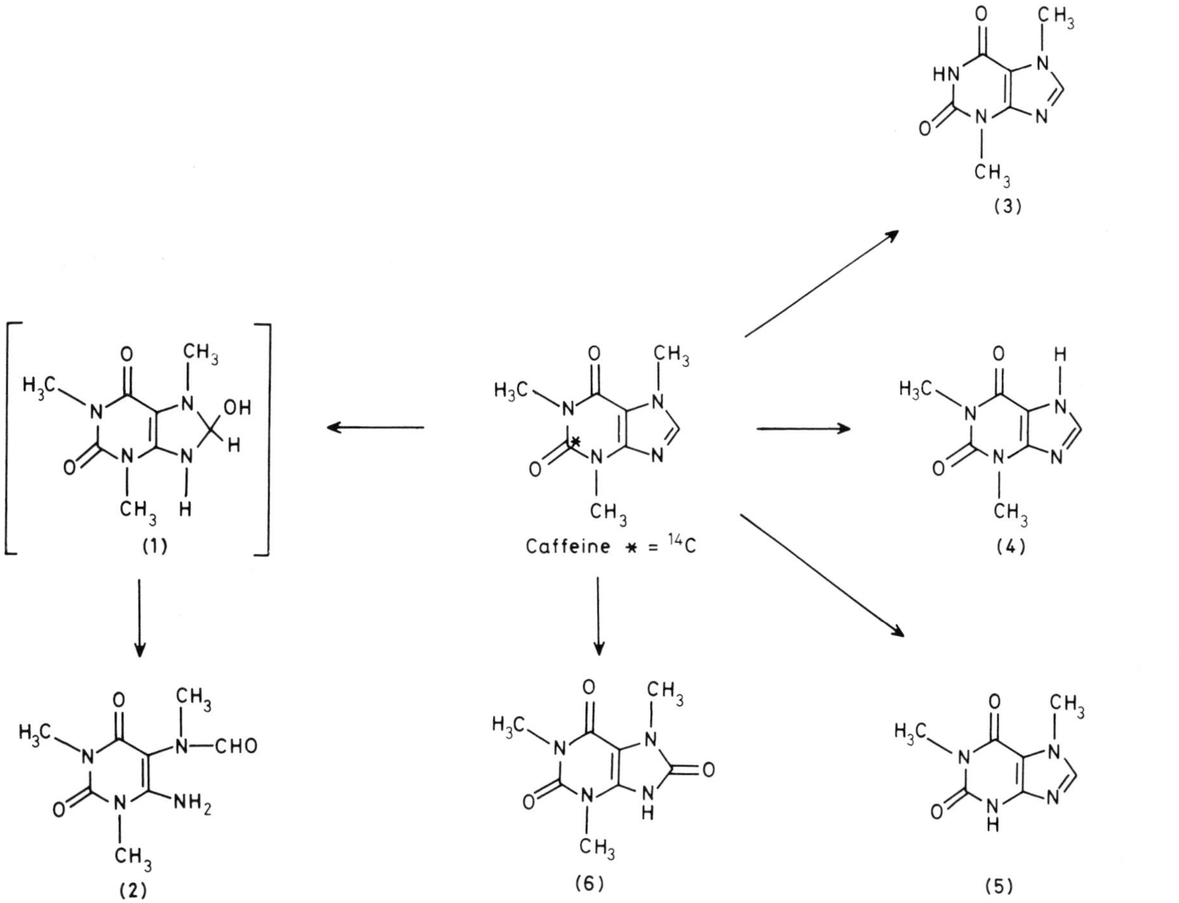

Caffeine * = ^{14}C

(1) (2) (3) (4) (5) (6)

The metabolism of caffeine to 6-amino-5-(*N*-methylformylamino)-1,3-dimethyluracil (2) was studied in the isolated perfused liver. Metabolite (2) was confirmed by MS. The kinetics of caffeine metabolism during 2 hours were measured in livers from control rats and from rats pretreated with phenobarbital, β-naphthoflavone, or 3-methylcholanthrene. Caffeine and (2) were determined by TLC under conditions where other metabolites (3)–(6) were well separated. Phenobarbital did not modify the rate of caffeine elimination or the extent of formation of (2). In contrast, there was a highly significant inducing effect on both caffeine elimination and formation of metabolite (2) in perfusions of livers from β-naphthoflavone and 3-methylcholanthrene pretreated rats.

Reference

A. Guaitani, R. Abbruzi, A. Bastone, M. Bianchi, M. Bonati, P. Catalani, R. Latini, C. Pantarotto, and K. Szczawinska, Metabolism of caffeine to 6-amino-5-(*N*-methylformylamino)-1,3-dimethyluracil in the isolated, perfused liver from control or phenobarbital-, β-naphthoflavone-, and 3-methylcholanthrene pretreated rats, *Toxicol. Lett.*, 1987, **38**, 55.

Caffeine, Paraxanthine, Theobromine, Theophylline

Use/occurrence:	Stimulants naturally occurring in chocolate, coffee, and tea
Key functional groups:	*N*-Methylpurine, xanthine
Test system:	Human, mouse, and rat liver microsomes

Structure and biotransformation pathway

(1)

(2)

(4)

(3)

The *in vitro* metabolisms of caffeine, paraxanthine, theophylline, and theobromine were found to be similar in respect of the NADPH dependence and response to P-450 inhibitors. There were, however, differences in the kinetic parameters for metabolism of the different substrates. At least two distinct enzymes with different substrate affinities had the potential to effect *N*-demethylation (4) and C-8 oxidation (2) and (3) *in vitro*, shown for paraxanthine (1). The high affinity enzyme responsible for the demethylations was identified as a polycyclic hydrocarbon inducible isozyme of cytochrome P-450. It was suggested that *in vivo* the activity of this isozyme would predominate owing to the low substrate concentrations. A possible *in vivo* test for the activity of the inducible P-450 isozyme was suggested based on a measure of the ratio of paraxanthine 7-demethylation to 8-hydroxylation products in urine after caffeine intake. The methylxanthines were measured using an HPLC assay.

Reference

M. E. Campbell, D. M. Grant, T. Inaba, and W. Kalow, Biotransformation of caffeine, paraxanthine, theophylline, and theobromine by polycyclic aromatic hydrocarbon inducible cytochrome(s) P-450 in human liver microsomes, *Drug Metab. Dispos.*, 1987, **15**, 237.

Caffeine

Use/occurrence:	Stimulant, naturally occurring in tea and coffee
Key functional groups:	N-Methylpurine, xanthine
Test system:	Human liver microsomes

Structure and biotransformation pathway

A lower rate of caffeine metabolism was found *in vitro* compared with that predicted from *in vivo* studies. The *in vitro* production of metabolites (1)–(4) was found to involve cytochrome P-450. Total percentage conversion of caffeine to its metabolites never exceeded 1% at a substrate concentration of 1 mM for any of the livers tested. The relatively high concentrations used *in vitro* were thought to be responsible for the relatively higher rate of production of (1) than predicted from *in vivo* studies.

Kinetic studies indicated that the formation of each of the N-demethylated metabolites (2)–(4) was mediated by both a high and a low affinity catalytic site over a concentration range of 0.05–80 mM. Only the high affinity component was thought to be important at concentrations encountered *in vivo*. The high affinity component was inhibited by the isozyme selective P-450 inhibitor α-naphthoflavone, while 1-phenylimidazole was found to inhibit the low affinity component.

Reference

D. M. Grant, M. E. Campbell, B. K. Tang, and W. Kalow, Biotransformation of caffeine by microsomes from human liver, Kinetics and inhibition studies, *Biochem. Pharmacol.*, 1987, **36**, 1251.

10-Ethyl-10-deaza-aminopterin

Use/occurrence:	Anti-cancer agent
Key functional groups:	Amino acid, aryl carboxamide, pteridine
Test system:	Rat (intravenous, 50 mg kg^{-1}), dog (intravenous, 0.25–5 mg kg^{-1})

Structure and biotransformation pathway

(1)

(2)

(3)

10-Ethyl-10-deaza-aminopterin (1) is an experimental anti-folate anti-tumour agent, which shows greater accumulation in tumour cells than that of the clinically used analogue methotrexate. Following intravenous administration of [^{3}H]-(1) to rats the drug was eliminated in three phases ($t_{1/2}$ 6.3 min) 0.8 hours, 18.5 hours), with 18.5% of the radioactivity being excreted in the urine and 66.4% in the faeces in 72 hours. HPLC analysis showed that faecal radioactivity was accounted for by 23% (1), 63% deglutamate metabolite (2), and 14% 7-hydroxy-10-ethyl-10-deaza-aminopterin (3). Polyglutamates of (1) were detected in kidney, liver, and small intestine by an HPLC-fluorescence assay. In the dog (1) was also eliminated in three phases (terminal $t_{1/2}$ 9.1 hours). In the urine there was less than 5% of the activity as (2) and no (3) could be detected.

Reference

M. P. Fanucchi, J. J. Kinahan, L. L. Samuels, C. Hancock, T.-C. Chou, D. Niedzwiecki, F. Farag, P. M. Vidal, J.-I. DeGraw, S. S. Sternberg, F. M. Sirotnak, and C. W. Leung, Toxicity, elimination, and metabolism of 10-ethyl-10-deaza-aminopterin in rats and dogs, *Cancer Res.*, 1987, **37**, 2334.

Methotrexate (MTX), γ-t-Butylmethotrexate (TBM)

Use/occurrence:	Anti-cancer drugs (folic acid antagonists)
Key functional groups:	Amino acid, aryl carboxamide, pteridine
Test system:	Human leukaemic cell lines in culture and rabbit hepatic aldehyde oxidase

Structure and biotransformation pathway

(1) X = H or $(CH_3)_3C$; * = ^3H

(2)

The rabbit hepatic aldehyde oxidase oxidized methotrexate (1; X = H) to its 7-hydroxy metabolite (2; X = H). The K_m of the aldehyde oxidase was 8-fold lower with TBM [1; X = $(CH_3)_3C$] as substrate compared with that with MTX as substrate. The V_{max} with TBM as substrate was also 8-fold greater than that with MTX as substrate. In cultured cells TBM was not metabolized, whereas MTX was polyglutamated at the γ-carboxy group [1; X = $(glu)_n$]. In addition TBM appeared to be effective in MTX resistant cell lines. Metabolites were identified by HPLC co-chromatography with standards.

Reference

J. E. Wright, A., Rosowsky, D. J. Waxman, D. Trites, C. A. Cucchi, J. Flatlow, and E. Frei, III, Metabolism of methotrexate and γ-t-butylmethotrexate by human leukemic cells in culture and by hepatic aldehyde oxidase *in vitro*, *Biochem. Pharmacol.*, 1987, **36**, 2209.

Nabilone

Use/occurrence:	Anti-emetic
Key functional groups:	Cyclohexanone, prochiral carbon
Test system:	Rat, dog, and monkey S-9 liver homogenates

Structure and biotransformation pathway

Nabilone is a racemic mixture of (1) and (2)

The metabolism of R,R-(1) and S,S-(2) ketones and the racemic mixture was investigated in liver homogenates with added NADP and NADPH as co-factors. The R,R,S-(S,S,R) (3) and S,S,S-(R,R,R) (4) carbinols were determined by GC–MS. In rat and dog liver homogenates incubation of (2) produced (4) only: no (3) was detected. However, (1) gave rise to mixtures of (3) and (4). Both (3) and (4) were formed from the racemic mixture but the amounts of (4) were higher. In contrast to rat and dog, the formation of carbinols by monkey liver homogenate was stereoselective from each ketone enantiomer, with over 80% being in the configuration of (4). Yields of carbinols were considerably greater in liver preparations from monkey than from the other two species. Accumulation of carbinol metabolites in brain has been implicated as the causative factor for CNS toxicity observed in dogs. Comparative pharmacokinetic studies in dog, human, and monkey have shown that the production of carbinol metabolites is unique to dog. On this basis, the monkey has been proposed as a more appropriate model than dog for toxico-

365

logical studies. The finding in the present study that carbinol production *in vitro* is greater for monkey than for dog suggests that the mechanism underlying the species dependent difference in accumulation of carbinol metabolites lies in their subsequent elimination rather than their formation.

Reference

H. R. Sullivan, G. K. Hanasono, W. M. Miller, and P. G. Wood, Species specificity in the metabolism of nabilone. Relationship between toxicity and metabolic routes, *Xenobiotica*, 1987, **17**, 459.

Harmine

Use/occurrence:	Model compound
Key functional groups:	Methoxyphenyl, pyridine, pyrrole
Test system:	Mouse liver microsomes

Structure and biotransformation pathway

(2)   (1)   (4)

(3)

Previous work had shown that metabolism of harmine (1) in control or phenobarbitone-induced microsomes proceeded via *O*-demethylation to yield harmol (4). In 3-methylcholanthrene induced microsomes formation of the novel metabolite (2) was the major pathway at an initial rate of 11 nmol min^{-1} mg^{-1} microsomal protein. A further novel metabolite (3) was produced at an initial rate of 3.8 nmol min^{-1} mg^{-1} protein, which was similar to the rate of formation of the phenol (4). Compounds (1) and (2) were characterized by NMR and MS, but the evidence did not allow the precise position of the hydroxy group in (2) to be assigned.

Reference

D. J. Tweedie and M. D. Burke, Metabolism of the β-carbolines, harmine and harmol, by liver microsomes from phenobarbitone- or 3-methylcholanthrene-treated mice, *Drug Metab. Dispos.*, 1987, **15**, 74.

Stobadine

Use/occurrence:	Anti-arrhythmic
Key functional groups:	Chiral carbon, methylindole, *N*-methyl-piperidine
Test system:	Rat liver microsomes

Structure and biotransformation pathway

(2)   (1)   (3)

   The biotransformation of [³H]stobadine (1) was investigated in rat liver microsomes. Stobadine was metabolized to *N*-desmethylstobadine (2) and two optically active stereoisomers of stobadine *N*-oxide (3). The metabolites were separated by TLC and their spectra compared with those of authentic compounds obtained by GC–MS.

Reference

M. Stefek, L. Benes, M. Jerglova, V. Scasnar, L. Turi-Nagy, and P. Kocis, Biotransformation of stobadine, a γ-carboline anti-arrhythmic and cardioprotective agent in rat liver microsomes, *Xenobiotica*, 1987, **17**, 1067.

Pentachlorodibenzofurans

Use/occurrence:	Environmental contaminants
Key functional groups:	Chlorophenyl, dibenzofuran
Test system:	Rat [oral, 0.36 mg, (1); 0.25 mg, (2); 1 mg (3)]

Structure and biotransformation pathway

	R^1	R^2	R^3
(1)	H	Cl	Cl
(2)	Cl	H	Cl
(3)	Cl	Cl	H

The metabolism of 2,3,4,7,8- (1), 1,2,3,7,8- (2), and 1,2,3,4,8- (3) penta-chlorodibenzofurans (penta-CDFs) was investigated in bile duct-cannulated rats. Bile was treated with enzyme or acid to hydrolyse possible conjugates, the metabolites were purified by solvent extraction and TLC, and phenolic compounds were identified as methoxy derivatives by GC–MS. All three compounds were excreted unchanged in bile but in small quantities and bio-transformation was presumed to be rate-limiting for elimination. Compound (3) was metabolized more effectively than the other two compounds, which is

369

consistent with the belief that the availability of vicinal unsubstituted carbon atoms facilitates metabolism via arene oxide intermediates. Dimethoxy-penta-CDF (4), thought to be the 6,7-dimethoxy isomer, was formed from (3) but not from the other two compounds. Cleavage of the ether bond to give the dimethoxypentachlorobiphenyl (5) was observed as a major metabolic route for (1). Small amounts of a methylthio-tetra-CDF (6) were formed from (1) but not from (2) or (3). Methoxy-penta-CDF (7) was formed from each penta-CDF and was a major metabolite of (1) and (3). Reductive dechlorination of (7) to produce methoxy-tetra-CDF (8) was observed for each compound but hydrolytic dechlorination of (8) to form dimethoxy-tri-CDF (10) also occurred with (1) and (2) only.

Reference

N. Pluess, H. Poiger, C. Schlatter, and H. R. Buser, The metabolism of some penta-chlorodibenzofurans in the rat, *Xenobiotica*, 1987, **17**, 209.

Etodolac

Use/occurrence:	Non-steroidal anti-inflammatory
Key functional groups:	Alkyl carboxylic acid, dihydropyran, indole
Test system:	Mouse (oral, 15 mg kg^{-1}), dog (oral, 50 mg kg^{-1}), rat (oral, 60 mg kg^{-1}), man (oral, 200 mg)

Structure and biotransformation pathway

(1) * = ^{14}C

(2)

Following single doses of [^{14}C]etodolac (1), urinary excretion was extensive in mouse (53% dose in 24 hours) and man (66% dose in 60 hours), but limited in rat and dog to 11% and 6% dose respectively after 72 hours. The ureide metabolite (2) was detected, as its methyl ester, in methylated extracts of urine of all four species. Identification was by MS, ^{1}H NMR, and HPLC comparison with a synthetic standard. The diastereomers of (2), which were well separated by HPLC, were formed in equal quantities. The metabolite (2) was relatively minor in all four species; the proportion of the dose associated with it ranged from 0.5% (dog) up to 6% (rat).

Reference

E. S. Ferdinandi, D. Cochran, and R. Gedamke, Identification of the etodolac metabolite 4-ureidoetodolac, in mouse, rat, dog, and man, *Drug Metab. Dispos.*, 1987, **15**, 921.

8-Methoxypsoralen

Use/occurrence:	Antipsoriasis drug
Key functional groups:	Furan, lactone, methoxyphenyl
Test system:	Rat (intravenous, 10 mg kg^{-1}), rat liver microsomes, rat liver 9000g supernatant

Structure and biotransformation pathway

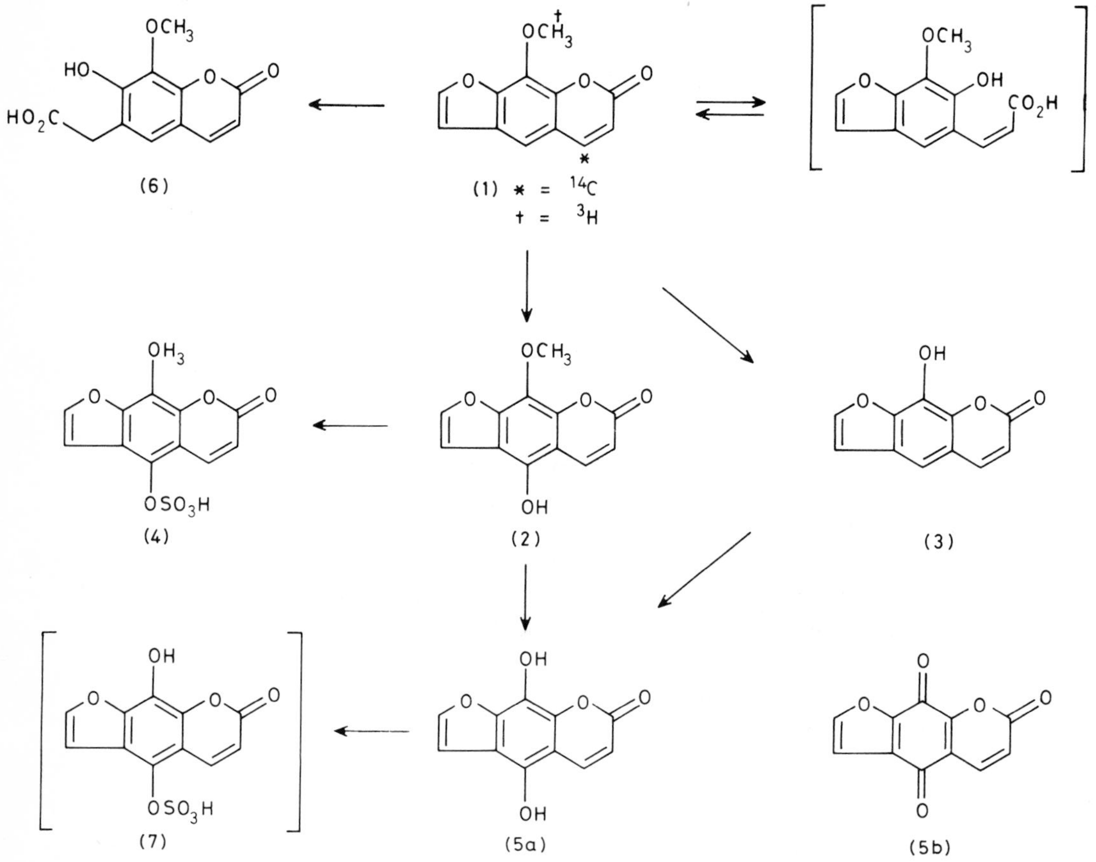

The exact position of the sulphate group in (7) is not known; it is shown here at the 5-position

Thirteen separated radioactive components were detected by HPLC in rat urine after intravenous administration of 8-methoxypsoralen. Compounds (1), (2), (3), (5), and (6) had been identified in a previous study. In the study under review further rats were co-administered [^{35}S]Na$_2$SO$_4$ and [^{3}H]-8-MOP and co-chromatography of the ^{35}S and ^{3}H radiolabels was used to provide evidence for the presence of sulphate conjugates. Metabolite (4) was identi-

372

fied as the sulphate conjugate of (2) by showing incorporation of ^{35}S and by thermospray LC–MS and FAB–MS. (7) was tentatively identified as a sulphate conjugate of (5a) on the basis of ^{35}S incorporation and its chemical decomposition to a mixture of (5a) and (5b) during isolation. Incorporation of ^{35}S indicated that a further component was also a sulphate conjugate but no further information was obtained on its structure. Five more polar metabolites were unidentified. The major metabolites (as % urinary radio-activity) were (4) (10%), (8) (15%), (7) (17%), and an unknown (10%). The proportion of (4) increased to 40% in phenobarbital pretreated rats, while pretreatment with β-naphthoflavone or (1) increased the proportion of (7) to 45% and 37% respectively. Rates of metabolism of (1) were measured in *in vitro* systems but no individual products were quantified or identified. It was suggested that (1) may be metabolized to a reactive product capable of binding to tissue macromolecules.

Reference

D. C. Mays, S. G. Hecht, S. E. Unger, C. M. Pacula, J. M. Climie, D. E. Sharp, and N. Gerber, Disposition of 8-methoxypsoralen in the rat, *Drug Metab. Dispos.*, 1987, **15**, 318.

Quinpirole

Use/occurrence: Antihypertensive drug

Key functional groups: Alkylamino, piperidine, pyrazole, tetrahydroindazole

Test system: Mouse, rat, rhesus monkey (oral, 2 mg kg^{-1}), dog (intravenous, 0.2 mg kg^{-1})

Structure and biotransformation pathway

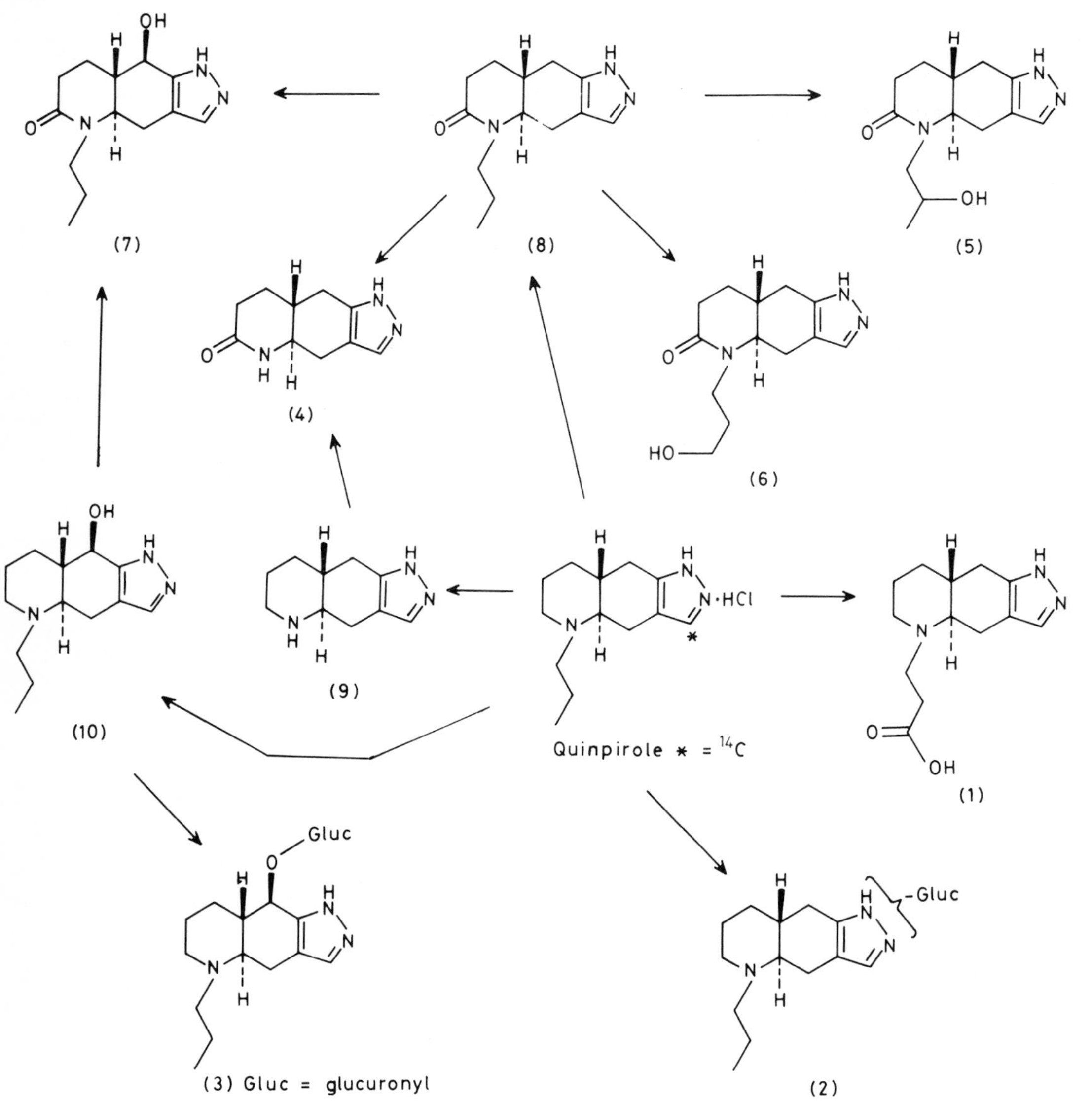

(7)

(8)

(5)

(4)

(6)

(10)

(9)

Quinpirole $* = {}^{14}C$

(1)

(3) Gluc = glucuronyl

(2)

374

Rats, mice, and monkeys excreted means of 65%, 85%, and 96% of a single oral dose via the urine during 0–48 hours. Dogs excreted a mean of 94% of an intravenous dose via the urine within 72 hours. Metabolites (1)–(10) were identified by MS, NMR, and IR and by comparison with synthetic standards in some cases. Unchanged quinpirole accounted for 57%, 43%, 3%, and 13% urinary radioactivity (UR) in rat, mouse, monkey, and dog respectively. The des-N-propyl compound (9) was important (10–44% UR) in all four species. The conjugated hydroxy compound (3) was the major metabolite in the dog (30% UR), followed by (9) and (10) (10% UR), but only small amounts of (3) and (10) were detected in the urine of other species. The majority of quinpirole metabolites in mouse urine contained the lactam grouping, $i.e.$ (4)–(8).

Reference

N. G. Gallick Whitaker and T. D. Lindstrom, Disposition and biotransformation of quinpirole, a new D-2 dopamine agonist antihypertensive agent, in mice, rats, dogs, and monkeys, *Drug Metab. Dispos.*, 1987, **15**, 107.

Amonafide

Use/occurrence:	Anti-cancer agent
Key functional groups:	Aryl amine, *N*-alkyl imide, dimethylaminoalkyl
Test system:	Human patients (intravenous, 400 mg m^{-2} day^{-1})

Structure and biotransformation pathway

* Position of hydroxy group uncertain

During 24 hours after dosing (30 minute infusion) renal excretion of unchanged amonafide (1) ranged from 5% to 45% dose in seven patients (mean 22%). Thermospray LC–MS was used to identify or partially identify eight metabolites of (1) in urine. Synthetic standards were available for metabolites (2) and (3) only. The MS data obtained did not allow the positions of hydroxylation in metabolites (7), (8), and (9) to be assigned. In addition to hydroxylation, routes of metabolism involved *N*-oxidation and *N*-demethylation of the aliphatic amine moiety, acetylation of the arylamine group, and formation of a glucuronide conjugate of the parent. Characterization of this conjugate was based on LC–MS data of the intact conjugate. Acetylated metabolites were absent from the urine of two of the patients. In an *in vitro* test the *N*-acetyl metabolite (3) was only slightly less cytotoxic than the parent compound. The *N*-oxide metabolite (2) was inactive in this test, demonstrating the importance of the dimethylaminoethyl group for anti-tumour activity.

Reference

T. B. Felder, M. A. McLean, M. L. Vestal, K. Lu, D. Farquhar, S. S. Legha, R. Shah, and R. A. Newman, Pharmacokinetics and metabolism of the anti-tumour drug amonafide (NSC-308847) in humans, *Drug Metab. Dispos.*, 1987, **15**, 773.

2-Amino-3-methylimidazo[4,5-*f*]quinoline (IQ), 2-Amino-3,4-dimethylimidazo-[4,5-*f*]quinoline (MeIQ)

Use/occurrence:	Food mutagens
Key functional groups:	Aminoimidazole, *N*-methylimidazole
Test system:	Rat (intraperitoneal, 5 mg kg^{-1}), 100 000g rat liver supernatant

---

Structure and biotransformation pathway

(1) R = H
(2) R = CH$_3$

(3) R = H
(4) R = CH$_3$

The aminoimidazoazaarenes 2-amino-3-methylimidazo[4,5-*f*]quinoline (IQ) (1) and 2-amino-3,4-dimethylimidazo[4,5-*f*]quinoline (MeIQ) (2) are examples of the group of highly mutagenic amines that have been isolated from cooked meat. Although it is known that radioactivity is virtually quantitatively excreted following administration of [^{14}C]-(1) or -(2), only 1.3% of the dose of (1) and 5.1% of (2) could be extracted from the urine and faeces into dichloromethane in this experiment. Metabolites in these extracts were characterized by low resolution MS after purification by TLC and HPLC. The urine of rats treated with (1) contained 0.6% of the dose as compound (1) and 0.14% as the *N*-acetylated metabolite (3). Rats treated with (2) excreted 0.93% of (2) and 0.37% of the *N*-acetylated metabolite (4) in their urine. The faeces of rats treated with (1) contained 0.56% of (1) but no (3), whereas treatment with (2) yielded 2.7% (2) and 1.1% (4). The bile contained both (2) and (4), however. The fluorescent spectra of (1)–(4) were also reported. *In vitro* studies demonstrated the presence of the *N*-acetylating enzyme in 100 000g rat liver homogenate. Rat liver cytosol deacetylated (3) and (4) to (1) and (2) respectively.

Reference

F. C. Størmer, J. Alexander, and G. Becher, Fluorometric detection of 2-amino-3-methylimidazo[4,5-*f*]quinoline, 2-amino-3,4-dimethylimidazo[4,5-*f*]quinoline, and their *N*-acetylated metabolites excreted by the rat, *Carcinogenesis*, 1987, **8**, 1277.

2-Amino-3,8-dimethylimidazo[4,5-*f*]-quinoxaline (MeIQx)

Use/occurrence:	Mutagen in cooked food
Key functional groups:	Aminoimidazole, *N*-methylimidazole, methylquinoxaline
Test system:	Mouse (intraperitoneal, 20 mg kg^{-1})

Structure and biotransformation pathway

$*$ = ^{14}C or ^{13}C
$\dagger$ = ^{15}N

After an intraperitoneal dose of MeIQx to mice about 24% was excreted in the urine during the first 24 hours with about 40% in the faeces. Four major radiolabelled components were detected by HPLC, one of which corresponded to the parent compound (9% dose). The three other more polar components were analysed by LC–MS. To assist with the detection and identification of metabolites a [^{13}C,^{15}N$_2$]-labelled analogue of MeIQx was used. One metabolite (5% dose) was assigned as a glucuronide of hydroxylated MeIQx although the position of hydroxylation was unknown. Mass spectra of the other two metabolites were similar to that of MeIQx, and some indication was obtained that one was an *N*-glucuronide of MeIQx whereas the other was uncharacterized.

Reference

N. J. Gooderham, D. Watson, J. C. Rice, S. Murray, G. W. Taylor, and D. S. Davies, Metabolism of the mutagen MeIQx *in vivo*: metabolite screening by liquid chromatography-thermospray mass spectrometry, *Biochem. Biophys. Res. Commun.*, 1987, **148**, 1377.

2-Amino-3,8-dimethylimidazo[4,5-*f*]quinoxaline (MeIQx)

Use/occurrence:	Mutagen in cooked meat
Key functional groups:	Aminoimidazole, *N*-methylimidazole, methylquinoxaline
Test system:	Rat (oral), rat liver S9 fraction

Structure and biotransformation pathway

MeIQx (1) is a mutagen found in cooked beef, which causes hepatocellular carcinomas in mice. In the present study rats were fed (1) and their urine and faeces examined for mutagenic metabolites. Three such metabolites were identified in urine: 2-amino-8-hydroxymethyl-3-methylimidazo[4,5-*f*]quinoxaline (2), 2-acetylamino-3,8-dimethylimidazo[4,5-*f*]quinoxaline (3), and 2-amino-8-methylimidazo[4,5-*f*]quinoxaline (4). Metabolites (2) and (4) were also found in faeces. Compounds (2) and (3) were as mutagenic as the parent compound in the Ames test, but (4) had much lower activity. In the presence of rat liver S9 (1) could be converted into (2) *in vitro*. Only a small proportion of (1) was converted into mutagenic metabolites and only ~ 5% was excreted unchanged in urine and faeces.

Reference

H. Hayatsu, H. Kasai, S. Yokoyama, T. Miyazawa, Z. Yamaizumi, S. Sato, S. Nishimura, S. Arimoto, T. Hayatsu, and Y. Ohara, Mutagenic metabolites in urine and faeces of rats fed with 2-amino-3,8-dimethylimidazo[4,5-*f*]quinoxaline, a carcinogenic mutagen present in cooked meat, *Cancer Res.*, 1987, **47**, 791.

2-Amino-3-methylimidazo[4,5-f]quinoline, 2-Amino-3,4-dimethylimidazo[4,5-f]quinoline

Use/occurrence:	Pyrolysis product of amino acids
Key functional groups:	Aminoimidazole, N-methylimidazole
Test system:	Rat liver hepatocytes

Structure and biotransformation pathway

(1) R = H
(4) R = CH$_3$

(2) R = H
(5) R = CH$_3$

(3) R = H
(6) R = CH$_3$

2-Acetylamino-3-methylimidazo[4,5-f]quinoline (2) was the major ethyl acetate-extractable metabolite found after incubation of the mutagen [2-^{14}C]-2-amino-3-methylimidazo[4,5-f]quinoline (1) (50 μM) with rat liver hepatocytes. The metabolite accounted for 2.4% of the parent substance added to the hepatocytes and was identified by HPLC, TLC, and UV comparison with a reference standard. A larger proportion of the radioactivity (36%) was converted into water-soluble metabolites. Attempted hydrolysis of these with either β-glucuronidase or aryl sulphatase released only small amounts (<5%) of the radioactivity as the parent (1) and the acetylated metabolite (2) together with an unidentified metabolite. Acid hydrolysis, however, converted 55% of the water-soluble radioactivity into (1), and as a result it was concluded that the sulphamate (3) was probably a metabolite of (1). An unidentified metabolite (1.6% of the water-soluble radioactivity) was also formed by acid hydrolysis. About 65% of the parent was metabolized by the hepatocytes. Qualitatively similar results were obtained using [2-^{14}C]-2-amino-3,4-dimethylimidazole[4,5-f]quinoline (4). The N-acetyl metabolite (5) was formed to a greater extent (9%), less of the radioactivity appeared in the water-soluble fraction (16%), and only 40% of the parent was metabolized. A smaller proportion (20%) of the water-soluble radioactivity was converted into the parent by acid hydrolysis. This same metabolite and the parent (5% of each) were also produced by treatment of the water-soluble fraction with β-glucuronidase. With both compounds there was some covalent binding of radioactivity to macromolecules and this was greater for the 3,4-dimethyl compound (4).

Reference

J. A. Holme, J. Alexander, G. Becher, and B. Trygg, Metabolism of 2-amino-3-methylimidazo[4,5-*f*]quinoline and 2-amino-3,4-dimethylimidazo[4,5-*f*]quinoline in suspensions of isolated rat-liver cells, *Toxicol. in vitro*, 1987, **1**, 175.

Functional Nitrogen Compounds

Hydrazine, Isoniazid, Iproniazid

Use/occurrence:	Antitubercular, antidepressant
Key functional groups:	Hydrazide, hydrazine
Test system:	Perfused rat liver

Structure and biotransformation pathway

Spin-trapping techniques were used to investigate formation of free radicals from bioactivation of hydrazine derivatives in perfused rat liver. The liver perfusion was carried out in the presence of phenyl-t-butylnitrone and ESR spectra were recorded to detect free-radical adducts in samples of perfusate taken at different times. Hydrazine (1), acetylhydrazine (2), and ioniazid (3) all formed the same adduct, which was confirmed as the acetyl radical by spectroscopic comparison with the authentic adduct produced chemically. Iproniazid similarly resulted in formation of the isopropyl radical. The results demonstrate that the hydrazine group was rapidly acetylated *in vivo* and results in formation of acetyl radicals which could be responsible for cellular damage known to occur with this class of compound.

Reference

B. K. Sinha, Activation of hydrazine derivatives to free radicals in the perfused rat liver: a spin-trapping study, *Biochim. Biophys. Acta*, 1987, **924**, 261.

Isoniazid

Use/occurrence:	Antitubercular
Key functional groups:	Hydrazide
Test system:	Mouse (oral 1.1 mg/ animal)

Structure and biotransformation pathway

(1)   (2)

Isoniazid (1), which is used as an antitubercular drug, has shown some carcinogenic properties in mice, and consequently its interaction with nucleic acids has been widely studied. The presently reported study involves the production and mutagenic effectiveness of cytosine adducts in DNA of mouse liver or lung. The DNA of mice was radiolabelled by subcutaneous administration of [^{3}H]deoxycytidine from day 2 to day 21 of life, and then isoniazid (1) was administered by gavage. Modified cytosine deoxynucleosides were separated by reversed phase HPLC. Three adducts were observed including a minor amount of the previously known product from *in vitro* experiments, 4-deamino-4-isoniazidocytosine (2). The identity of the other two products is unknown; they were eliminated more slowly from the carcinogenic target tissue, lung, than from the liver. Adduct (2) was shown to be weakly promutagenic during *in vitro* synthesis of DNA.

Reference

G. B. Maru, S. Bhide, R. Saffhill, and P. J. O'Connor, Formation and persistence of isoniazid-DNA adducts in mouse tissues, *Human Toxicol.*, 1987, **6**, 153.

Phenelzine

Use/occurrence:	Antidepressant
Key functional groups:	Alkyl hydrazine
Test system:	Rat liver microsomes

Structure and biotransformation pathway

Six metabolites of phenelzine (1) have been identified on incubation with liver microsomes from phenobarbital pretreated rats. Metabolites were separated, identified, and quantified by GC and comparison with authentic reference standards. Besides phenylacetaldehyde (2), benzaldehyde (3), and the corresponding alcohols (4) and (5) ethylbenzene and toluene were also detected as metabolites. The mechanism of formation was investigated by the use of deuterated phenelzine, $^{18}O_2$ labelling experiments, and EPR spectroscopy for the detection of radical intermediates. It was shown that the microsomal metabolism of 2-phenylethyl hydroperoxide also produced the aldehyde and alcohol metabolites. Results indicated that the 2-phenylethyl radical was the precursor of all the microsomal metabolites including the intermediate hydroperoxides.

Reference

P. R. Ortiz de Montellano and M. D. Watanabe, Free radical pathways in the *in vitro* hepatic metabolism of phenelzine, *Mol. Pharmacol.*, 1987, **31**, 213.

Phenylhydrazine

Use/occurrence: Haemolytic

Key functional groups: Aryl hydrazine

Test system: Rat (subcutaneous, 500 μmol kg^{-1})

Structure and biotransformation pathway

Phenylhydrazine (1) is known to interact with haemoglobin in erythrocytes to precipitate Heinz body and to induce haemolysis. The *in vitro* reaction of (1) with oxyhaemoglobin yields *N*-phenylprotoporphyrin and the currently reported work is an investigation of the nature and extent of this reaction *in vivo*. Compound (1) was injected subcutaneously into the back of male Wistar rats at a dose of 500 μmol kg^{-1}. Blood and spleen were collected at 3, 24, and 48 hours, and treated with acidic methanol and zinc acetate. Two green pigments were then separated by TLC, and shown by FAB–MS and NMR to be zinc complexes of *N*-phenylated porphyrin dimethyl ester (2). Each of the resolved pigments was a mixture of two isomers; one pair of isomers had phenylated vinyl-substituted pyrrole rings of the porphyrin and the other had phenylated propionic acid substituted pyrrole rings. The same products were produced by the chemical reaction of oxyhaemoglobin with (1). The pigments could be detected 3 hours after the injection of (1) and doubled in amount to 0.4% of the dose of (1) after 24 hours. It was postulated that the mechanism of the reaction involved the initial formation of phenyldiazene which decomposes to a phenyl radical.

Reference

K. Hirota, T. Hatanaka, and T. Hirota, Isolation of *N*-phenylprotoporphyrin IX from the red cells and spleen of the phenylhydrazine-treated rat, *Arch. Biochem. Biophys.*, 1987, **255**, 42.

4-Aminobiphenyl

Use/occurrence: Environmental carcinogen

Key functional groups: Aryl amine, biphenyl

Test system: Human cigarette smokers

Structure and biotransformation pathway

Earlier studies in rats had indicated that 4-aminobiphenyl (1) forms an acid-labile sulphinamide adduct (2) with a cysteine residue in haemoglobin. Quantitative methods have now been developed to investigate the possible formation of this metabolite-adduct in human cigarette smokers. The haemoglobin was subjected to mild basic hydrolysis which liberated the 4-aminobiphenyl from the cysteinyl residue. The amine was then extracted, derivatized to the pentafluoropropionamide, and analysed by negative ion chemical ionization MS with selected ion recording. Higher levels of the adduct (mean 154 pg g^{-1} protein) were found in the haemoglobin of smokers compared to non-smokers (28 pg g^{-1}). Adduct levels in five individuals who gave up smoking decreased over 6–8 weeks to non-smoker levels.

Reference

M. S. Bryant, P. L. Skipper, S. R. Tannenbaum, and M. Maclure, Haemoglobin adducts of 4-aminobiphenyl in smokers and non-smokers, *Cancer Res.*, 1987, **47**, 602.

4-Aminobiphenyl, 1-Naphthylamine, 2-Naphthylamine

Use/occurrence:	Environmental carcinogens
Key functional groups:	Aryl amine, biphenyl
Test system:	Rat liver microsomes, rat liver UDP-glucuronosyltransferase

Structure and biotransformation pathway

N-Glucuronidation of carcinogenic aromatic amines may be considered as a competition pathway for the bioactivating pathway of N-hydroxylation. The N-glucuronidation of three amines, 4-aminobiphenyl (1), 1-naphthylamine (2), and 2-naphthylamine (3) has now been studied using rat liver microsomes and five purified UDP-glucuronosyltransferases (UDPGT). For microsomes from Sprague–Dawley rats, N-glucuronidation (determined fluorimetrically) was much more extensive for (2) (60.3 nmol conjugated min^{-1} mg^{-1} protein) than for (3) (6.6 nmol min^{-1} mg^{-1} protein) and (1) (3.4 nmol min^{-1} mg^{-1} protein). For Wistar rat microsomes the extent of N-glucuronidation showed individual variation according to the level of one isozyme of UDPGT (3α-hydroxy-steroid UDPGT).

Purified 3α-hydroxy-steroid UDPGT catalysed the conjugation of all three amines, but (1) was not a substrate for the four other UDPGTs examined (p-nitrophenyl, 17β-hydroxy-steroid, morphine, and digitoxigenin monodigitoxoside UDPGT). Amines (2) and (3) were, however, also conjugated by p-nitrophenol (3-methylcholanthrene inducible) UDPGT and 17β-hydroxy-steroid UDPGT.

Reference

M. D. Green and T. R. Tephly, N-Glucuronidation of carcinogenic aromatic amines catalysed by rat hepatic microsomal preparations and purified rat liver uridine 5'-diphosphate-glucuronasyl-transferases, *Cancer Res.*, 1987, **47**, 2028.

2-, 3-, and 4-Aminobiphenyl, 2-, 3-, and 4-Acetamidobiphenyl

Use/occurrence:	Model compounds
Key functional groups:	*N*-acetyl aryl amine, aryl amine, biphenyl
Test system:	Rat liver microsomes

Structure and biotransformation pathway

(a) Hydroxylation

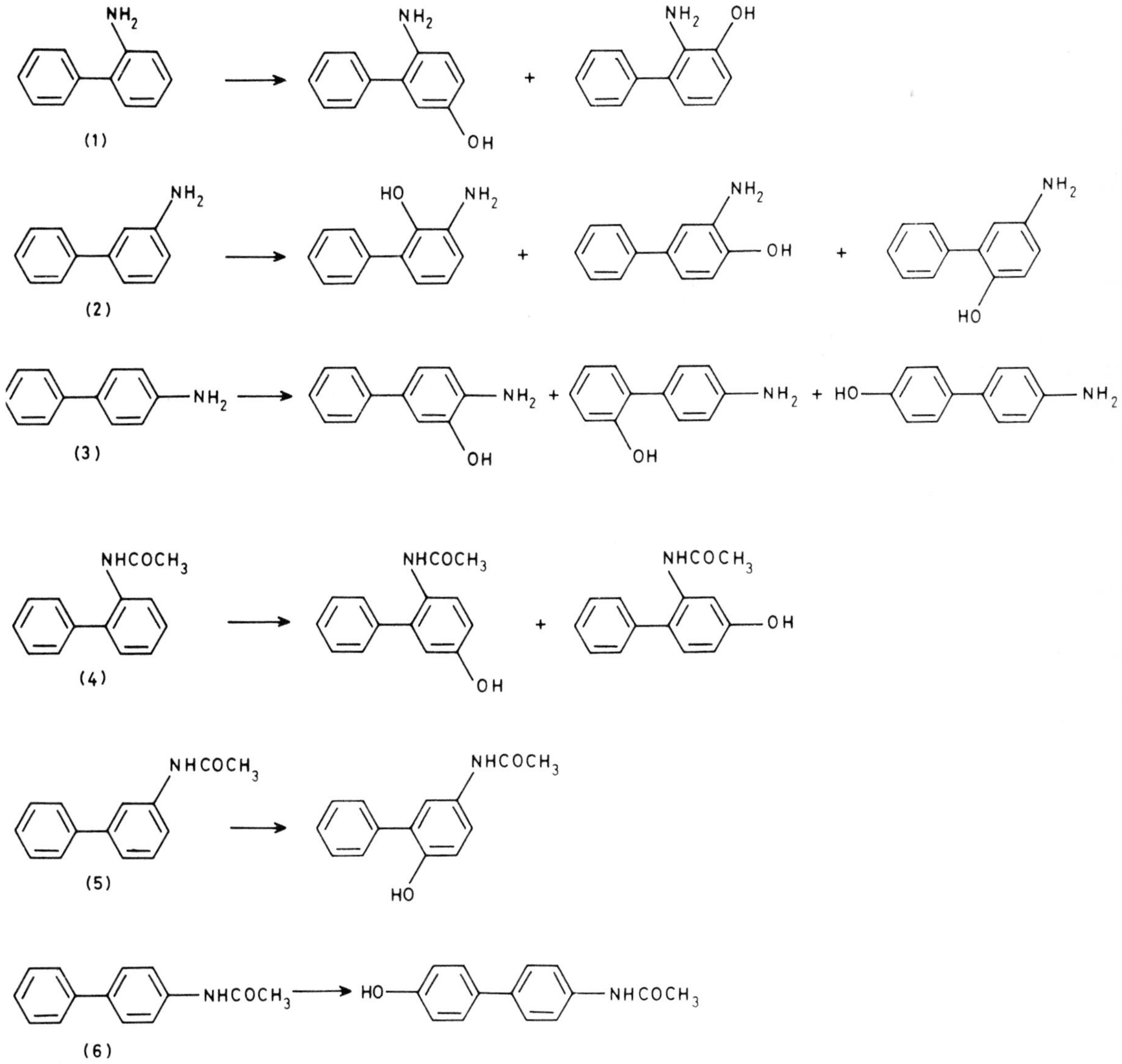

393

(b) Nitrogen oxidation

(1) → (? non-enzymic artefact)

(2)

(3)

4-Aminobiphenyl (3) is carcinogenic in animals, whereas 3-aminobiphenyl (2) is only weakly carcinogenic and 2-aminobiphenyl (1) does not show carcinogenic activity. The metabolism of these three aromatic amines and of their potential metabolites, the acetamide derivatives (4), (5), and (6), were therefore compared for the purpose of establishing some correlation with tumour induction. Rat liver microsomal metabolism of all amines (1)–(3) gave hydroxy-4-aminobiphenyls with the hydroxy group *ortho-* or *para-* to the amino function. Identity of products was confirmed by comparison of their TLC behaviour with that of authentic standards. Metabolism of the acetamidobiphenyls (4)–(6), in the presence of a deacetylation inhibitor, sodium fluoride, led largely to derivatives with a hydroxyl group *para-* to the acetamide group. 3-Aminobiphenyl (2) and 4-aminobiphenyl (3) were also metabolized by nitrogen oxidation to hydroxylamines and nitroso compounds, but 2-aminobiphenyl (1) only produced a minor amount of 2-nitrobiphenyl. However, production of nitro compounds by non-enzymic processes in heat-inactivated microsomes was also observed. It was concluded that the lack of carcinogenicity and mutagenicity for 2-aminobiphenyl (1) is due to its resistance to nitrogen oxidation.

Reference

N. Bayraktar, M. Kajbaf, S. D. Jatoe, and J. W. Gorrod, The oxidation of isomeric amino and acetamido-biphenyls by rat hepatic microsomal preparations, *Arch. Toxicol.*, 1987, **60**, 91.

4-Aminobiphenyl

Use/occurrence:	Environmental carcinogen
Key functional groups:	Aryl amine, biphenyl
Test system:	Mouse, hamster, rat, guinea pig, rabbit (intraperitoneal, 1 mg kg^{-1}); hepatocytes from the above species

Structure and biotransformation pathway

The urinary metabolite profile of 4-aminobiphenyl (1) in five species (mouse, hamster, rat, guinea pig, and rabbit) was compared with that from hepatocytes from the same species, in order to validate the use of the latter as an *in vitro* model for aromatic amine metabolism. 4'-Hydroxy-4-acetamido-biphenyl (2) was the major metabolite for all the species except the mouse. Other metabolites detected (HPLC) in all species were 4-acetamidobiphenyl (3), 3-hydroxy-4-aminobiphenyl (4), and 4'-hydroxy-4-aminobiphenyl (5). Mouse, rat, and guinea pig also excreted 2'-hydroxy-4-aminobiphenyl (6) and 2'-hydroxy-4-acetamidobiphenyl (7). Oxidation products of the amino group [4-hydroxylaminobiphenyl (8), 4-nitrosobiphenyl (9), and 4-nitrobiphenyl (10)] were also found as minor components from all species. The metabolite profile for the urinary excretion products was in general comparable with the profile from hepatocytes, supporting the use of the latter as a screening system for human metabolism of (1).

Reference

S. D. Jatoe and J. W. Gorrod, The *in vitro/in vivo* comparative metabolism of 4-aminobiphenyl using isolated hepatocytes, *Arch. Toxicol.*, 1987, **60**, 65.

4-Nitrosophenetol

Use/occurrence:	Model compound[f]
Key functional groups:	Alkyl aryl ether, arylnitroso
Test system:	Human erythrocytes

Structure and biotransformation pathway

When 4-nitrosophenetol (1) was incubated with human red blood cells it was rapidly metabolized. Lipophilic metabolites were investigated by HPLC of ether extracts. A major metabolite was 4-phenetidine (2) but also present were the dimers 4-ethoxy-4'-nitrosodiphenylamine (3) and 4-amino-4'-ethoxydiphenylamine (4). It was demonstrated that the nitroso compound was rapidly converted into the amine by red cells. Identity of the metabolites was confirmed by chromatographic comparison with authentic reference compounds and by UV spectroscopy of the isolated components. It was shown that 4-ethoxy-4'-nitrosodiphenylamine was produced as one of the products formed by reaction of 4-nitrosophenetol with an equimolar amount of glutathione. The corresponding amine was synthesized by reduction with titanous chloride. The mechanism of formation of the dimers is not known but is thought to involve nucleophilic *para*-substitution of an activated 4-phenetidine.

Reference

H. Klehr, P. Eyer, and W. Schafer, Formation of 4-ethoxy-4-nitrosodiphenylamine in the reaction of the phenacetin metabolite 4-nitrosophenetol with glutathione, *Biol. Chem. Hoppe-Seyler*, 1987, **368**, 895.

2-Naphthylamine

Use/occurrence:	Carcinogen, dye manufacture
Key functional groups:	Aryl amine
Test system:	Ram seminal vesicles microsomes, purified prostaglandin H synthase, horseradish peroxidase

Structure and biotransformation pathway

Prostaglandin H synthase is known to convert arylamines into compounds which bind to nucleic acids. The details of this metabolic activation have now been explored using 2-naphthylamine (1) as substrate, and prostaglandin H synthase (crude or purified) or horse radish peroxidase as oxidizing systems.

[^{3}H]-(1) (50 μM) was incubated with microsomal prostaglandin H synthase from ram seminal vesicles with initiation by H_2O_2 or ascorbic acid. Metabolites were extracted and analysed by reverse phase HPLC. Structures were determined by UV, MS, and comparison with authentic standards. The three main products were 2-aminodinaphthylamine (2), 2-amino-1,4-naphtho-quinone-N^4-2-naphthylimine (3) and dibenzo[a,h]phenazine (4). The structures of two minor products were not proved, but they were suggested to be 1,4-dinaphthylamino-2-naphthylamine (5) and 2,2'-azobis(1,1'-dihydroxy-naphthalene) (6). Purified prostaglandin H synthase produced smaller amounts of similar products. Horseradish peroxidase gave major amounts of metabolites (2) and (4).

Thus (1) appears to be metabolized by a one-electron oxidative system, dependent upon the peroxidase activity of prostaglandin H synthase. A similar profile of metabolites was obtained with a chemical one-electron oxidizing system (potassium ferricyanide). The binding of (1) to DNA was also shown to be dependent on the peroxidase activity of prostaglandin H synthase. Species responsible for this binding may include 2-amino-1-naphthol (7) (oxidized to 2-imino-1-naphthoquinone) and also a free-radical species.

Reference

J. A. Boyd and T. E. Eling, Prostaglandin H synthase-catalysed metabolism and DNA binding of 2-naphthylamine, *Cancer Res.*, 1987, **47,** 4007.

N,N,N′,N′-Tetramethylbenzidine

Use/occurrence:	Model substrate
Key functional groups:	Aryl amine, benzidine, N-methyl aryl amine
Test system:	Horseradish peroxidase (*in vitro*)

Structure and biotransformation pathway

(CH$_3$)$_2$N — ⟨benzidine⟩ — N(CH$_3$)$_2$

(1)

GS — ⟨aryl⟩ — N(CH$_3$)$_2$ ... (CH$_3$)$_2$N — ⟨biphenyl⟩ — N(CH$_3$)$_2$, SG

(2)

GS = glutathionyl

(3)

N,N,N′,N′-Tetramethylbenzidine (1) is known to be dealkylated in a peroxidase system. It is possible to inhibit this reaction by the addition of glutathione, and the current study involved the investigation of the mechanism of this protection. N,N,N′,N′-Tetramethylbenzidine was incubated with horseradish peroxidase, hydrogen peroxide, and glutathione, and the water-soluble metabolites were separated by reverse phase HPLC. The two main products were identified by ^{1}H NMR and FAB–MS as the diglutathionyl adducts substituted at the 2,2′- (2) and 3,3′- (3) positions. A further product was shown to be a complex mixture, possibly containing monosubstituted glutathione conjugates. When N,N,N′,N′-tetramethylbenzidine was incubated with horseradish peroxidase and hydrogen peroxide in the absence of glutathione, the N_4-tetramethylbenzoquinone di-imine was first formed. Addition of glutathione removed the di-imine and formed the same two water soluble metabolites.

Reference

L. C. McGirr and P. J. O'Brien, Glutathione conjugate formation without N-demethylation during the peroxidase catalysed N-oxidation of N,N′,N,N′-tetramethylbenzidine, *Chem. Biol. Interact.*, 1987, **61**, 61.

N-Methyl-4-aminoazobenzene

Use/occurrence: Experimental carcinogen, dye

Key functional groups: Aryl amine, azobenzene, N-methyl aryl amine

Test system: Rat (oral, 0.06% in diet)

Structure and biotransformation pathway

The formation of adducts of metabolites of N-methyl-4-aminoazobenzene (1) with DNA has been studied in rats fed a diet containing 0.06% of this carcinogenic aminoazo dye (labelled with ^3H at the 3'-position). DNA was isolated from liver, kidney, and spleen and hydrolysed enzymatically. The nucleoside adducts with metabolites of (1) were separated by HPLC. The major products, each of which accounted for nearly half of the binding of (1) to DNA, were adducts at the N^2 and C-8 positions of guanine (2) and (3). Smaller amounts of the N^6 adduct of adenine (4) and the demethylated guanine C-8 adduct (5) were also detected. As dosing continued up to five weeks the concentrations of the adducts in the liver increased 2–3-fold and were considerably greater than those in the kidney or spleen. Several other minor adducts were also observed, including N-(guanin-8-yl)-N-methyl-4-aminoazobenzene (6) [resulting from depurination of (3)], N-guanosin-8-yl-N-methyl-4-aminoazobenzene (7) (from RNA contamination), the cis-isomers of (3) and (7) (from photolytic activity), and the imidazole ring-opened form of (3). 16/30 animals in a parallel experiment with dosing of unlabelled (1) developed hepatocellular carcinoma, and it was postulated by the authors that levels of the N^2 adduct with guanine (2) best correlated with the tissue carcinogenic specificity of (1).

Reference

D. L. Tullis, K. L. Dooley, D. W. Miller, K. P. Baetcke, and F. F. Kadlubar, Characterization and properties of the DNA adducts formed from N-methyl-4-aminoazobenzene in rats during a carcinogenic treatment regime, *Carcinogenesis*, 1987, **8,** 577.

Carmoisine

Use/occurrence: Food colouring

Key functional groups: Azobenzene, naphthalene, phenol, sulphonic acid

Test system: Rat, mouse, guinea pig (oral, 0.5 and 50 mg kg^{-1})

Structure and biotransformation pathway

(2) (1) * = ^{14}C (4)

(3)

The major constituent of the azo dye carmoisine (1) is the disodium salt of 4-hydroxy-3-(4-sulpho-1-naphthylazo)naphthalene-1-sulphonic acid. After a single oral dose of [^{14}C]-(1) to rats, mice, or guinea pigs, most of the radio-activity was excreted, mainly in the faeces, within 72 hours. The proportion of radioactivity in guinea pig urine was greater than in the other two species. No significant accumulation of radioactivity was found in any tissue. Pregnant rats eliminated an oral dose at a similar rate to non-pregnant rats, and the level of radioactivity in foetuses was similar to that in maternal tissues. The major urinary metabolite was naphthionic acid (2). Rat and mouse also excreted 2-amino-1-naphthol-4-sulphonic acid (3), and the guinea pig excreted 1,2-naphthoquinone-4-sulphonate (4) in addition to (3). Of the above metabolites only (2) was also found in the faeces of all three species, but at least five additional metabolites were present, two of which were hydrolysed by β-glucuronidase and sulphatase.

Reference

J. C. Phillips, C. Bex, D. G. Walters, and I. F. Gaunt, Metabolic disposition of ^{14}C-labelled carmoisine in the rat, mouse, and guinea pig, *Food Chem. Toxicol.*, 1987, **25**, 927.

Quinoneimines and Quinonedi-imines, including 2-amino-1,4-naphthoquinone-imine, 2,3′,6-trichloroindophenol

Use/occurrence:	Experimental anti-cancer agents
Key functional groups:	Quinoneimine, quinonedi-imine
Test system:	Rat liver microsomes

Structure and biotransformation pathway

The activity of the quinoneimine actinomycin D against human cancer has prompted an interest in quinoneimines and quinonedi-imines as potential anti-tumour agents. The ability of a wide series of quinone(di)imines to liberate reactive oxygen species following metabolic reduction has therefore now been studied. The compounds (22) were incubated with rat liver microsomes and the rates of oxidation of NADPH or NADH and the formation of super-oxide were determined. A wide range of activities was observed, with quinoneimines generally being metabolized faster than quinonedi-imines. The greatest superoxide formation was seen with 2-amino-1,4-naphtho-quinoneimine (1) (K_m 5.2 μM with NADPH, 19.6 μM with NADH) and N,N'-diacetyl-2-amino-1,4-naphthoquinoneimine (2), and the greatest oxida-tion of reduced pyridine nucleotide was with these two compounds and with N,N-dimethylindoaniline (3), indophenol (4), and 2,3′,6-trichloroindophenol

(5). Superoxide anion radical formation was correlated with extent of metabolism for most compounds although several compounds which were well metabolized did not produce superoxide. ESR was used to study the formation of free radicals; that produced by (5) in air was stable for 24 hours, but that from (1) was short-lived and only detectable under anaerobic conditions. The quinone(di)imines also reacted *in vitro* with reduced glutathione although the extent of reaction was not related to their toxicity to hepatocytes or Chinese hamster ovary cells.

Reference

G. Powis, E. M. Hodnett, K. S. Santone, K. L. See, and D. C. Melder, Role of metabolism and oxidation–reduction cycling in the cytotoxicity of antitumour quinone-imines and quinonediimines, *Cancer Res.*, 1987, **47**, 2363.

Acetophenone oxime, Salicylaldoxime, *d*-Camphor oxime, Benzamidoxime

Use/occurrence:	Model compounds
Key functional groups:	Amidoxime, oxime
Test system:	Liver cytosol (hog, guinea pig, hamster, rabbit, mouse), rabbit liver aldehyde oxidase

Structure and biotransformation pathway

Liver cytosols from hog, guinea pig, hamster, rabbit, rat, and mouse metabolized acetophenone oxime (1) to acetophenone, (2) in the presence of 2-hydroxypyrimidine (an electron donor of aldehyde oxidase), and under anaerobic conditions.

Guinea pig and hamster liver cytosols were most active followed by mouse, rabbit, hog, and rat. Rabbit liver cytosol also metabolized three other oximes to varying extents [salicylaldoxime (3) to salicylaldehyde (4), *d*-camphor oxime (5) to *d*-camphor (6) and *d*-camphor imine (7), benzamidoxime (8) to benzamidine (9)]. This metabolizing activity was inhibited by menadione, an inhibitor of aldehyde oxidase, and purified rabbit liver oxidase was also capable of metabolizing the four oximes under anaerobic conditions, in the presence of an electron donor (*e.g.* 2-hydroxypyrimidine, N'-methylnicotin-amide). The identity of the oxo- and imino-products was confirmed by TLC, UV, and mass spectrometry. Ammonia was also detected as a product, in equimolar amounts to the oxo-compound produced. The mechanism proposed for the reaction involved reduction of the oximes by liver oxidase to ketimines followed by the hydrolysis of the latter to the oxo-compounds and ammonia.

Reference

K. Taksumi and M. Ishigai, Oxime-metabolizing activity of liver aldehyde oxidase, *Arch. Biochem. Biophys.*, 1987, **253**, 413.

N-Methylformamide

Use/occurrence:	Solvent with antineoplastic activity
Key functional groups:	Formamide
Test system:	Mouse (intraperitoneal, 6.8 mml kg^{-1})

Structure and biotransformation pathway

The origin of several known metabolites of N-methylformamide (1) was investigated in mice. Samples of bile taken from mice 4 hours after dosing with N-methylformamide were found to contain S-(N-methylcarbamoyl)-glutathione (2). This biliary metabolite is the precursor to the corresponding mercapturic acid (3), which was found in urine. The formation of these metabolites was found to be subject to a primary isotope effect when the formyl C—H was replaced by C—D. The amount of methylamine (4) found in urine was also subject to a primary isotope effect. These data suggest that methylamine (4) does not arise from simple enzymic hydrolysis but is associated with a primary oxidative process. The proposed metabolic scheme involves a primary oxidative step to an unknown, followed by conjugation with glutathione to produce (2), which is further metabolized to (3). Methyl isocyanate was postulated as a candidate for the unknown reactive intermediate. Methylamine could arise from (2) or (3) or the proposed intermediates. The deutero-analogue of N-methylformamide was found to be less toxic.

Reference

M. D. Threadgill, D. B. Axworthy, T. A. Baillie, P. B. Farmer, K. C. Farrow, A. Gescher, P. Kestell, P. G. Pearson, and A. J. Shaw, Metabolism of N-methylform-amide in mice: primary kinetic deuterium isotope effect and identification of S-(N-methylcarbamoyl)glutathione as a metabolite, *J. Pharmacol. Exp. Ther.*, 1987, **242**, 312.

Cyanamide

Use/occurrence: Alcohol-deterrent agent

Key functional groups: Nitrile

Test system: Bovine liver catalase

Structure and biotransformation pathway

$$CH_3\overset{\overset{O}{\|}}{C}NH\!-\!C\!\equiv\!N \quad \longleftarrow \quad H_2N\!-\!C\!\equiv\!N \quad \longrightarrow \quad \left[\begin{matrix} H \\ \diagdown \\ HO^{\diagup}N\!-\!C\!\equiv\!N \end{matrix}\right] \quad \longrightarrow \quad {}^-C\!\equiv\!N \; + \; [H\!-\!N\!=\!O]$$

$$(2) \qquad\qquad\qquad (1) \qquad\qquad\qquad (3) \qquad\qquad (4) \qquad\qquad (5)$$

Acetylcyanamide (2) had been previously identified as the major urinary metabolite of cyanamide (1) in several species. In this study, cyanide was detected, by a specific assay method, as a product of the *in vitro* enzymic oxidation of cyanamide (1) under conditions that also produced an active aldehyde dehydrogenase inhibitor. The reaction was inhibited by ethanol and required the presence of a glucose/glucose oxidase system to generate a steady supply of hydrogen peroxide. It was postulated that the reaction proceeded via the N-hydroxycyanamide intermediate (3) but attempts to synthesize this compound were unsuccessful. It was suggested that cyanide formation, combined with limited N-acetylation capacity, is responsible for the exceptional susceptibility of the dog to the toxic effects of cyanamide. It was also suggested that the potential stoichiometric product from the decomposition of (3), nitroxyl (5), might be the active aldehyde dehydrogenase inhibitor but no evidence was presented for the formation of such a species.

Reference

F. N. Shirota, E. G. Demaster, and H. T. Nagasawa, Cyanide is a product of the catalase-mediated oxidation of the alcohol deterrent agent, cyanamide, *Toxicol. Lett.*, 1987, **37**, 7.

Sulfamidine hydrochloride

Use/occurrence:	Insecticide
Key functional groups:	Amidine, *N*-arylimine, chlorophenyl, dialkylamino, methylphenyl, methyl thioether
Test system:	Rat (oral, 130 mg kg^{-1}), *in vitro* (liver 20 000g supernatant)

Structure and biotransformation pathway

The metabolism of a new insecticide sulfamidine was studied in both *in vivo* and *in vitro* test systems and compared with that of chlordimeform. During seven days following a single oral dose of [^{14}C]sulphamidine to rats 80% of the dose was recovered in the urine and 11% in the faeces. Chloroform extracts of urine were found to contain 30% of the radioactivity dosed, and a number of metabolites were identified in these extracts. Metabolites identified were 4-chloro-*o*-toluidine (5) (2.8% of the dose), *N*-desmethyl-chlordimeform (2) (1.1%), *N*-formyl-4-chloro-*o*-toluidine (3) (3%), sulphamidine sulphoxide (1) (0.3%), *N,N*-didesmethylchlordimeform (4) (0.9%), and 5-chloroanthranilic acid (6) (6.4%). Other metabolites not identified in the chloroform extracts accounted for 14.1% of the dose.

In the *in vitro* experiment sulphamidine (25 μM) was almost completely metabolized after 2 hours incubation in the presence of 2 mg ml^{-1} rat liver 20 000g supernatant protein with appropriate co-factors. The products obtained were 4-chloro-*o*-toluidine (5) (4.5%), *N*-desmethylchlordimeform (2) (19.7%), *N*-formyl-4-chloro-*o*-toluidine (3) (25%), sulfamidine sulphoxide (1) (22.4%), and *N,N*-didesmethylchlordimeform (4) (1.9%), with others not identified (4.8%). In the presence of SK525A, sulfamidine sulphoxide (1) was the major product.

Metabolites were identified by TLC comparison with standards and by GC–MS analysis. The primary rat metabolites, sulphamidine sulphoxide (1) and *N*-desmethylchlordimeform (2), have been shown to have insecticidal activity. *N*-Desmethylchlordimeform is also the primary metabolite of chlordimeform, and this metabolite, together with *N,N*-didesmethylchlordimeform, was thought to be responsible for the mammalian toxicity of chlordimeform and sulfamidine.

Reference

Y. Watanabe and F. Matsumara, Comparative metabolism of sulfamidine and chlordimeform in rats, *J. Agric. Food Chem.*, 1987, **35**, 379.

Diethylamine, Dibutylamine, Methylbenzylamine, Morpholine

Use/occurrence:	Model compounds
Key functional groups:	Dialkylamine
Test system:	Immunostimulated macrophages

Structure and biotransformation pathway

Macrophages are known to be able to synthesize nitrite and nitrate upon immunostimulation. The present study explored the capacity of such cells to nitrosate secondary amines. Macrophages (from immortalized cell lines or freshly isolated from mice) were stimulated by treatment with *Escherichia coli* lipopolysaccharide (LPS) and then incubated with the secondary amines, diethylamine (1), dibutylamine (2), methylbenzylamine (3), and morpholine (4). Products were extracted into dichloromethane and determined by GC with thermal energy analysis, using N-nitrosopropylamine as internal standard. The cell lines all produced N-nitrosomorpholine (8) (114–940 nM) after incubation with 15 mM morpholine (4). Addition of interferon-γ enhanced (3–4-fold) the effect of LPS, and similar effects were also seen on the production of nitrite by the cells. The comparative metabolism of the four amines (1)–(4) at a concentration of 5 mM was studied in one cell line, after stimulation with LPS and interferon-γ. The product yields were (5) 4 nM, (6) 23 nM, (7) 255 nM, and (8) 1680 nM.

The mechanism of this metabolism is unknown, although it was suggested by the authors on the basis of the time courses of production of nitrosamine and of nitrite that the amine was trapping a nitrosating species formed during the cells' synthesis of nitrite and nitrate.

Reference

M. Miwa, D. J. Stuehr, M. A. Marletta, J. S. Wishnok, and S. R. Tannenbaum, Nitrosation of amines by stimulated macrophages, *Carcinogenesis*, 1987, **8**, 955.

N-Methylbenzamidine

Use/occurrence:	Model compound
Key functional groups:	*N*-Methylamidine
Test system:	Rabbit liver fractions

Structure and biotransformation pathway

(1)   (2)   + CH_2O

The metabolism of (1) was investigated in rabbit liver fractions with and without added co-factors. Formation of the dealkylated amidine (2) was determined by a novel HPLC method. Production of formaldehyde was also monitored. *N*-Dealkylation of (1) was measurable in 12 000g supernatant and microsomes only. The reaction required NADPH and O_2 and was inhibited by SKF 525-A, KCN, metyrapone, and carbon monoxide. The direct involvement of cytochrome P-450 is therefore implicated although pretreatment of animals with phenobarbitone, 3-methylcholanthrene, and (1) did not significantly increase the rate of reaction. From the results of this study it is argued that benzamidines with hydrogen atoms in the α-position to the amidine nitrogen are *N*-dealkylated rather than *N*-oxygenated.

Reference

B. Clement and M. Zimmermann, Hepatic microsomal *N*-demethylation of *N*-methylbenzamidine, *Biochem. Pharmacol.*, 1987, **36**, 3127.

Nitrosamines

[^{2}H$_6$]-N-Nitrosodimethylamine

Use/occurrence:	Model compound
Key functional groups:	N-Methylnitrosamine, nitrosamine
Test system:	Rat liver microsomes

Structure and biotransformation pathway

N-Nitrosodimethylamine (1) is metabolized by two pathways, demethylation and denitrosation. The former pathway leads to a methylating agent which may be responsible for the carcinogenicity of (1). The fully deuterated analogue of (1), [^{2}H$_6$]-N-nitrosodimethylamine (2), has lower carcinogenicity and methylating ability than (1). The kinetics of the metabolism of (2) have therefore now been studied in an attempt to explain these phenomena.

Compound (2) was incubated with acetone-induced rat liver microsomes at concentrations of 0.04–1.5 mM. Nitrite and formaldehyde were measured colorimetrically as estimates of the denitrosation and demethylation pathways. Additionally amines were determined as their dinitrophenyl derivatives by GC–MS, using N-methyl-[N-^{2}H$_3$]methylamine as internal standard. Parallel experiments were carried out with the non-deuterated substrate (1). The K_m values for demethylation and denitrosation of (1) were both 0.06 mM; these were increased to 0.30 and 0.19 mM respectively, although the difference between these latter values was claimed to be of doubtful significance. The V_{max} values were 7.9 and 0.83 nmol min^{-1} mg^{-1} for demethylation and denitrosation of (1) and these were unchanged for (2).

The isotope effect on K_m for denitrosation of (2) was determined in competitive incubations containing 0.3–1 mM [^{15}N]-(1) and equal amounts of (2), by GC–MS determinations of the [^{15}N]- and [^{2}H$_3$]-methylamine produced. The mean K_m ratio was 5.7 ± 1.0, in line with the demethylation K_m ratio that had been determined colorimetrically. It was concluded that deuteration of (1) did not lead to any switching of metabolism from demethylation to denitrosation, and that no explanation had been found for the reduced carcinogenicity of (2).

Reference

D. Wade, C. S. Yang, C. J. Metral, J. M. Roman, J. A. Hrabie, C. W. Riggs, T. Anjo, L. K. Keefer, and B. A. Mico, Deuterium isotope effect on denitrosation and demethylation of *N*-nitrosodimethylamine by rat liver microsomes, *Cancer Res.*, 1987, **47**, 3373.

N-Nitrosodimethylamine, *N*-Nitrosomethylurea

Use/occurrence:	Environmental chemicals
Key functional groups:	*N*-Methylnitrosamine, nitrosamine
Test system:	Trout liver microsomes

Structure and biotransformation pathway

$$(CH_3)_2N—NO \longrightarrow NO_2^-$$

$$CH_3NH\overset{\overset{\displaystyle O}{\|}}{C}NHNO \longrightarrow NO_2^-$$

Incubation of the two nitrosamines with trout liver microsomes together with NADPH and molecular oxygen showed the formation of nitrite and formaldehyde. The extent of formation was inhibited by common inhibitors of microsomal enzymes although formation of nitrite was affected more than that of formaldehyde, indicating that they were produced by separate pathways.

Reference

A. Arillo and F. Tosetti, Denitrosation of *N*-nitrosodimethylamine and *N*-nitrosomethylurea by liver microsomes from trout, *Environ. Res.*, 1987, **42**, 366.

N-Nitrosodimethylamine

Use/occurrence:	Environmental carcinogen
Key functional groups:	N-Methylnitrosamine, nitrosamine
Test system:	Rat liver microsomes

Structure and biotransformation pathway

$(CH_3)_2N—NO \longrightarrow$
$\begin{array}{c} CH_3 \\ \\ H \end{array} \Big\rangle N—H$ + NO_2^- (+ ? HCHO)

N-Nitrosodimethylamine is known to be metabolically activated to a methylating agent, which is thought to be responsible for its mutagenic and carcinogenic activity. This metabolism involves α-hydroxylation followed by decomposition of the resulting methylol to formaldehyde and a methyl diazonium ion. An alternative metabolic pathway, denitrosation, has now been demonstrated to occur in rat liver microsomal systems. It was postulated that denitrosation could occur by a reductive process, yielding dimethylamine, or by an oxidative process via the imine $CH_3N=CH_2$, which would hydrolyse to methylamine and formaldehyde. Incubation of N-nitrosodimethylamine (4 mM) with ethanol-induced microsomes produced methylamine (19.8 μM) and nitrite (20.3 μM) but only insignificant amounts of dimethylamine. Incubation of N-nitrosodimethylamine, labelled with ^{15}N at the amino nitrogen, showed that the methylamine nitrogen was derived from the substrate. Dimethylamine is probably not an intermediate in the production of methylamine as its demethylation was shown not to be a significant metabolic pathway. Methylamine and dimethylamine were determined by GC–MS of their 2,4-dinitrophenyl derivatives; quantification was carried out by selected ion recording using deuterated internal standards.

Reference

L. K. Keefer, T. Anjo, D. Wade, T. Wang, and C. S. Yang, Concurrent generation of methylamine and nitrite during the denitrosation of N-nitrosodimethylamine by rat liver microsomes, *Cancer Res.*, 1987, **47**, 447.

Nitrosocimetidine

Use/occurrence:	Potential metabolite of anti-ulcer drug cimetidine
Key functional groups:	Guanidine, nitrosamine
Test system:	Blood (*in vitro*: rat, mouse, guinea pig, hamster, human), blood (*in vivo*: hamster)

Structure and biotransformation pathway

(1)   →   (2)

Nitrosocimetidine (1) is a DNA methylating agent and gives positive results in several short-term genotoxicity tests. It is consequently a potentially hazardous possible metabolite of the widely used anti-ulcer drug cimetidine (2). However, whole blood and haemoglobin are capable of denitrosating nitrosocimetidine. This deactivating activity is species-dependent (rat > mouse/guinea pig > human/hamster). In rat blood at least 75% of the nitrosocimetidine decomposition products consists of cimetidine; the half-life is *ca.* 2 min at 37 °C. In hamster blood 40% of the product is cimetidine, with a half-life of 27 min. Other degradation products are of unknown structure. The denitrosation pathway appears to be dependent on haemoglobin cysteine residues; modification of the latter with iodoacetamide abolished the degradation rate enhancement caused by human haemoglobin. In hamster blood nitrosocimetidine administered intravenously was almost totally converted into cimetidine with a half-life less than 5 min. The compounds were identified by reverse phase HPLC in comparison with authentic compounds.

Reference

D. E. Jensen, G. J. Stelman, and A. Spiegel, Species difference in blood-mediated nitrosocimetidine denitrosation, *Cancer Res.*, 1987, **47**, 353.

N-Nitrosobis(2-oxopropyl)amine

Use/occurrence:	Experimental carcinogen
Key functional groups:	Alkyl ketone, nitrosamine
Test system:	Rat microsomes, hepatocytes

Structure and biotransformation pathway

N-Nitrosobis(2-oxopropyl)amine (1) is known to be metabolized in rats and hamsters to CO_2, *N*-nitroso(2-hydroxypropyl)(2-oxopropyl)amine (2), and *N*-nitrosobis(2-hydroxypropyl)amine (3). *N*-Nitroso-*N*-methyl-2-oxopropylamine (4) has also been detected in hamster urine, which could possibly be the metabolite responsible for the methylation of DNA caused by (1). The current study investigated the biotransformation of $[1\text{-}^{14}C]\text{-}(1)$ by microsomes and hepatocytes from uninduced male Fischer-344 rats. Metabolites were separated and identified by HPLC on reverse phase and silica columns, or by GC.

No metabolites could be detected after incubation of (1) with microsomes. However, hepatocytes converted (1) into $^{14}CO_2$ and into several non-volatile radiolabelled metabolites. Two of these materials were of unknown identity, although (2) and (3) were both identified from the HPLC elution profiles. No (4) was found, however, by HPLC. A small proportion of the acetol (6) derived from the postulated diazo intermediate (5) was also detected by GC.

Reference

T. G. Farrelly, J. E. Saavedra, R. J. Kupper, and M. L. Stewart, The metabolism of *N*-nitrosobis(2-oxopropyl)amine by microsomes and hepatocytes from Fischer-344 rats, *Carcinogenesis*, 1987, **8**, 1095.

N-Nitroso(2-hydroxypropyl)(2-oxopropyl)-amine

Use/occurrence:	Model compound
Key functional groups:	Secondary alkyl alcohol, alkyl ketone, nitrosamine
Test system:	Hamster and rat (subcutaneous, 10–150 mg kg^{-1}), hamster and rat liver microsomes

Structure and biotransformation pathway

(2)   (1) * = ^{14}C   (3)

(4) Gluc = glucuronyl

Hamsters treated with one of a number of nitrosamines, including *N*-nitroso(2-hydroxypropyl)(2-oxopropyl)amine, HPOP (1), develop ductal adenocarcinomas of the pancreas, the yield depending on carcinogen, dose, route of administration, and sex. This is not the case in rats.

Hamsters and rats metabolize (1) quite differently as judged by urinary metabolites excreted 6 hours after administration. Hamsters form the sulphate ester of HPOP (2) as a major metabolite and also reduce it to *N*-nitrosobis(2-hydroxypropyl)amine (3). In contrast rats form the glucuronide conjugate (4) of (1) and also excrete more of it unchanged. Conjugation by sulphotransferases or glucuronyl transferases occurs in the liver. Glucuronyl transferase activity towards phenolic compounds was comparable in the two species, but glucuronidation of (1) was three times greater in rat than in hamster. Sulphation of (1) on the other hand was 10 times faster in hamster than in rat, a reaction not inhibited by classical phenol sulphotransferase inhibitors. This suggests that *β*-hydroxynitrosamines are most efficiently sulphated by the aliphatic (hydroxy-steroid) sulphotransferase isozymes. *In vitro* only one isomer of (1), where the nitroso group is *syn* to the free keto group, was sulphated to an appreciable extent.

Reference

D. M. Kokkinakis, D. G. Scarpelli, V. Subbarao, and P. F. Hollenberg, Species differences in the metabolism of *N*-nitroso(2-hydroxypropyl)(2-oxopropyl)amine, *Carcinogenesis*, 1987, **8**, 295.

N-Butyl-N-(4-hydroxybutyl)nitrosamine

Use/occurrence:	Experimental carcinogen
Key functional groups:	Alkyl alcohol, nitrosamine
Test system:	Rat hepatocytes, rat (intraperitoneal, 25 mg kg^{-1})

Structure and biotransformation pathway

$$n\text{-}C_4H_9\text{—}\overset{\overset{\displaystyle NO}{|}}{N}\text{—}CH_2CH_2CH_2CH_2OH \longrightarrow n\text{-}C_4H_9\text{—}\overset{\overset{\displaystyle NO}{|}}{N}\text{—}CH_2CH_2CH_2CO_2H \longrightarrow n\text{-}C_4H_9\text{—}\overset{\overset{\displaystyle NO}{|}}{N}\text{—}CH_2\underset{\underset{\displaystyle OH}{|}}{C}HCH_2CO_2H$$

$$(1) \qquad\qquad\qquad (2) \qquad\qquad\qquad (3)$$

This work extends previous well reported studies on the *in vivo* metabolism of N-butyl-N-(4-hydroxybutyl)nitrosamine (1) with investigations of the effects of disulfiram on the metabolism of (1) *in vivo* and in isolated hepatocytes.

$[^{14}C]$-(1) was incubated with rat hepatocytes and the products were analysed on a reverse phase HPLC column. Rapid metabolism of (1) occurred (2.1 μmol h^{-1}/5 million cells) with more than 98% of (1) being metabolized in 4 hours. The major metabolite was N-butyl-N-(3-carboxypropyl)nitrosamine (2) and N-butyl-N-(2-hydroxy-3-carboxypropyl)nitrosamine (3) was a minor metabolite. Disulfiram, which inhibits the carcinogenic action of (1), did not alter significantly the rate of metabolism of (1) by hepatocytes, but did increase the amount of (3) formed (*ca.* 3.5-fold), suggesting that disulfiram inhibits further metabolism of (3). *In vivo* studies in rats treated intraperitoneally with (1) were consistent with earlier literature; pre-administration of disulfiram (0.5% diet, 2 weeks) did not alter the amount of urinary excretion of radioactivity or of (2). However, excretion of (3) was increased from 11% to 22%, supporting the postulate that further metabolism of (3) is inhibited by disulfiram.

Reference

C. C. Irving and D. S. Daniel, Influence of disulfiram on the metabolism of the urinary bladder carcinogen N-butyl-N-(4-hydroxybutyl)nitrosamine in the rat, *Carcinogenesis*, 1987, **8**, 1309.

N-Nitrosoproline

Use/occurrence:	Nitrosated amino acid
Key functional groups:	Amino acid, nitrosamine, pyrrolidine
Test system:	Rat, *in vivo*, DNA, *in vitro* ± microsomes

Structure and biotransformation pathway

(1)   (2)

In most studies *N*-nitrosoproline (1) was found to undergo minimal metabolism, to be excreted unchanged, and to be non-carcinogenic in several long-term animal studies. However, results from one *in vitro* enzyme-free study suggested that (1) was activated to a strong alkylating species. In the present study biotransformation of (1) was investigated in nephrectomized rats to see whether impaired urinary excretion would increase its metabolism. *In vitro* metabolism and DNA binding with an enzyme-mediated and enzyme-free system were also studied.

When L-[U-^{14}C]-(1) was administered to nephrectomized rats about 1% of the radioactivity appeared as CO_2, but metabolism of (1) was not significantly increased compared with that in sham-operated animals. Urinary excretion of (1) in sham-operated rats was 98%, while in rats after unilateral nephrectomy it was 59%. In both *in vitro* enzyme-mediated and enzyme-free systems where rapid elimination is not possible covalent binding of [2,3,4,5-^{3}H]-(1) to calf thymus DNA was demonstrated, possibly through the decarboxylation product of (1), nitrosopyrrolidine (2), which is a powerful carcinogen. The authors conclude that there is evidence for very low levels of *in vivo* metabolism of (1).

Reference

L. Y. Y. Fong, D. E. Jensen, and P. N. Magee, Evidence for metabolism of *N*-nitrosoproline, *Chem. Biol. Interact.*, 1987, **64**, 115.

Use/occurrence:	Tobacco-specific carcinogens
Key functional groups:	Aryl ketone, *N*-methylnitrosamine, nitrosamine, pyrrolidine
Test system:	Rat (intraperitoneal, 0.03–3.9 μmol kg^{-1})

Structure and biotransformation pathway

4-(Methylnitrosamino)-1-(3-pyridyl)butan-1-one (1) and *N'*-nitroso-nornicotine (2) are two carcinogens present in tobacco and tobacco smoke. The initiating stage in the carcinogenic process for nitrosamines is thought to be alkylation of DNA by a reactive species produced by metabolic α-hydroxylation of the compounds. Alkylation of proteins also occurs and, although this is thought to be of less significance with respect to the carcinogenic pathway, it has been suggested that protein alkylation products may be useful as an internal dosimeter for carcinogen exposure. For this reason the reaction products of metabolically activated (1) and (2) with haemoglobin have now been investigated.

[5-^{3}H]-(1) was administered intraperitoneally to rats at doses of 0.03–3.9 μmol kg^{-1}, and a linear dose–response relationship was observed for radioactivity bound to globin with 0.1% of the dose being bound. Treatment of the globin with dilute NaOH or HCl released 10–15% of the radioactivity; the main constituent was shown by HPLC analysis to be 4-hydroxy-1-(3-pyridyl)butan-1-one (3). The structure of this was confirmed by GC–MS analysis of the product formed after administration of a larger dose of (1) (50 mg kg^{-1} day^{-1} for 3 days). The same adduct was produced to a smaller extent after administration of (2) or 4-(ethoxycarbonylnitrosamino)-1-(3-

427

pyridyl)butan-1-one (4). The metabolic scheme that was therefore postulated involves α-hydroxylation of (1) and (2) followed by breakdown of the products to the diazohydroxide (5), which forms adducts with nucleophilic centres in haemoglobin. Monitoring of the keto-alcohol (3) released hydrolytically from the protein may thus be a valuable indication of the received dose of (1) or (2).

Reference

S. G. Carnella and S. S. Hecht, Formation of haemoglobin adducts upon treatment of F344 rats with the tobacco-specific nitrosamines 4-(methylnitrosamino)-1-(3-pyridyl)butan-1-one and N'-nitrosonornicotine, *Cancer Res.*, 1987, **47**, 2626.

4-(*N*-Methyl-*N*-nitrosamino)-1-(3-pyridyl)butan-1-one

Use/occurrence:	Tobacco-specific carcinogen
Key functional groups:	*N*-methylnitrosamine, nitrosamine
Test system:	Rat (intraperitoneal, 0.1–100 mg kg^{-1} day^{-1}, 1–12 days)

Structure and biotransformation pathway

4-(*N*-Methyl-*N*-nitrosamino)-1-(3-pyridyl)butan-1-one (1) is one of the two major tobacco-specific nitrosamines derived from the nitrosation of nicotine. Compound (1) has been shown to be a carcinogen in rodents and is thought to exert its genotoxic effects after being metabolically activated via α-hydroxylation to a methylating agent. The present study investigated the dose–response relationships for production of a promutagenic methylated adduct of DNA (O^6-methylguanine) (2) following intraperitoneal administration of (1) at doses from 0.1 to 100 mg kg^{-1} day^{-1} to Fischer-344 rats. Compound (2) was determined at the higher doses (10–100 mg kg^{-1} day^{-1}) of (1) by fluorescence-linked HPLC, after acidic depurination of DNA isolated from whole lung cells and specific lung cell populations.

For the lower doses of (1) a radioimmunoassay procedure was used to estimate O^6-methyldeoxyguanosine in an enzymic hydrolysate of DNA. The comparability of results from the two analytical methods was confirmed at a dose of (1) of 3 mg kg^{-1} day^{-1}. The relationship between dose of (1) and the production of (2) in lung was not linear, with an increased ratio of amount of (2) to dose of (1) at lower doses of (1). Specific lung cell populations showed differences in their concentrations of (2), with Clara cells (which are known for their localization of cytochrome P-450 activity) showing the higher concentrations. The efficiency of formation of (2) in Clara cells was particularly notable at low doses of (1), being 38-fold greater at the dose of 0.3 mg kg^{-1} day^{-1} compared with 100 mg kg^{-1} day^{-1} of (1). This observation was supported by autoradiographic studies. Nitrosodimethylamine at a dose of 0.4 mg kg^{-1} (18 times over 6 weeks) gave about half of the alkylation of O^6 of guanine in lung DNA compared with that from an equimolar dose of (1).

Reference

S. A. Belinsky, C. M. White, T. R. Devereux, J. A. Swenberg, and M. W. Anderson, Cell selective alkylation of DNA in rat lung following low dose exposure to the tobacco-specific carcinogen 4-(*N*-methyl-*N*-nitrosamino)-1-(3-pyridyl)butan-1-one, *Cancer Res.*, 1987, **47**, 1143.

N-Nitroso-N-methylaniline

Use/occurrence: Environmental carcinogen

Key functional groups: Aryl amine, N-methyl-nitrosamine, nitrosamine

Test system: Rat hepatic and oesophageal preparations, Ames test in *S. typhimurium* TA 1537

Structure and biotransformation pathway

A number of non-symmetrical N-nitrosamines, including N-nitroso-N-methylaniline (1), are oesophageal carcinogens in the rat. It was postulated that metabolism of (1) proceeded via α-oxidation, which would give N-nitroso-N-hydroxymethylaniline (2), which in aqueous solution would yield the benzenediazonium ion (3), both of these compounds being positive in the Ames test. Previous work showed that chemicals which induced microsomal demethylation of (1) as measured by formaldehyde formation also induced denitrosation resulting in nitrite. These were thought to be independent parallel pathways using the same cytochrome P-450 isozyme.

The present study examines the metabolic activation of (1) by rat liver and oesophageal S9 fractions prepared after induction by phenobarbital (PB) or pyrazole. The latter is known to induce metabolism of other N-nitrosamines. Using these preparations the major metabolite was aniline, with formation of a smaller amount of N-methylaniline (4). In the presence of PB-induced S9 only some phenol was generated. These metabolites are not consistent with the originally proposed metabolic pathway. The authors therefore suggest (and present evidence for) an initial denitrosation to give (4) and NO. The latter is rapidly oxidized to nitrite, while (4) is demethylated to yield formaldehyde and aniline (5). Compound (4) is converted into aniline (5) five times faster than (1), thus explaining the small amount of this metabolite found during metabolism of (1). Thus denitrosation and demethylation are sequential steps. Compound (1) is converted in a dose dependent manner into a mutagen positive in the Ames test only by pyrazole-induced S9. Only metabolism by PB-induced S9 generated phenol, showing that direct demethylation of (1) via (3) was also an important pathway in the presence of this particular inducer. Neither (4) nor aniline (5) was metabolized to phenol by either induced system. Results using S9 from induced oesophageal tissue were qualitatively similar.

Reference

B. Gold, J. Farber, and E. Rogan, An investigation of the metabolism of *N*-nitroso-*N*-methylaniline by phenobarbital- and pyrazole-induced Sprague–Dawley rat liver and esophagus derived S9, *Chem. Biol. Interact.*, 1987, **61**, 215.

Diphenylnitrosamine

Use/occurrence:	Accelerator in vulcanizing rubber
Key functional groups:	Aryl amine, *N*-aryl nitrosamine
Test system:	Mouse liver microsomes

Structure and biotransformation pathway

The generally accepted mechanism for bioactivation of carcinogenic nitrosamines involves hydroxylation α to the *N*-nitroso group. Diphenylnitrosamine (1) (which has been shown to be carcinogenic after painting on the skin of hairless mice) has no α-hydrogen to allow this bioactivation to occur and its metabolism has therefore now been studied. Compound (1) was incubated with phenobarbital-induced mouse liver microsomes and the products were analysed by HPLC. Three metabolites were found, diphenylamine (2), 4-hydroxydiphenylamine (3), and the quinoneimine derivative of (3) [which was reduced to (3) to enable it to be analysed by HPLC]. Diphenylhydroxylamine, which is a possible metabolite that could account for the observed carcinogenicity and toxicity of (1), was not found as a metabolite.

Reference

K. E. Appel, S. Görsdorf, T. Scheper, M. Bauszus, and A. G. Hildebrandt, Enzymatic denitrosation of diphenylnitrosamine: activation or inactivation?, *Arch. Toxicol.*, 1987, **60**, 204.

433

Amino Acids and Peptides

Dopamine

Use/occurrence:	Biogenic amine
Key functional groups:	Amino acid
Test system:	Sheep (intravenous)

Structure and biotransformation pathway

The metabolites of dopamine (1), (2)–(4), in sheep blood were separated and analysed by HPLC using available reference compounds. After an intravenous dose the aldehyde (2) was detected as a major metabolite in plasma and its concentration increased as the dopamine concentration decreased. Smaller concentrations of the alcohol (4) and the acid (3) were also detected. *In vitro* experiments indicated that the aldehyde was produced by plasma amine oxidase and then partly converted into the alcohol by aldehyde reductases in red blood cells.

Reference

D. F. Sharman, The metabolism of dopamine in the blood of ruminant animals: formation of 3,4-dihydroxyphenylacetaldehyde, *Comp. Biochem. Physiol.*, 1987, **86C**, 151.

Benzylcysteine

Use/occurrence:	Model compound
Key functional groups:	Alkyl aryl thioether, amino acid
Test system:	Normal and analbunemic rats (oral and intravenous, 20 μmol kg^{-1})

Structure and biotransformation pathway

$* = {}^{14}C$

(1)

The excretion of benzylcysteine was studied in normal and analbunemic rats following both oral and intravenous administration. Experiments were repeated in rats which had been surgically nephrectomized and biliary cannulated. Benzylcysteine was metabolized to the mercapturic acid (1). The rate of urinary excretion was rapid; within 2 hours 53–67% of an oral dose of benzylcysteine was recovered as the mercapturate (1). It was suggested that the major part of an oral dose of benzylcysteine was transferred to the liver where it was rapidly acetylated. The absence of albumin in analbunemic rats did not prevent the mercapturate from being transferred to the kidney for rapid elimination via the urine. However, following intravenous administration of benzylcysteine the urinary excretion of the mercapturate was significantly less in analbunemic rats compared with normal rats. Thus it was suggested that albumin is important in the final elimination of mercapturic acids when animals are extraorally challenged with a large dose of toxic electrophiles.

Reference

M. Inoue, K. Okajima, S. Nagase, and Y. Morino, Inter-organ metabolism and transport of a cysteine-*S*-conjugate of xenobiotics in normal and mutant analbunemic rats, *Biochem. Pharmacol.*, 1987, **36**, 2145.

L-Canavanine

Use/occurrence:	Natural product with anti-tumour activity
Key functional groups:	Amino acid, guanidine, oximino
Test system:	Rat (oral, intravenous, subcutaneous, 2 g kg^{-1})

Structure and biotransformation pathway

Following the oral, intravenous, or subcutaneous administration of [^{14}C]-L-canavanine to rats most of the radioactivity was recovered in the urine (83, 68, and 61% respectively). Expired ^{14}CO$_2$ accounted for 5–8% of the dose. The major radioactive metabolite was urea (1), accounting for 88% of the radio-activity in the urine following an intravenous dose, and 75 and 50% respectively following subcutaneous and oral doses. Formation of urea was believed to be due to the action of arginase, producing the corresponding non-radioactive product L-canaline (2). The alternative pathway leading to guanidine (3) and L-homoserine (4) was a less important route regardless of route of administration. Guanidine (3) itself accounted for *ca.* 5% of urinary radioactivity. Methylguanidine (5) accounted for 8% of urinary radioactivity following intravenous administration but only 1–2% following oral or subcutaneous administration. Guanidoacetic acid (6), a product of the transamidation of glycine, accounted for between 1 and 2% of urinary radioactivity. [^{1}C]Urea was determined by a radioenzymic technique while ^{14}CO$_2$ was trapped and determined by liquid scintillation counting. Urinary guanidine compounds were assayed using an amino acid autoanalyser.

Reference

D. A. Thomas and G. A. Rosenthal, Metabolism of L-[*guanidooxy*-^{14}C]canavanine in the rat, *Toxicol. Appl. Pharmacol.*, 1987, **91**, 406.

N^G,N^G-Dimethylarginine, N^G,N'^G-Dimethylarginine

Use/occurrence:	Marker of *in vivo* protein breakdown
Key functional groups:	Amino acid, guanidine, methylamino
Test system:	Rat (intraperitoneal, 0.5 μmol)

Structure and biotransformation pathway

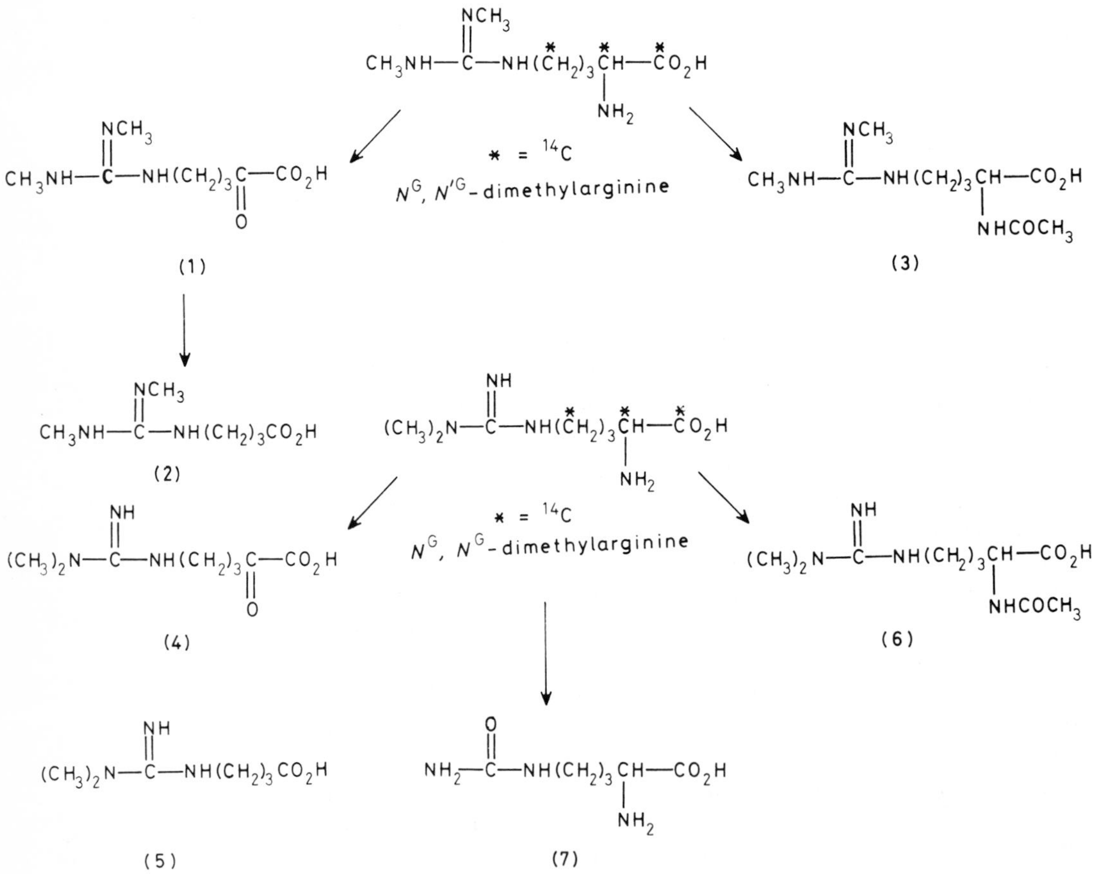

N^G,N'^G-dimethylarginine

$* = {}^{14}C$

N^G,N^G-dimethylarginine

(1) (2) (3) (4) (5) (6) (7)

N^G,N^G-Dimethylarginine (DMA) and N^G,N'^G-dimethylarginine (DM'A) are naturally occurring in proteins and are excreted in the urine following protein breakdown. Consequently they have been proposed as index compounds for protein degradation. However, in rats following intraperitoneal administration of [^{14}C]-DMA or [^{14}C]-DM'A, extensive metabolism occurs

440

leading both to the *N*-acetyl conjugates (6) and (3) and also via pathways to the α-keto-acids (4) and (1), and subsequent degradation products. DMA is also converted into citrulline (7) and its metabolic products ornithine, arginine, and glutamic acid.

Injection of [^{14}C]-DM'A (0.5 μmol) led to 75% of the radioactivity being excreted in the first 12 hours, with 23.7% of this being unchanged DM'A. The metabolites (1), (2), and (3) accounted for 20.2%, 9.6%, and 48.4% respectively of urinary radioactivity. Tissues and plasma contained mostly unchanged DM'A. For rats treated with [^{14}C]-DMA (0.5 μmol) 13% of the radioactivity was excreted in 12 hours, containing 35.2% unchanged DMA, 16.4% metabolite (4), 18.4% (5), and 8.5% (6). In tissues and plasma radioactivity was associated mainly with citrulline (7), with smaller amounts of DMA, ornithine, arginine, and glutamic acid.

Metabolites (1) and (4) were identified by absorption spectra and mass spectra of their 2,4-dinitrophenylhydrazones, and by comparison of their IR spectra with authentic compounds. Upon reductive amination they yielded DMA and DM'A. Metabolites (2) and (5) were identified by GC–MS after conversion into *N*-dimethylamine-methylene methyl esters, and comparison with authentic compounds.

Reference

T. Ogawa, M. Kimoto, H. Watanabe, and K. Sasaoka, Metabolism of $N^{\mathrm{G}},N^{\mathrm{G}}$- and $N^{\mathrm{G}},N'^{\mathrm{G}}$-dimethylarginine in rats, *Arch. Biochem. Biophys.*, 1987, **252**, 526.

Enalapril, Perindopril, Ramipril

Use/occurrence:	Antihypertensive agents
Key functional groups:	Amino acid, alkyl carboxamide, alkyl ester, pyrrolidine amide
Test system:	Rat (oral and intravenous, 10 mg kg^{-1})

Structure and biotransformation pathway

This study was conducted using non-radioactive methodology. Extracts of urine collected during 48 hours after single doses of enalapril (1), perindopril (2), or ramipril (3) were analysed, after methylation, by GC–MS. All three compounds underwent ester hydrolysis, and mean 0–24 hour urinary excretion (24–48 hour excretion appeared to be insignificant) of the resulting diacids amounted to 28%, 22%, and 15% of orally administered (1), (2), and (3) respectively, and 35% of intravenously administered (1). Quantification for each compound was achieved by using the authentic diacid of one of the other compounds as an internal standard. Enzymic deconjugation experiments indicated that the potential acyl glucuronides were not formed from any of the three compounds. In urine from rats receiving enalapril (1), a further metabolite (4) was detected resulting from hydrolysis of the amide linkage. An estimated 9% of orally administered (1) was excreted as (4) in urine during 0–24 hours, but only trace amounts were detected after intravenous administration. The corresponding hydrolytic metabolites of (2) and (3), in which the amide nitrogen was incorporated into the bulkier and more lipophilic bicyclic groupings, were not detected. It was suggested that liver enzymes might be responsible for the formation of (4) from (1) since (4) was formed in much higher amounts after oral administration.

Reference

O. H. Drummer and S. Kourtis, Biotransformation studies of diacid angiotensin converting enzyme inhibitors, *Arzneim.-Forsch.*, 1987, **37** (II), 1225.

N-Ethylmaleimide-*S*-glutathione

Use/occurrence:	Model compound
Key functional groups:	Glutathione
Test system:	Nephrectomized rats (intravenous, 4.14 mg kg^{-1}), perfused rat liver

Structure and biotransformation pathway

(1) * = ^{14}C (2) (3)

GS = glutathionyl

After intravenous administration of the glutathione conjugate (1) the cysteine conjugate (3) was detected in both normal and nephrectomized rats although the ratio of this metabolite to unchanged compound was higher in normal animals. The rapid appearance of the metabolite in nephrectomized rats indicated that its formation occurred in extrarenal tissues. This was demonstrated by experiments with the perfused liver where the cysteinyl-glycine (2) and cysteine conjugate (3) were detected in the perfusate. In addition it was shown that (2) could be converted into the cysteine conjugate in plasma.

Reference

T. Hirota and T. Komai, Extrarenal metabolism of *N*-ethyl[2,3-^{14}C]maleimide-*S*-glutathione, a model compound of glutathione conjugate, *J. Pharmacobio-Dyn.*, 1987, **10**, 336.

Bialaphos

Use/occurrence:	Herbicide (produced by a micro-organism)
Key functional groups:	Alanine, methylphosphinoyl, peptide
Test system:	Mouse (oral, 1.85 mg kg^{-1})

Structure and biotransformation pathway

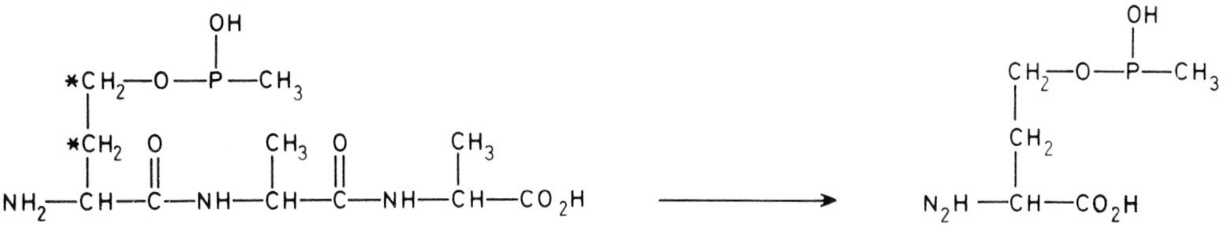

(1) * = ^{14}C

(2)

Bialaphos (1) has a unique tripeptide structure and it is synthesized by a micro-organism *Streptomyces hygroscopicus*. Approximately 90% of an oral dose of [^{14}C]bialaphos was excreted in the faeces and 10% via the urine. Four metabolites were found in the urine and three in faeces. 2-Amino-4[(hydroxy)(methyl)phosphinyl]butyric acid (2) was found to be the major metabolite present in faeces, accounting for *ca.* 50% of the dose. The metabolite was identified by co-chromatography and ninhydrin colour reaction.

Reference

A. Suzuki, K. Nishide, M. Shimaru, and I. Yamamoto, Metabolism of bialaphos in mice, *J. Pesticide Sci.*, 1987, **12**, 105.

445

Cyclosporin A (CsA cyclosporine)

Use/occurrence:	Fungal metabolite, immunosuppressant drug
Key functional groups:	Alkene, isobutyl, leucine, peptide, *N*-methylamide
Test system:	Rat (oral, intravenous, refs. 1, 2), rabbit hepatocytes (ref. 3), rabbit liver microsomes (ref. 4)

Structure and biotransformation pathway

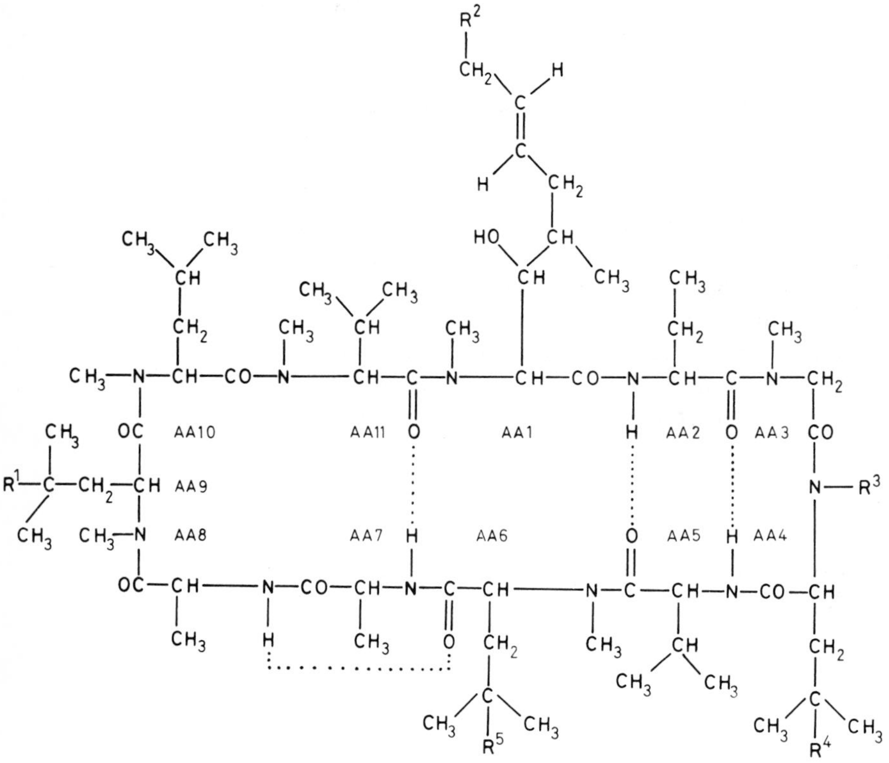

AA = Amino acid

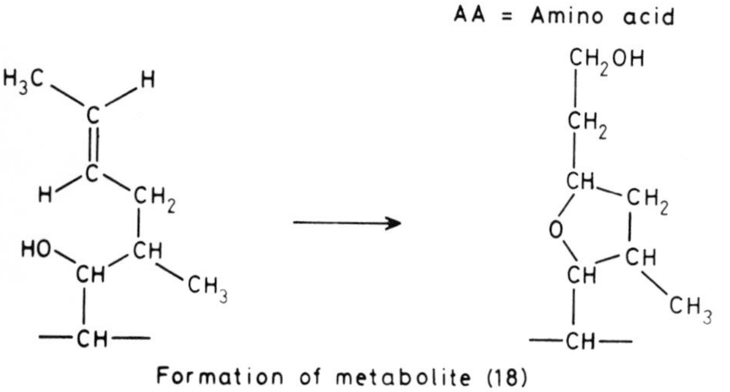

Formation of metabolite (18)

446

Metabolite No.	R^1	R^2	R^3	R^4	R^5	Other modification
Cyclosporin A	H	H	CH$_3$	H	H	
(1)	OH	H	CH$_3$	H	H	
(8)	OH	OH	CH$_3$	H	H	
(9)	OH	H	H	H	OH	
(10)	OH	H	CH$_3$	OH	H	
(16)	OH	H	CH$_3$	H	OH	
(17)	H	OH	CH$_3$	H	H	
(18)	H	OH	CH$_3$	H	H	Intramolecular cyclization of amino acid (1) side-chain (see below)
(21)	H	H	H	H	H	
(13)	hydroxylated and *N*-demethylated derivative of cyclosporin A (positions not definitely assigned)					

Cyclosporin A is a cyclic undecapeptide. Its α-amino acid sub-units have the natural *S*-configuration with the exception of amino acid 8 (D-alanine) which has the *R*-configuration. The ^{3}H-labelled compound was prepared biosynthetically and the positions of labelling were determined by ^{3}H NMR. Tritium was located in the methyl group at position 4 of amino acid 1 and in the seven *N*-methyl groups. The isolation and characterization of nine ether-soluble metabolites from the urine of dog and man and rat bile and faeces are described in ref. 1. Structural assignments were mainly based on ^{1}H and ^{13}C NMR, FD–MS, and the results of amino acid analysis after hydrolysis with hydrochloric acid.

The structures of several further metabolites could not be completely determined. All the identified metabolites retained the intact cyclic oligo-peptide structure of the parent drug. Transformation processes principally involved hydroxylation at the terminal carbon atom of amino acid 1 and at the γ-position of the *N*-methyl-leucines 4, 6, and 9 and *N*-desmethylation of the *N*-methyl-leucine 4.

Following oral administration (10 or 30 mg kg^{-1}) to rats (ref. 2) more than 70% dose was excreted in faeces and up to 15% in urine. Elimination in bile accounted for 10% and 60% of oral and intravenous (3 mg kg^{-1}) doses respectively. Identified metabolites accounted for 53% of the radioactivity in the 0–24 hour urine. The monohydroxylated metabolites (1) and (17) accounted for 1.5% and 0.8% dose respectively and the dihydroxylated metabolite (8) accounted for 0.7% dose. In bile (oral dose) the major identi-fied components were (8) (2.6% dose) and (1) (1.7% dose). Metabolites (17), (10) (dihydroxy), and (13) (2-monohydroxy-*N*-desmethyl) each accounted for 1.0–1.5% dose. Parent drug was the major radioactive component in blood and tissues but amounts in urine and bile were negligible.

In the study of the metabolism of CsA in rabbit hepatocytes (ref. 3), metabolite assignations were based on HPLC comparison with known stan-dards. Monohydroxylated (first generation) metabolites appeared rapidly in the intracellular medium of the hepatocytes and achieved a maximum intra-cellular concentration after 20 minutes. Dihydroxylated and dihydroxylated-*N*-demethylated metabolites (second generation) achieved steady-state intracellular concentrations between 30 and 60 minutes. The potential first generation non-hydroxylated-*N*-demethylated metabolite (21) was not formed in significant amounts. This latter metabolite was seen in studies of the metabolism of CsA in rabbit liver microsomes, together with the other metabolites already mentioned (ref. 4). Microsomes from animals which had received a variety of pretreatments were investigated, but only microsomes from animals treated with macrolide antibiotics exhibited a type I binding spectrum upon CsA addition. This led to the conclusion that macrolide antibiotic-inducible cytochrome P-450 3c (or a related form) was responsible for the major part of CsA metabolism by rabbit liver microsomes.

References

[1] G. Maurer, H. R. Loosli, E. Schreier, and B. Keller, Disposition of cyclosporine in several animal species and man, 1. Structural elucidation of its metabolites, *Drug Metab. Dispos.*, 1984, **12**, 120.

[2] O. Wagner, E. Schreier, F. Heitz, and G. Maurer, Tissue distribution, disposition and metabolism of cyclosporine in rats, *Drug Metab. Dispos.*, 1987, **15**, 377.

[3] G. Fabre, P. Bertault-Peres, I. Fabre, P. Maurel, S. Just, and J.-P. Cano, Metabolism of cyclosporin A, I. Study in freshly isolated rabbit hepatocytes, *Drug Metab. Dispos.*, 1987, **15**, 384.

[4] P. Bertault-Peres, C. Bonfils, G. Fabre, S. Just, J.-P. Cano, and P. Maurel, Metabolism of cyclosporin A, II. Implication of the macrolide antibiotic inducible cytochrome P-450 3c from rabbit liver microsomes, *Drug Metab. Dispos.*, 1987, **15**, 391.

Steroids

Pregnenolone

Use/occurrence:	Progestational hormone
Key functional groups:	Pregnene, steroid
Test system:	Mouse adrenal cells

Structure and biotransformation pathway

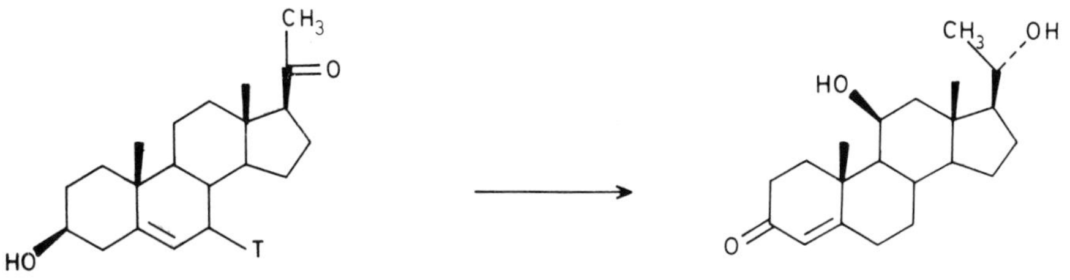

(1) T = ³H   (2)

A major metabolite of [³H]pregnenolone (1) was detected in the culture medium after incubation with Y-1 mouse adrenocortical tumour cells. The metabolite was analysed by HPLC and characterized by GC–MS. Identification as 11β,20α-dihydroxypregn-4-en-3-one was confirmed by comparison with the authentic reference compound. Other intermediate metabolites detected were progesterone and 20α-dihydroxyprogesterone.

Reference

M. Matsuguchi, L. Dehennin, G. Habrioux, L. Matsuguchi-Moreau, and H. Degrelle, Pregnenolone metabolism in Y-1 mouse adrenal cells: HPLC analysis and identification of metabolites by gas chromatography and mass spectrometry, *J. Steroid Biochem.*, 1987, **28**, 311.

5α-Cholest-8(14)-en-3β-ol-15-one

Use/occurrence:	Antilipemic
Key functional groups:	Cholestenone, steroid
Test system:	Baboon (oral, 75 mg kg^{-1})

Structure and biotransformation pathway

(1) $*$ = ^{14}C

(2)

A mixture of [2,4-^{3}H]- and [4-^{14}C]-(1) was administered orally to a male baboon which had been administered the non-radiolabelled drug for 85 days. Plasma samples taken at 4–24 hours after dosing were analysed by silicic acid column chromatography for the unchanged drug and cholesterol (2) and their palmitic acid esters which were available as authentic reference standards. Esters of (1) represented the major metabolites in plasma at early times but amounts of components corresponding to cholesterol and its esters increased throughout to reach maximum concentrations at 24 hours. The identity of cholesterol as a metabolite was confirmed by isolation and co-chromatographic analysis of the acetate with reference compounds. The cholesterol ester was similarly analysed after hydrolysis and also purified as the di-bromide derivative. There was no indication of the mechanism of formation of the metabolite.

Reference

G. J. Schroepfer, T. N. Pajewski, M. Hylarides, and A. Kisic, 5α-Cholest-8(14)-en-3β-ol-15-one. *In vivo* conversion to cholesterol upon oral administration to a non-human primate, *Biochem. Biophys. Res. Commun.*, 1987, **146**, 1027.

Homoursodeoxycholic acid

Use/occurrence:	Gallstone dissolution
Key functional groups:	Cholanic acid, steroid
Test system:	Rat (intraduodenal, 5 mg, intracaecumal, 1 mg)

Structure and biotransformation pathway

Following intraduodenal administration of (1) to biliary fistulated rats more than 80% of the radioactivity was excreted in bile within 0–24 hours. Bile was chromatographed on LH-20 columns to separate non-conjugated from material conjugated with glycine, or glucuronic acid or taurine. Analysis of the non-conjugated fraction by TLC indicated the presence of norursodeoxycholic acid (2) and β-muricholic acid (3). The identity of these compounds was confirmed by GC–MS after HPLC separation as bromophenacyl ester derivatives. The three acids were primarily converted into the glucuronide conjugates rather than to glycine or taurine conjugates. Following intracaecumal injection 50% of the radioactivity was excreted in bile largely as unconjugated material. In addition to (2) and (3) homolithocholic acid (4) was identified.

Reference

T. Kuramoto, S. Moriwaki, K. Kawamoto, and T. Hoshita, Intestinal absorption and metabolism of homoursodeoxycholic acid in rats, *J. Pharmacobio-Dyn.*, 1987, **10**, 309.

Cinobufagin

Use/occurrence:	Cardiotonic
Key functional groups:	Epoxide, lactone, steroid
Test system:	Male albino rats, male dogs, male and female cats (oral, 0.6 or 5 mg kg^{-1}, intravenous, 0.05 mg or 0.25 mg kg^{-1})

Structure and biotransformation pathway

	R^1	R^2
Cinobufagin (1)	OH	OCOCH$_3$
Desacetylcinobufagin (2)	OH	OH
3-Epidesacetylcinobufagin (3)	OH	OH

Samples of serum from rats, dogs, or cats dosed with cinobufagin (1) either intravenously or orally were analysed by alumina column chromatography and HPLC for the parent drug, desacetylcinobufagin (2), and 3-epidesacetyl-cinobufagin (3). After an intravenous dose to rats all three compounds were present in rat plasma, but after an oral dose only (3) and traces of cinobufagin were detected. Following an intravenous dose to cats or dogs only cinobufagin was measured in plasma, and after an oral dose none of the compounds could be detected in plasma. Cinobufagin was metabolized to (2) in rabbit serum, and to (2) and (3) in minced rat liver. NMR, IR, and GC–MS were used to confirm the identities of the metabolites isolated and purified by HPLC.

Reference

S. Toma, S. Marishita, K. Kuronuma, Y. Mishima, Y. Hirai, and M. Kawakami, Metabolism and pharmacokinetics of cinobufagin, *Xenobiotica*, 1987, **17**, 1195.

Dehydrocholic acid

Use/occurrence:	Choleretic
Key functional groups:	Alkyl ketone, steroid
Test system:	Rat (intravenous infusion, $10\ \mu$mol min^{-1} kg^{-1}), human patient (intravenous, 300 mg)

Structure and biotransformation pathway

Bile was collected at intervals before and after intravenous administration of dehydrocholic acid (1). Bile acids were isolated by solvent extraction after enzymatic hydrolysis. Components were derivatized by formation of O-methyloximes (keto groups) and dimethylethylsilyl ethers. Metabolites were identified by GC–MS and comparison with authentic reference compounds. The $3\alpha,7\beta$-hydroxy metabolite (2) was identified in both rat and human bile and the trihydroxylated metabolite (3) in rat bile. Previous work had shown that 7α-hydroxy metabolites were formed in animals and man.

Reference

Y. Miyazaki, H. Ichimya, H. Miyazaki, F. Nakayama, and M. Nakagaki, Identification of 7β-hydroxy metabolites of dehydrocholate in man and rat, *Chem. Pharm. Bull.*, 1987, **35**, 3955.

Norethindrone

Use/occurrence:	Contraceptive steroid
Key functional groups:	Ethynyl, oestren-3-one, steroid
Test system:	Lactating women (oral, 1 mg day^{-1})

Structure and biotransformation pathway

(1)

(2)

(3) R^1 = SO$_3^-$, R^2 = H
(4) R^1 = H , R^2 = SO$_3^-$
(5) R^1 = R^2 = SO$_3^-$

Ethynyl steroids were analysed in milk samples taken from a lactating woman administered 1 mg norethindrone per day for 13 days. Steroids were isolated by extraction and column chromatography and analysed by GC–MS as o-methyloxime trimethylsilyl ether derivatives. Sulphates were identified by acetylation followed by hydrolysis of the sulphate groups prior to derivatization as above. Besides norethindrone four sulphated tetrahydro derivatives were identified by comparison of GC retention times and partial mass spectra with those of reference compounds. The metabolites included the 3-sulphate of 5β-oestrane-3α,17β-diol (2) and the two monosulphates and a disulphate of 5α-oestrane-3α,17β-diol (3)–(5). The 3α,5α-disulphate was the major metabolite in the day 13 sample, representing about 200 pg equivalents ml^{-1}.

Reference

B. L. Sahlberg, The characterization of sulphated metabolites of norethindrone in human milk after oral administration of contraceptive steroids, *J. Steroid. Biochem.*, 1987, **26**, 481.

Testosterone

Use/occurrence:	Endogenous steroid
Key functional groups:	Androsten-3-one, steroid
Test system:	Rat skin microsomes

Structure and biotransformation pathway

(1) * = ^{14}C (2)   (3)   (4)

[^{14}C]Testosterone was incubated separately with microsomes from whole skin, dermis, and epidermis of neonatal rats. Metabolites were extracted with ethyl acetate and identified by TLC comparison with reference standards. Three hydroxylated compounds were formed from all microsomal samples, namely 16α- (2), 6β- (3), and 7α-hydroxytestosterone (4). Maximum hydroxylation occurred in the epidermal microsomes and NADPH was a necessary co-factor. Comparison of metabolism by hepatic and epidermal microsomes revealed that the latter possessed 3–5% of the hepatic activity in adult male rats.

Reference

H. Muktar, M. Athar, and D. R. Bickers, Cytochrome P-450 dependent metabolism of testosterone in rat skin, *Biochem. Biophys. Res. Commun.*, 1987, **145**, 749.

4-Hydroxyoestradiol, 2-Hydroxyoestradiol

Use/occurrence:	Oestrogen metabolites
Key functional groups:	Oestradiol, steroid
Test system:	Male human (intravenous infusion, 90 or 150 μg during 90 min)

Structure and biotransformation pathway

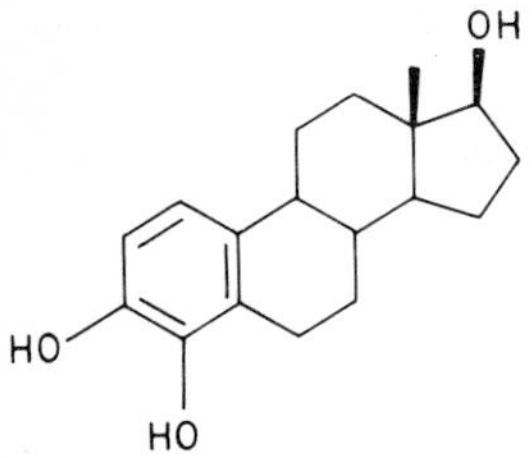

(1) 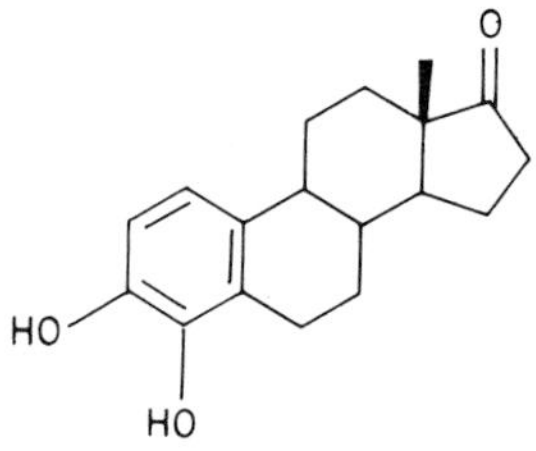 (2) 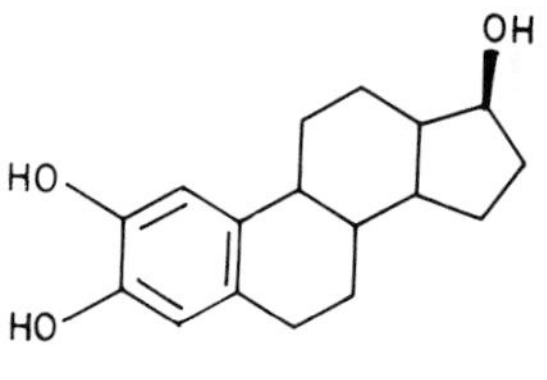 (3)

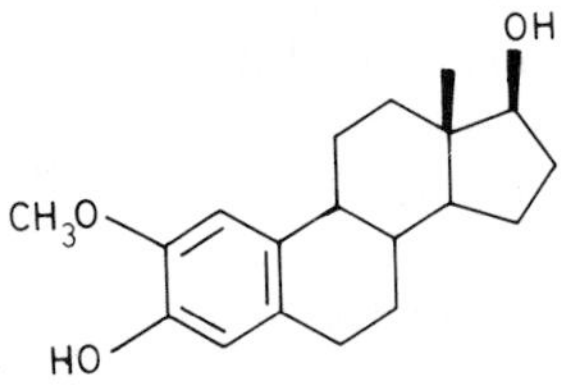

(4) 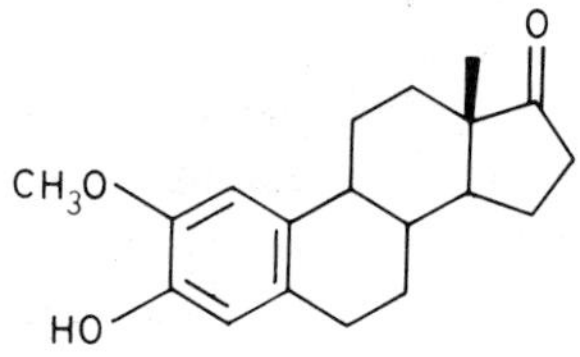 (5)

During infusion of 4-hydroxyoestradiol (1) significant plasma concentrations were measured whereas the 2-hydroxyoestradiol was so rapidly cleared that levels were almost undetectable. Metabolites of both compounds were monitored in plasma and urine. 4-Hydroxyoestradiol was eliminated mainly as glucuronide or sulphate conjugates of itself or the hydroxyoestrone (2). 2-Hydroxyoestradiol was extensively converted into the 2-methylethers (4) and (5) which were excreted in urine as conjugates.

Reference

G. Emons, G. R. Merriam, D. Pfeiffer, D. L. Loriaux, P. Ball, and R. Knuppen, Metabolism of exogenous 4- and 2-hydroxyestradiol in the human male, *J. Steroid Biochem.*, 1987, **28**, 499.

Oestrone, Dehydroepiandrosterone

Use/occurrence:	Model compounds
Key functional groups:	Oestrone, steroid
Test system:	Rat and guinea pig perfused liver

Structure and biotransformation pathway

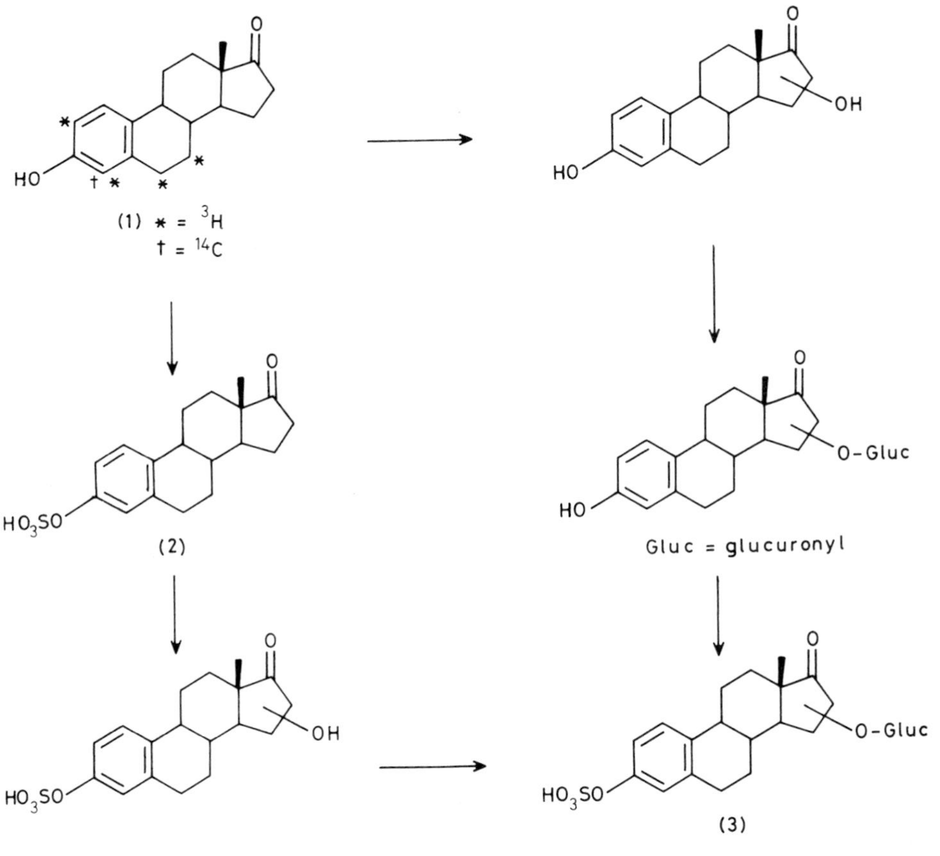

(1) * = ^{3}H
† = ^{14}C

Gluc = glucuronyl

Species differences in the metabolism of $[^3H]/[^{14}C]$oestrone (1) and other steroids were investigated using the perfused rat and guinea pig liver. $^{35}SO_4$ was added to the perfusion system to monitor the production of sulphate metabolites. Sulphate metabolites of oestrone (2) were found to accumulate in the liver within 10 minutes but were converted into sulphoglucuronides [glucuronic acid in the D ring (3)] and partially hydrolysed to be reconjugated as glucuronides. The major biliary metabolites of oestrone were glucuronides and sulphoglucuronides, with unconjugated steroid, sulphate conjugates, and other polar metabolites being excreted. In the perfused guinea pig liver oestrone was rapidly taken up to form a single glucuronide which was

459

excreted in bile and secreted in the medium. Sulphoglucuronides were not formed. The rat liver also formed sulphoglucuronides of dehydroepiandrosterone but the guinea pig liver did not. This was attributed to a restricted ability to hydroxylate the D-ring of steroids.

Reference

A. B. Roy, C. G. Curtis, and G. M. Powell, The metabolism of oestrone and some other steroids in isolated perfused rat and guinea pig livers, *Xenobiotica*, 1987, **17**, 1299.

Budesonide

Use/occurrence:	Glucocorticoid
Key functional groups:	Acetal, prednisolone, steroid
Test system:	Rat and mouse 9000g liver supernatant

Structure and biotransformation pathway

The major metabolites of budesonide (1) are known to be 6β-hydroxybudesonide (2) and 16α-hydroxyprednisolone (3). These metabolites were identified by MS of fractions from an HPLC separation. A moving belt HPLC–MS interface has now been used to investigate the metabolism of (1) and this has revealed the presence of novel metabolites.

9000g supernatants from the livers of rats and mice were incubated with an equimolar mixture of (1) and [^{2}H$_8$]-(1) at a concentration of 10–20 μM. Both epimers (22R and 22S) were studied separately. Metabolites were purified by SepPak extraction followed by reverse phase HPLC, linked by a moving belt interface to a mass spectrometer.

461

In addition to hydroxylation at the 6β-position a further hydroxylated metabolite (4) was identified. This must be hydroxylated on the acetal group as one deuterium atom was lost in its formation from $[^2H_8]$-(1). (A deuterium isotope effect k_H/k_D of ca. 6–7 was also observed.) The mechanism of formation of (3) was studied by carrying out incubations under $^{18}O_2$. No ^{18}O was incorporated into (3) although it was present in (2). Evidence was also obtained for the formation of Δ^6-budesonide (5).

Reference

C. Lindberg, J. Paulson, and S. Edsbäcker, The use of on-line liquid chromatography/mass spectrometry and stable isotope techniques for the identification of budesonide metabolites, *Biomed. Environ. Mass Spectrom.*, 1987, **14**, 535.

Budesonide

Use/occurrence:	Glucocorticoid
Key functional groups:	Acetal, prednisolone, steroid
Test system:	9000g supernatant fraction from livers of rat, mouse, and man

Structure and biotransformation pathway

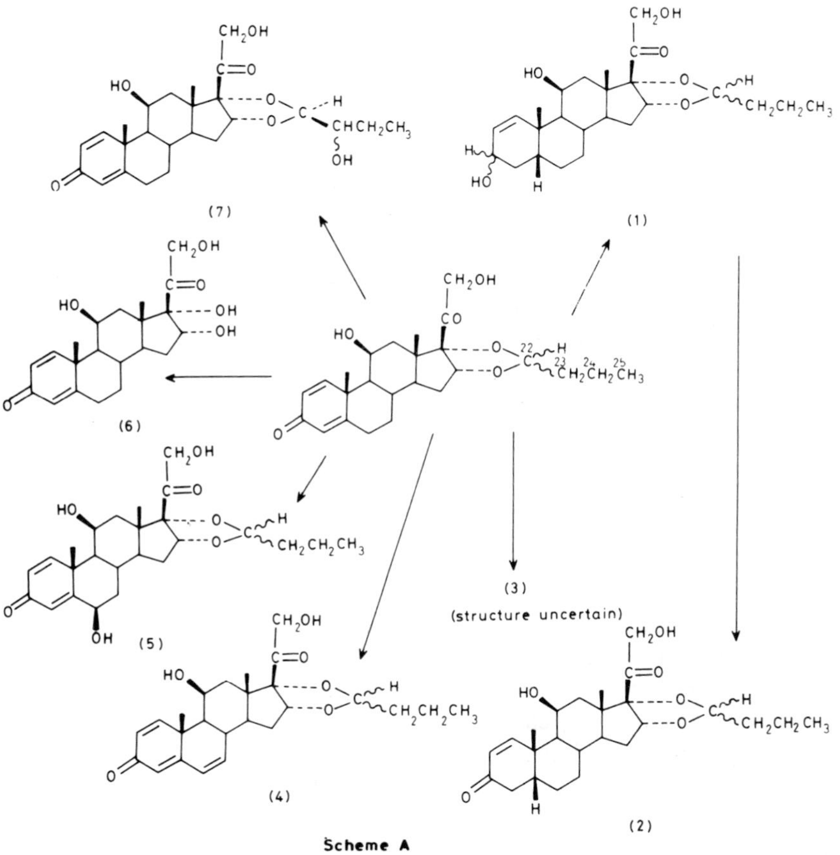

Scheme A

Budenoside is a C-22 1:1 epimeric mixture, the $(22R)$-epimer having 2–3 times higher topical glucocorticoid potency than the $(22S)$-epimer. In this study (Scheme A, ref. 1) the C-22 epimers were separately studied and metabolites were characterized by MS and ^{1}H NMR. Incubations for analysis by MS utilized a 1:1 mixture of budesonide and [^{2}H$_8$]budesonide while [1,2-^{3}H]-labelled budesonide was added in trace amounts in incubations for subsequent analysis by NMR. There were significant differences in the metabolism of the two epimers and also species differences in the metabolism of each

463

Scheme B

separate epimer. Oxidative metabolism predominated and the 6β-hydroxy metabolite (5) was formed from both epimers in all three species. However, of the other oxidative metabolites, (6) was formed from the $(22R)$-epimer, whereas (7) was produced from the $(22S)$-epimer [the structural assignation of (7) as the 23-hydroxy compound was tentative]. Compound (4) appeared to be a minor metabolite except in the case of the $(22S)$-epimer in the mouse. For each epimer, rates and routes of metabolism in liver were more similar to man in the mouse than in the rat. For the $(22R)$-epimer (5) was the major oxidative metabolite in the male rat, whereas (6) predominated in mouse and man (both sexes). In the female rat liver budesonide metabolism was slow and reductive pathways were more important. For the $(22S)$-epimer [which was metabolized more slowly than the $(22R)$-epimer except in the rat] relative proportions of oxidative metabolites (5) and (7) were generally similar, (5) being the more important in all cases except the female rat. A-Ring reduction to yield metabolites (1) and (2) was quantitatively more important in the rat: metabolite (1) was not detected in incubations of mouse or human liver and (2) only as a minor metabolite in human samples. In female rat liver, reductive enzymes were located in the 100 000g supernatant fraction. Subsequent incubation of isolated (1) and (2) in the 9000g supernatant fraction showed that (1) could be transformed to (2) but the reverse reaction did not occur. Metabolite (3) could not be sufficiently purified to allow complete identification but UV and MS data indicated that the A-ring was reduced. Metabolite (3) was formed in incubations from livers of mouse and man but not rat. It was concluded that the mouse was a more relevant species than the rat in studies of the pharmacology and toxicology of budesonide. Substrate-selective oxidation of the non-symmetric $16\alpha,17\alpha$-acetal group was proposed as the reason for the differences in biotransformation of the two budesonide epimers.

The mechanism of the acetal splitting of the $(22R)$-epimer to yield (6) was further investigated (ref. 2), using the compound labelled with deuterium or tritium in the 22-, 23-, 24- and 25-positions, and by performing incubations under $^{18}O_2$. On the basis of results obtained, including the identification of butyric acid as a product and the non-detection of butanol or butyraldehyde,

464

a mechanism (Scheme B) was proposed. This mechanism involved initial hydroxylation of the acetal carbon, followed by rearrangement to an intermediary ester, which was then hydrolysed to the final product.

References

[1] S. Edsbäcker, P. Andersson, C. Lindberg, J. Paulson, A. Ryrfeldt, and A. Thalén, Liver metabolism of budesonide in rat, mouse and man, *Drug Metab. Dispos.*, 1987, **15**, 403.

[2] S. Edsbäcker, P. Andersson, C. Lindberg, A. Ryrfeldt, and A. Thalén, Metabolic acetal splitting of budesonide, *Drug Metab. Dispos.*, 1987, **15**, 412.

9α-Fluorobudenoside

Use/occurrence:	Glucocorticosteroid
Key functional groups:	Acetal, fluoroalkyl, steroid
Test system:	Rat and human liver 9000g supernatant fraction

Structure and biotransformation pathway

$$
\begin{array}{ll}
(1) & R = \quad H \\
(2) & R = \alpha\text{-}F
\end{array}
\qquad\qquad
\begin{array}{ll}
(3) & R = \quad H \\
(4) & R = \alpha\text{-}F
\end{array}
$$

9α-Fluorobudenoside (1) and $6\alpha,9\alpha$-difluorobudenoside (2) (5 μM) were metabolized by rat and human liver 9000g supernatant fraction to triamcinolone [9α-fluoro-16α-hydroxyprednisolone, (3)] and fluocinolone [$6\alpha,9\alpha$-difluoro-16α-hydroxyprednisolone (4)] respectively. Metabolism was stereospecific, occurring with the $(22R)$- but not the $(22S)$-epimers. Metabolites were identified by retention times on HPLC but were not quantified.

Reference

P. Anderson, M. Lihne, A. Thalen, and A. Ryrfeldt, Effect of structural alterations on the biotransformation rate of glucocorticosteroids in rat and human liver, *Xenobiotica*, 1987, **17**, 35.

Spironolactone

Use/occurrence:	Diuretic
Key functional groups:	Acetylthio, steroid
Test system:	Guinea pig hepatic and renal microsomes, guinea pig (intraperitoneal, 25 mg kg^{-1})

Structure and biotransformation pathway

Analysis of (1) and metabolites formed by microsomes was performed by HPLC. Analysis of metabolites in plasma was carried out by GC–MS. Incubation of (1) with microsomes resulted in the formation of a single metabolite, the 7α-thio derivative (2). The rate of formation of (2) was higher in liver microsomes than in kidney microsomes. The addition of S-adenosylmethionine (SAM) brought about the formation of the 7α-thiomethyl metabolite (3) in both liver and kidney preparations. Confirmation of this metabolic route was obtained from the formation of (3) when (2) was added to microsomal preparations containing SAM. Small amounts of canrenone (4) were also produced independently of SAM addition.

The major metabolite in plasma after intraperitoneal administration of (1) to guinea pigs was (3). Lower amounts of (4) were also detected. In contrast, in guinea pigs receiving (4), (4) was the major compound in plasma and (3)

467

could not be detected. These results indicate that (3) is not produced from conjugates of (4) by the C–S lyase enzyme system.

Reference

L. B. La Cagnin, P. Lutsie, and H. D. Colby, Conversion of spironolactone to 7α-thiomethyl-spironolactone by hepatic and renal microsomes, *Biochem. Pharmacol.*, 1987, **36**, 3439.

Gomphoside

Use/occurrence:	Cardiotonic agent
Key functional groups:	Acetal, digitoxigenin, steroid
Test system:	Rat (intraperitoneal, 2 mg kg^{-1}), rat liver microsomes

Structure and biotransformation pathway

[G-^{3}H]Gomphoside (1) was dosed to rats with cannulated bile ducts and metabolites were isolated from 0–8 hour bile samples which contained about 68% of the administered dose, but only about 10% of this material was chloroform-extractable. Liver microsomes from phenobarbitone-induced rats were also used to investigate formation of gomphoside metabolites when a maximum conversion of 16% was achieved during a one hour incubation. Similar non-polar metabolites were detected in bile and microsomal incubations consisting of one major and four minor components. Metabolites were isolated and purified by a combination of TLC and HPLC and identified by MS and NMR. The major metabolite was identified as gomphosgenin (3) derived by cleavage of the acetal-linked carbohydrate moiety. Identification was confirmed by comparison with the authentic compound obtained by hydrolysis of gomphoside. A minor metabolite (7) was shown to be an isomer

469

of (3) resulting from epimerization of one of the hydroxy groups. This metabolite could be formed via the intermediate ketone (6). Two further metabolites were shown to be the $3'\alpha$-hydroxy- and $3'\beta$-hydroxy-19-oxo compounds (4), confirmed by comparison with authentic reference standards. The metabolite (2) was derived by opening of the $2'$-hemiacetal and metabolic reduction of the resulting keto group which was confirmed by comparison with the compound synthesized by sodium cyanoborohydride reduction of gomphoside. A single major polar metabolite in bile was shown to be a glucuronide of gomphoside (5). The position of conjugation was assigned as the $3'$-hydroxy on the basis of mass spectrometry and ^{1}H and ^{13}C NMR.

Reference

A. E. Mutlib, H. T. A. Cheung, and T. R. Watson, *In vivo* and *in vitro* metabolism of gomphoside, a cardiotonic steroid with doubly-linked sugar, *J. Steroid Biochem.*, 1987, **65**, 28.

Mespirenone

Use/occurrence:	Aldosterone antagonist
Key functional groups:	Acetylthio, lactone, pregnadiene, steroid
Test system:	Rat (oral, 200 mg kg^{-1})

Structure and biotransformation pathway

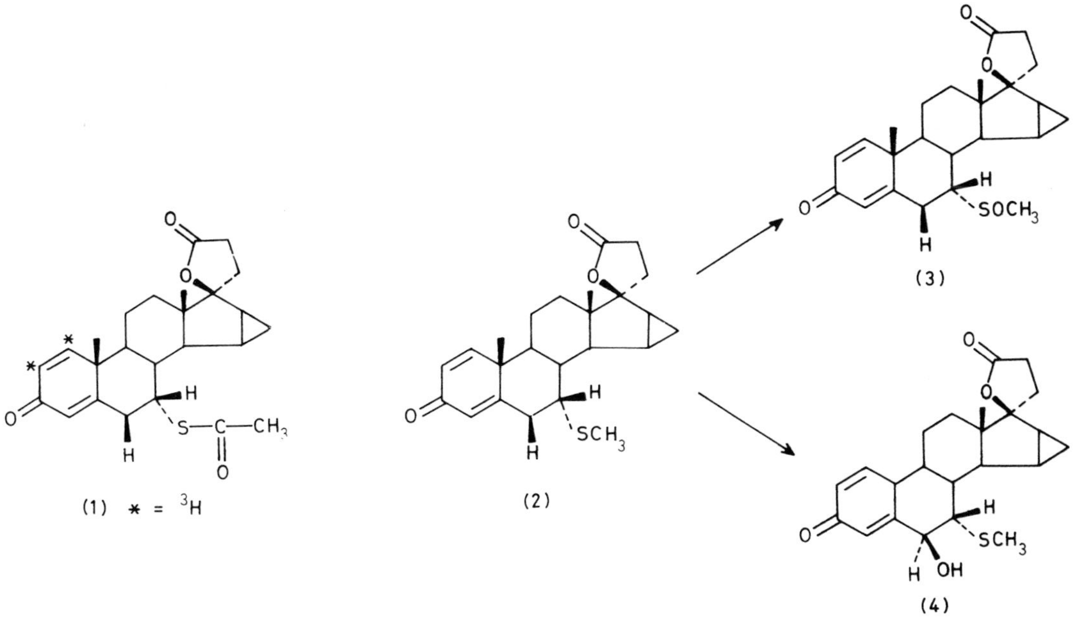

(1) ∗ = ^{3}H

(2)

(3)

(4)

Urinary metabolites of mespirenone (1) were extracted into organic solvent, separated by HPLC, and subsequently identified by MS and NMR. 12% of the radioactivity was excreted in urine, only 20% of which was extractable. HPLC revealed six components, four of which were characterized. Unchanged (1) in urine represented 0.1% of the dose. The 7α-thiomethyl metabolite (2; 0.3% dose) probably arises through deacetylation and subsequent S-methylation. Further metabolism produced the sulphoxide (3; 1.0% dose) and hydroxy derivative (4; 0.4% dose).

Reference

M. Hildebrand, W. Krause, G. Kuhne, and G.-A. Hoyer, Pharmacokinetics and metabolism of mespirenone, a new aldosterone antagonist, in rat and cynomolgus monkey, *Xenobiotica*, 1987, **17**, 623.

Miscellaneous

Selenocyanate

Use/occurrence:	Possible metabolite of selenium
Key functional groups:	Selenocyanate
Test system:	Rat (subcutaneous, 2 mg kg^{-1})

Structure and biotransformation pathway

$$(CH_3)_2Se \quad \longleftarrow \quad SeCN^- \quad \longrightarrow \quad (CH_3)_3Se^+$$
$$(2) \qquad\qquad\qquad (1) \qquad\qquad\qquad (3)$$

This metabolic study of selenocyanate (1) was initiated because of the compound's possible formation as a metabolite of selenium. [^{75}Se]-(1) and [^{14}C,^{75}Se]-(1) were synthesized and injected subcutaneously into male rats. $26.8 \pm 8.1\%$ of the dose was excreted in the breath as dimethyl selenide (2), which was trapped by nitric acid or benzyl chloride. There was no ^{14}C in (2). The major urinary metabolite, accounting for $14.5 \pm 5.1\%$ of the total dose, was the trimethylselenonium ion (3), which was determined by cation-exchange HPLC. Again no ^{14}C was incorporated into (3). Only about 2% of the dose of (1) was recovered from the urine as unchanged material. The uptake of selenium was highest in the kidney ($1.89 \pm 0.2\%$ of the dose g^{-1}) followed by the liver, blood, adrenal, and heart.

Reference

S. Vadhanavikit, R. J. Kraus, and H. E. Ganther, Metabolism of selenocyanate in the rat, *Arch. Biochem. Biophys.*, 1987, **258**, 1.

Dimethylarsinic acid

Use/occurrence: Herbicide

Key functional groups: Organoarsenic

Test system: Mouse, hamster (oral, 40 mg As kg^{-1}), man (oral, 0.1 mg As)

Structure and biotransformation pathway

$$(CH_3)_2\overset{\overset{\displaystyle O}{\|}}{As}\text{—OH} \longrightarrow (CH_3)_3As=O$$

$$(1) \qquad\qquad\qquad (2)$$

Following administration of [^{74}As]dimethylarsinic acid (1) to hamsters and mice 57% and 69% dose respectively were excreted in urine and 42% and 29% dose respectively were excreted in faeces. Ion exchange chromatography, paper electrophoresis, thin-layer chromatography and arsine generation–gas chromatography combined with atomic absorption spectrophotometry or mass spectrometry were used to characterize arsenic compounds. In mice and hamsters, 3.5% and 6.4% dose respectively were excreted in urine as trimethylarsine oxide (2). 76% dose (hamsters) or 80.7% dose (mice) were eliminated in urine and faeces as unchanged (1), while a further 13–15% dose was excreted as an unidentified complex of (1), mainly in urine. No demethylation of (1) was observed. In a human subject about 4% dose was excreted in urine as trimethylarsine oxide (2) and about 80% as unchanged (1).

Reference

E. Marafante, M. Vahter, H. Norin, J. Envall, M. Sandström, A. Christakopoulos, and R. Ryhage, Biotransformation of dimethylarsinic acid in mouse, hamster and man, *J. Appl. Toxicol.*, 1987, **7**, 111.

Sodium arsenite

Use/occurrence:	Model compound
Key functional groups:	Inorganic arsenic
Test system:	Rat (intraperitoneal and oral, $0.25-5$ mg kg^{-1})

Structure and biotransformation pathway

$$NaAsO_2 \longrightarrow (CH_3)AsO \longrightarrow (CH_3)_2AsOH$$

$$(1) \qquad\qquad (3)$$

Like man the rat inactivates inorganic arsenic by methylation to mono-methylarsonic (1; MMA) and dimethylarsinic (2; DMA) acids which are excreted in the urine along with inorganic arsenic. In this study the effect of a variety of agents on the biotransformation of inorganic arsenic has been studied *in vivo*. Reduction of hepatic glutathione levels by greater than 90% following phorone pretreatment greatly modified the *in vivo* metabolism of inorganic arsenic leading to decreased urinary excretion of MMA (1) and DMA (2). This was also associated with an increased accumulation of in-organic arsenic in the liver. Thus a drastic reduction of hepatic glutathione not only impairs the methylation of inorganic arsenic but also impairs its biliary excretion. When glutathione depletion was less severe the total amount of arsenic excreted was similar to that in non-pretreated rats. How-ever, the proportions of MMA (1), DMA (2), and inorganic arsenic were different: MMA was reduced whereas DMA and inorganic acid were increased.

Reference

J. P. Buchet and R. Lauwerys, Study of factors influencing the *in vivo* methylation of inorganic arsenic in the rat, *Toxicol. Appl. Pharmacol.*, 1987, **91**, 65.

Potassium bromate

Use/occurrence:	Food additive
Key functional groups:	Bromate
Test system:	Rat (oral)

Structure and biotransformation pathway

KBrO$_3$
(1)

(2)

Potassium bromate (1), a strong oxidant which has been used as a food additive, is a renal carcinogen in rats. Many such compounds can form oxygen radicals which interact with DNA to form 8-hydroxydeoxyguanosine (8-OH-dG) (2). Potassium bromate and two non-carcinogenic oxidants, NaClO and NaClO$_2$, were administered orally to rats, and DNA prepared from liver and kidney was analysed by HPLC for (2). A significant increase in this altered base was found in kidney DNA only after treatment with (1). Also the lesion was not evident in liver, a non-target tissue. These results appear to suggest that formation of (2) is closely linked to the carcinogenic effect of (1).

Reference

H. Kasai, S. Nishimura, Y. Kurokawa, and Y. Hayashi, Oral administration of the renal carcinogen, potassium bromate, specifically produces 8-hydroxydeoxyguanosine in rat target organ DNA, *Carcinogenesis.*, 1987, **8**, 1959.

Cypermethrin, 3-(2,2-Dichlorovinyl)-2,2-dimethylcyclopropanecarboxylic acid

Use/occurrence:	Insecticide and one of its metabolites
Key functional groups:	Benzyl nitrile, chloroalkene, cyclopropyl carboxylate, dimethylcyclopropyl
Test system:	Hens (oral, 3–6 mg kg^{-1})

Structure and biotransformation pathway

479

Hens eliminated 75–86% of an oral dose of either [^{14}C]cypermethrin, [^{14}C]-*cis*-3-(2,2-dichlorovinyl)-2,2-dimethylcyclopropanecarboxylic acid (1), or [^{14}C]-*trans*-3-(2,2-dichlorovinyl)-2,2-dimethylcyclopropanecarboxylic acid (8) within 24 hours. Methanol extracts of excreta from hens given the *cis*-acid (1) contained four major components, which were present both free and as conjugates (mainly glucuronides). These were identified by GC and GC–MS. The parent acid (1) accounted for 40% of the radioactivity in the extract, metabolite (5) for 29%, and metabolites (6) and (7), which were not separately quantified, together accounted for 11%. A small amount (3%) of the lactone (3) and possibly the diacid (4) were also formed. A larger proportion (65%) of the *trans*-acid (8) was eliminated, either free or in a conjugated form. The *cis*-hydroxy metabolite (9) accounted for 9%, and the diacids (10) and (13), together with the hyroxy-acid (12), accounted for 4% of the radioactivity in the extract. These aglycones were also excreted as conjugates (probably glucuronides). The lactone (11) was also present (4%). In both the *cis*- and *trans*-acids metabolism occurred preferentially in the methyl group *trans* to the dichlorovinyl moiety.

Cypermethrin gave metabolites which were formed from both the *cis* (1) and *trans*- (8) acids and only small amounts of unchanged parent (4%) were excreted. There was no evidence for oxidation of either the methyl group or the aromatic ring before hydrolysis of the ester. *cis*- and *trans*-3-(2,2-dichlorovinyl)-2,2-dimethylcyclopropanecarboxylic acid (1) and (8) accounted for 14 and 46% of the radioactivity in methanol extracts of excreta. Minor metabolites present were the lactones (4) and (11) (3%), the hydroxy- and di-acids (5), (10), (12), and (13) (8%), and the hydroxy- and di-acids (6), (7), and (9) (10%).

Reference

M. H. Akhtar, R. M. G. Hamilton, and H. L. Trenholm, Excretion, distribution and depletion of [^{14}C]cypermethrin and *cis*-and *trans*-isomers of 3-(2,2-dichlorovinyl)-2,2-dimethylcyclopropanecarboxylic acid administered orally to laying hens, *Pestic. Sci.*, 1987, **20**, 53.

Cypermethrin

Use/occurrence:	Insecticide
Key functional groups:	Benzylnitrile, chloroalkene, cyclopropyl carboxylate, dimethylcyclopropyl
Test system:	Chicken (oral, 1.5 mg)

Structure and biotransformation pathway

Chickens were given daily oral doses of $[^{14}C]$cypermethrin for up to 14 days. After 10 days the radioactivity in eggs (principally in the yolk) consisted mainly of unchanged cypermethrin. *trans*-Hydroxycypermethrin (1) was tentatively identified as a minor metabolite (4% of the residue) together with 3-phenoxybenzoic acid (3) and 4'-hydroxycypermethrin (2) (each about 1%). 3-Phenoxybenzoic was also identified in the liver (*ca.* 3% of the residue). Metabolites were identified by TLC comparison with authentic reference compounds.

Reference

D. H. Hutson and G. Stoydin, Excretion and residues of the pyrethroid insecticide cypermethrin in laying hens, *Pestic. Sci.*, 1987, **18**, 157.

Cypermethrin (*cis* and *trans*)

Use/occurrence: Insecticide

Key functional groups: Benzyl nitrile, chloroalkene, cyclopropyl carboxylate, dimethylcyclopropyl

Test system: Trout (oral, 6.6 mg or in water)

Structure and biotransformation pathway

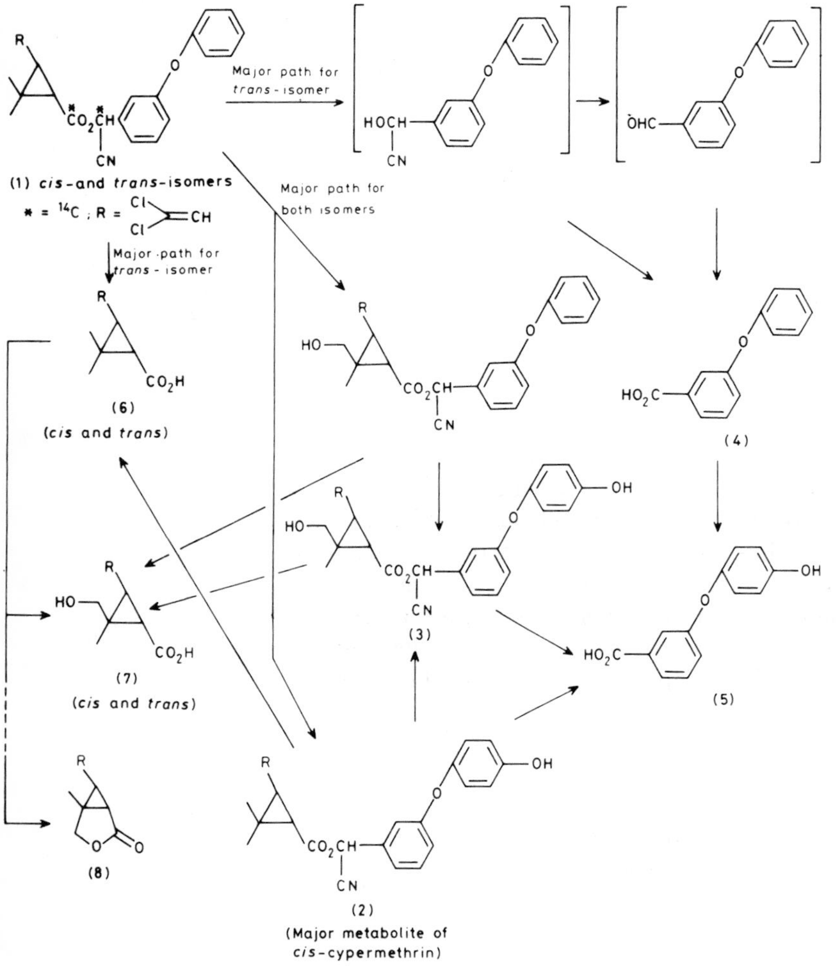

(1) *cis-* and *trans-*isomers
* = ^{14}C ; R = Cl₂C=CH—

(6) (*cis* and *trans*)

(7) (*cis* and *trans*)

(8)

(2) (Major metabolite of *cis*-cypermethrin)

(3)

(4)

(5)

cis- or *trans-*Cypermethrin (1) was administered to rainbow trout either orally (food) or in water. Orally administered cypermethrin was poorly absorbed from the intestine but was excreted from the intestine into water and via the gills. The principal route of elimination *in vivo* was via the bile,

482

with 20–28% of the dose absorbed excreted as biliary metabolites in 24 hours. The glucuronide of 4'-hydroxycypermethrin (2) was the major metabolite (80% of total bile radioactivity for the *cis*-isomer, but only 30% for the *trans*-isomer). Dichlorovinyldimethylcyclopropanecarboxylic acid (6) and its glucuronide, 3-(4-hydroxyphenoxy)benzoic acid (5), its sulphate and ester and ether glucuronides, and 3-phenoxybenzoic acid (4) glucuronide and taurine conjugates were detected by TLC, GLC, and HPLC.

Reference

R. Edwards, P. Millburn, and D. H. Hutson, The toxicity and metabolism of the pyrethroids *cis*- and *trans*-cypermethrin in rainbow trout, *Salmo gairdneri, Xenobiotica*, 1987, **17**, 1175.

Fenitrothion

Use:	Insecticide
Key functional groups:	*O*-Methyl-phosphorothioate, nitrophenyl
Test system:	Killifish, mullet (0.1 p.p.m. in water)

Structure and biotransformation pathway

$$(1) \ast = {}^{14}C \qquad (2) \qquad (3) \qquad (4) \qquad (5)$$

Killifish and mullet were exposed to 0.1 p.p.m. [^{14}C]fenitrothion (1) for up to 7 and 3 days respectively. Killifish were exposed in freshwater at 15 and 25 °C and in seawater at 25 °C. Under all three conditions the glucuronide conjugate of 3-methyl-4-nitrophenol (2) was the major metabolite found in the fish and accounted for between 16 and 58% of the radioactivity present. Desmethylfenitrothion (4) (1–7% of the radioactivity) was also present together with small amounts (< 10%) of the metabolites (2), (3), and (5). Fenitrothion accounted for greater than 88% of the radioactivity in the water, but metabolites (2)–(5) were also present.

Mullet were exposed to treated fresh- and sea-water at 25 °C. Desmethylfenitrothion (4) was the major metabolite found in the fish (45–70% of the radioactivity) together with the glucuronide of (2) (14–21%). Small amounts of metabolites (2), (3), and (5) were also identified. The fenitrothion content of the water was low (42–63% of the ^{14}C), and metabolites (2)–(5) were present in larger amounts than in the water to which the killifish were exposed. All metabolites were identified by TLC comparison with authentic reference compounds.

Reference

Y. Takimoto, M. Ohshima, and J. Miyamoto, Comparative metabolism of fenitrothion in aquatic organisms, I. Metabolism in the euryhaline fish *Oryzias latipes* and *Mugil cephalus*, *Ecotoxicol. Environ. Safety*, 1987, **13**, 104.

Fenitrothion

Use:	Insecticide
Key functional groups:	*O*-Methyl-phosphorothioate, nitrophenyl
Test system:	Freshwater snails (0.1 p.p.m. in water)

Structure and biotransformation pathway

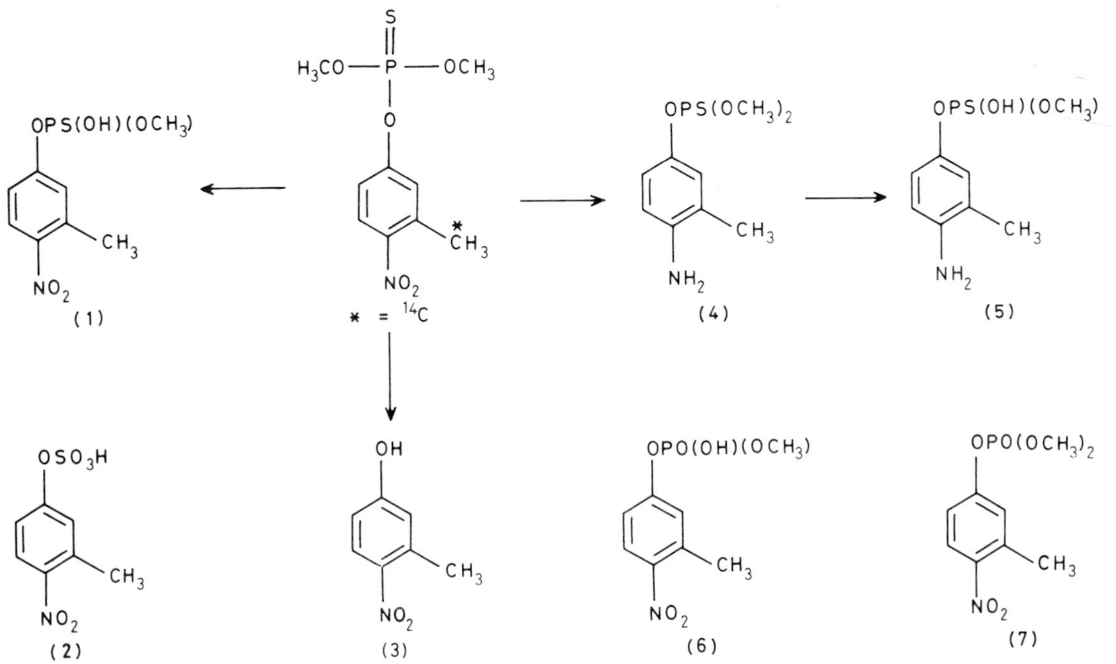

Two species of freshwater snails, *C. japonica* (pond snails) and *P. acuta*, were exposed to 0.1 p.p.m. [^{14}C]fenitrothion for 3 days using a dynamic flow system at 25 °C.

In the pond snail the concentration of fenitrothion reached equilibrium (1.13 p.p.m.) within one day of exposure and the parent compound constituted 52% of the radioactivity. The major metabolites found in the snail were desmethyfenitrothion (1) (19–23%) and 3-methyl-4-nitrophenyl sulphate (2) (8–10%). 3-Methyl-4-nitrophenol (3), aminofenitrothion (4), desmethyl-aminofenitrothion (5), and desmethylfenitrooxon (6) were the minor metabolites (<8%) found in the snails. 3-Methyl-4-nitrophenol (3) was the major metabolite present in the water, accounting for 7–10% of the radioactivity; (1), (2), (4), (6), and fenitrooxon (7) were also present as minor metabolites. Fenitrothion constituted more than 70% of the radioactivity in the water. Approximately 90% of the radioactivity was eliminated from the snails within

485

3 days of their transfer to fresh water. In the day 3 water fenitrothion and 3-methyl-4-nitrophenol (3) were the major components and each accounted for 20% of the radioactivity.

The concentration of fenitrothion also reached equilibrium (4.53 p.p.m.) within one day in *P. acuta*, and it accounted for 61–66% of the radioactivity present in the snail. Desmethylfenitrothion (1) and the glucoside of 3-methyl-4-nitrophenol (3) were the major metabolites (18–21% and 5–7% respectively of the radioactivity present). Metabolites (2)–(7) were also present but in small amounts (< 4%). In the exposure water fenitrothion constituted more than 88% of the radioactivity; small quantities (< 6%) of metabolites (1), (3), (4), (6), and (7) were present. When the snails were transferred to fresh water *ca.* 88% of the radioactivity was excreted into the recovery water in one day. Fenitrothion and 3-methyl-4-nitrophenol (3) accounted for 50% and 15% of the radioactivity respectively. Other components, except (3), were identified as minor metabolites (< 4%).

Reference

Y. Takimoto, M. Ohshima, and J. Miyamoto, Comparative metabolism of fenitrothion in aquatic organisms. II. Metabolism in the freshwater snails, *Cipangopaludina japonica* and *Physa acuta, Ecotoxicol. Environ. Safety,* 1987, **13,** 118.

Fenitrothion

Use:	Insecticide
Key functional groups:	*O*-Methyl-phosphorothioate, nitrophenyl
Test system:	Crustaceans (1 p.p.b. in water)

Structure and biotransformation pathway

Waterfleas (*D. pulex*) and shrimps (*P. paucidens*) were exposed to 1 p.p.b. [^{14}C]fenitrothion using a flow through system at either 18 °C (waterfleas) or 25 °C (shrimps).

The concentration of both fenitrothion (68 p.p.b.) and radioactivity (90 p.p.b.) reached equilibrium in the waterfleas after 8 hours exposure. The metabolites present in the waterfleas were desmethylfenitrooxon (1), 3-methyl-4-nitrophenyl sulphate (2), desmethylfenitrothion (3), 3-methyl-4-nitrophenol (4), and fenitrooxon (5), none of which accounted for more than 9% of the radioactivity. More than 90% of the radioactivity in the exposure water at 24 hours consisted of fenitrothion. The major metabolite in the water was 3-methyl-4-nitrophenol but this only constituted, at most, 2–4% of the radioactivity. Compounds (1), (2), (3), and (5) were present as minor metabolites. On transferring the waterfleas to freshwater, 95% of the radioactivity was eliminated within one day. Fenitrothion accounted for 64% of the radioactivity in the water together with smaller amounts ($<9\%$) of the metabolites (1)–(5).

In the shrimp fenitrothion reached equilibrium on day 1, the concentration being 5 p.p.b., and it comprised 44–53% of the radioactivity. 3-Methyl-4-nitrophenol (4) and its glucoside (6) each accounted for 8–11% of the radioactivity in the shrimp, and metabolites (1), (3), and (5) were also detected in small amounts ($<5\%$ of the radioactivity). The exposure water contained fenitrothion (85% of the radioactivity) together with metabolites (4) (2–10%), (1), (3), (5), and (6) (each less than 1%). When they were transferred to freshwater, 80% of the radioactivity in the shrimps was excreted into the water in 8 hours. The major products eliminated were 3-methyl-4-nitrophenol (4) (38% of the radioactivity originally in the shrimp), 3-methyl-4-nitrophenyl-β-glucoside (6) (22%), and fenitrothion (8%). Small amounts of metabolites (1), (3), and (5) were also detected.

Reference

Y. Takimoto, M. Ohshima, and J. Miyamoto, Comparative metabolism of fenitrothion in aquatic organisms. III. Metabolism in the crustaceans, *Daphnia pulex* and *Palaemon paucidens*, *Ecotoxicol. Environ. Safety*, 1987, **13**, 126.

Methylparathion

Use/occurrence:	Insecticide
Key functional groups:	*O*-Methyl-phosphorothioate, nitrophenyl
Test system:	Partially purified human foetal liver glutathione S-transferase

Structure and biotransformation pathway

(1) * = ^{14}C   (2)   GS = glutathionyl

Partially purified human foetal liver glutathione S-transferase catalysed the metabolism of methylparathion (1) exclusively to desmethylparathion (2) via *O*-dealkylation. HPLC, radiometric analysis of the enzyme reaction, and co-chromatography with a reference standard were used to characterize desmethylparathion.

Reference

L. L. Radulovic, A. P. Kulkarni, and W. C. Dauterman, Biotransformation of methylparathion by human foetal liver glutathione S-transferases: an *in vitro* study, *Xenobiotica*, 1987, **17**, 105.

Thiovamidithion, Vamidithion

Use/occurrence: Organophosphorus insecticides

Key functional groups: Dialkyl thioether, methyl amide, *O*-methyl-phosphorodithioate, *O*-methylphosphorothioate

Test system: Rat, mouse liver fractions

Structure and biotransformation pathway

The *in vitro* metabolism of vamidithion and its thio analogue thiovamidithion was investigated using rat and mouse subcellular fractions. Thiovamidithion was extensively oxidized to thiovamidithion sulphoxide (4), vamidithion (1), and vamidithion sulphoxide (3) in addition to oxidative

490

hydrolysis products (6)–(9). Vamidithion sulphoxide (3) was the principal metabolite of vamidithion and only small amounts of hydrolysis products were detected. The cytochrome P-450 mono-oxygenase system was found to be responsible for most of the *in vitro* metabolism of vamidithion and its thio analogue. However, glutathione S-transferase showed a moderate activity toward thiovamidithion producing metabolites (2), (6), (7), (8), and (9). There was no evidence that either vamidithion or its thio analogue were substrates for amidase. Metabolites were identified by TLC co-chromatography and by FT-IR and FT-NMR.

Reference

M. A. El-Oshar, N. Motoyama, and W. C. Dauterman, *In vitro* metabolism of vamidithion and its thio analogue by rat and mouse liver, *J. Agric. Food Chem.*, 1987, **35**, 138.

Isofenophos

Use/occurrence:	Organophosphorus insecticide
Key functional groups:	Isoalkylamino, isoalkyl ester, aryl carboxylate phosphoroamidothioate
Test system:	Rat (oral, 5 mg kg^{-1}), *in vitro* liver microsomes

Structure and biotransformation pathway

The *in vivo* metabolism of isofenophos in the rat was studied using both [*O-ethyl*-1-^{14}C]- and [*ring*-^{14}C]-isofenophos. Approximately 80% of the administered dose was excreted in the urine following a single oral dose of [*O-ethyl*-^{14}C]isofenophos with 3% in the expired air and 10% in the faeces.

492

The major metabolite found following oral administration of [*ring-*¹⁴C]isofenophos was (1), which was present in both free and conjugated forms while metabolite (5) was present as a minor metabolite. Metabolites (3) and (6) were identified in benzene extracts but these metabolites represented only a small proportion of the dose. In the day 1 urine of rats dosed [*O-ethyl-*¹⁴C]isofenophos metabolites (10), (2), (7), and (4) accounted for 24%, 15%, 12%, and 6% of the dosed radioactivity. Other metabolites found were (8) and (9), each accounting for less than 3% of the dosed radioactivity. The results suggest that cleavage of the P–O-aryl linkage predominated over cleavage of the P–N linkage. Metabolites were identified by TLC co-chromatography with reference standards.

In the *in vitro* experiment the microsomal metabolism of the separated chiral isomers of [*O-ethyl-*¹⁴C]isofenophos was studied. Isofenophos oxon (3) was formed in greater amounts (4-fold) from the (−)-isofenophos than from the (+)-isofenophos isomer. No other quantitative differences in the extent of isofenophos metabolism appeared to be significant.

References

M. Ueji and C. Tomizawa, Metabolism of the insecticide isofenophos in rats, *J. Pestic. Sci.*, 1987, **12**, 245.

M. Ueji and C. Tomizawa, Metabolism of chiral isomers of isofenophos in the rat liver microsomal system, *J. Pestic. Sci.*, 1987, **12**, 269.

Cyclophosphamide

Use/occurrence:	Anti-cancer agent
Key functional groups:	*N*-Chloroethyl, phosphoramide
Test system:	Rat (intravenous, oral, intraperitoneal, 20–40 mg kg^{-1})

Structure and biotransformation pathway

Cyclophosphamide (1) is known to be activated via metabolism to 4-hydroxycyclophosphamide (2). This is in tautomeric equilibrium with its ring-opened form, aldophosphamide (3), and (3) spontaneously breaks down to the active alkylating agent, phosphoramide mustard (4). Alcophosphamide (5) [which could be formed by phosphorodiamide esterase action on (1) or by reduction of (3)] has been observed as a minor metabolite, but no detailed characterization of its production had been carried out until the present study. An equimolar mixture of (1) and a deuterated analogue [β-^{2}H$_4$]-(1) (20 mg kg^{-1} each) was administered intravenously to male Sprague–Dawley rats. The first 12 hour urine was extracted and the product was trimethylsilylated and examined by ammonia CI MS. Protonated molecular ion doublets (^{2}H$_0$/^{2}H$_4$) were observed for derivatized (1), (5), and carboxyphosphamide (6). GC of derivatized (5) was shown (by MS) to result in its dehydrochlorination. Compound (5) was quantitated in plasma (using [^{2}H$_8$]-(1) as internal standard) after an intravenous dose of 20 mg kg^{-1} (1) to rats; its terminal $t_{1/2}$

was 76.2 ± 13.7 min. The comparative production of (5) after oral or intra-peritoneal administration of (1) was also studied using co-administration via different routes of (1) and $[^2H_4]$-(1). The terminal $t_{1/2}$ was 106.7 ± 25.2 min for the oral route and 73.9 ± 5.2 min for intraperitoneal injection. Following intravenous administration of (5) at 20 mg kg^{-1}, phosphoramide mustard (4) was found to be a major circulating and urinary metabolite, which suggests that (5) could be of significance for the anti-cancer activity of (1).

Reference

P. S. Hong and K. K. Chan, Identification and quantification of alcophosphamide, a metabolite of cyclophosphamide, in the rat using chemical ionization mass spectrometry, *Biomed. Environ. Mass Spectrom.*, 1987, **14**, 167.

Zinc ethylenebisdithiocarbamate (zineb)

Use/occurrence:	Fungicide
Key functional groups:	Dithiocarbamate
Test system:	Rat and marmoset (oral, 50 mg kg^{-1})

Structure and biotransformation pathway

Metabolites of zineb (1) in ethyl acetate and methanol extracts of urine and faeces were investigated by TLC. The major metabolites consisted of ethylenethiourea (2) and undifferentiated polar material.

Ethylenethiourea was present mainly in urine and represented about 20% of the administered dose in both species. Smaller amounts of ethyleneurea (3) were also detected in urine, representing 2% and 5% dose in marmoset and rat respectively. There was some evidence that ethylenethiourea in excreta was photolabile.

Reference

A. J. F. Searle, A. C. Stewart, and M. Paul, The measurement of ethylenethiourea and ethyleneurea in the rat and common marmoset *Callithrix jacchus* after zineb (zinc ethylenebisdithiocarbamate) dosing, *Xenobiotica.*, 1987, **17**, 733.

Carbon disulphide

Use/occurrence:	Chemical/solvent
Key functional groups:	Thiocarbonyl
Test system:	Rat hepatocytes and liver microsomes

Structure and biotransformation pathway

Carbon disulphide (1) is metabolized in rat liver predominantly by the cytochrome P-450 containing mono-oxygenase system. The product, whose structure remains to be firmly established, is probably an unstable intermediate (2) which reacts with water to form monothiocarbonate (5) and a reactive sulphur species. Monothiocarbonate exists in solution in equilibrium with, and can be converted into, carbonyl sulphide (6). In intact hepatocytes monothiocarbonate is further metabolized to carbon dioxide and hydrosulphide (7) and thence to thiosulphate (8) and sulphate (4). Carbonyl sulphide was characterized by GC–MS. [^{14}C]Carbon dioxide was determined by gas radiochromatography.

Reference

C. P. Chengelis and R. A. Neal, Oxidative metabolism of carbon disulphide by isolated rat hepatocytes and microsomes, *Biochem. Pharmacol.*, 1987, **36**, 363.

Compound Index

Key Functional Group Index

Reaction Type Index

Defluorination
fluoroacetate, 135
fluorocytosine, 313
Deformylation, *N*-formyl, 410
Dehydrochlorination, trichloromethyl, 124
N-Demethylation
tert-alkyl amine, 227, 281
aryl amine, 336
dimethylaminoalkyl, 201, 208, 214, 219, 227, 229, 231, 235, 376
N-methyl alkyl amine, 267
N-methyl amide, 329, 446
methylamino/dimethylamino, 43, 233, 334, 410
methylguanidine, 440
N-methylimidazole, 380
N-methylpiperidine, 207, 368
N-methylpyrazolinone, 284
N-methylpyrrolidone, 282
methylxanthine, 357, 359, 361, 362
nitrosamine, 417, 420, 421
O-Demethylation
methoxyalkyl, 244, 295, 297
methyoxyphenyl, 99, 167, 168, 208, 210, 267, 332, 344, 350, 367, 372
methylphosphorothioate, 484, 485, 487, 490
methylthiophosphate, 489
Denitration, nitrophenyl, 99
Denitrilation, 136
Denitrosation, nitrosamine, 417, 419, 420, 431, 433
Desaturation
alkane to alkene, 143, 144, 172
aklylthiomethyl to alkene, 467
chloroalkane to alkene, 122
cyclopentyl to cyclopentenyl, 242
dihydropyridine, 303
imidazoline to imidazole, 287
isoalkyl carboxylic acid, 143, 144
propylphenyl, 172
pyrrolidine to pyrrole, 338
tetrahydropyridine, 300
Desulphuration
alkylthiourea, 496
phosphorothioate, 484, 485, 487, 490, 492
Dihydrodiol formation
benzhydrol, 223
naphthyl/naphthalene, 45, 199
phenyl, 225

polycyclic aromatic hydrocarbon, 51, 54, 60, 63, 71, 74, 78, 80, 82, 93
Dimerization
aryl amine, 397, 398
aryl hydrazine, 354
aryl nitroso, 397
nitrofuran, 276

Epoxidation
alkene/alkadiene, 103
arylethylene, 104
cycloalkene, 131
dihydrodiol, 76
halogenophenyl, 163
polycyclic aromatic hydrocarbon, 50, 52, 54, 84

Formylation, aryl amine, 92
Free radical formation
N-acetylhydrazide, 387
N-isopropylhydrazide, 387

Glucose conjugate, phenol, 181
Glucuronic acid conjugates
sec-alcohol, 242, 337, 374, 423
alicyclic alcohol, 469
alkyl alcohol, 139, 149, 238
alkyl carboxylic acid, 40
aminoimidazole, 379
aryl amine, 321, 376, 392
aryl acetic acid, 290
benzoic acid, substituted, 197
benzylic alcohol, 92, 199
cytosine, 313
hydroxypyrimidine, 319
phenol, 67, 174, 175, 184, 207, 210 261, 349
pyrazole, 374
steroid, 459
Glutamine conjugates, aryl acetic acid, 189, 191
Glutathione conjugates
acrylonitrile, 116
aldehyde, α,β-unsaturated, 111
alkene, 113
alkyl epoxide, 106, 108, 246
alkyl sulphonate, 147
aryl amine, 202, 400
aryl epoxide, 95, 163
benzyl sulphate, 56
bromoacetyl, 119
bromoalkyl/bromoalkane, 118, 120